TANKS

OF THE WORLD

Taschenbuch der Panzer

Edited by Wolfgang Schneider

8 th Edition

Bernard & Graefe Verlag

Designed by Walter Amann, München
Lithographies and Set by: FIBO Druck- und Verlags GmbH,
Neuried b. München
Printed by Konkordia GmbH, Bühl
Printed in Germany

ISBN-3-7637-5984-0

List of contents

CHILE

CHINA

CROATIA

CZECH REPUBLIC/SLOVAKIA

EGYPT

FINLAND

FRANCE

GERMANY

GREAT BRITAIN

ITALY

JAPAN

KOREA, NORTH

KOREA, SOUTH

(Vehicles marked with (*) do not represent a seperate entry.)

Foreword

It is a pleasure for me to comply with the request of the editor and to contribute a foreword to what is now the 8th edition of the well known »TANKS OF THE WORLD« (»Taschenbuch der Panzer«). The development of the German Army, like that of armed forces in general, being based on the missions it is assigned to, is influenced to a major extent by operational requirements, openings and trends in technology in the foreseeable future and, of course, to an increasing extend by the financial resources available. NATO's new Strategic Concept and the task spectrum it covers define the scope of action of the German armed forces. National defence in cooperation with our allies remains our core mission. It will in future be considered an integral part of collective Alliance defence conducted outside Germany.

The functional capabilities which the Alliance demands of its members mean that the German Army must essentially be capable of conducting

+ mechanised operations + peace missions + special operations

We must assume that in future operations German forces will very likely be deployed outside Germany. New structures must therefore be geared towards the requirements of these operational forces for their missions abroad.

To accomplish its future tasks, the Army needs forces which can basically be employed throughout the task spectrum. Mechanised combat troops which are specially versatile and adaptable and offer optimum protection will therefore continue to form the core of the Army. They are perfectly suited both to conduct combined arms operations and to manage composite force operations in the context of peace missions.

The current operational setting, as far as mechanised operations are concerned, clearly shows that Germany's land forces, together with those of the NATO partners, will no longer be able to cover the given space entirely with forces – as it used to be in the past when the concept of forward defence applied in Central Europe. What is required today is the capability to operate freely in terms of space. The decisive feature of future operations will therefore be the way in which space is used. The aim being to allow an unexpected and superior response to be made at the precise point where the decision can be brought about.

The developments in technology lend modern land forces a new dimension in operations in the use of space by means of highly mobile forces and long-range reconnaissance and weapon systems. An integrated modern long-range artillery system, backed up by SMART ammunition, reconnaissance and combat drones and highly mobile air-mechanised forces, will in the future give the German Army a true deep-battle capacity and due to the effect of fire assist it in it's operational endeavour to attain a favourable local force ratio for tactical engagements involving armoured forces. Their primary function is to defeat enemy armoured forces in continuous battle by quickly alternating between the different types of combat, even when terrain and vision are poor. To do so, they must also be able to conduct mobile operations without the assistance of neighbours and while manoueuvring.

When it comes to enhancing protection, survivability and firepower of armoured vehicles, we are compelled to proceed step by step. Our focus is first on improving the protection components of both armoured and non-armoured vehicles. At this point, I would like to take a look on a few armoured systems as examples.

In the Leopard 2 MBT, especially in the improved version, the armour corps of the German Army has an appropriate high-powered weapon system that will stand it in good stead for quite some time to come. The primary weapon system of the mechanised infantry will remain, also in the foreseeable future, the improved MARDER 1 A 3 infantry fighting vehicle. The Panzerhaubitze 2000 combines with the intelligent

sensor-fused ammunition SMART, the drones CL 289, KZO and TAIFUN, the artillery radar system COBRA and the MLRS multiple launch rocket system – co-ordinated by the ADLER C2 and weapons direction system, an artillery system of the highest international standard.
Against the backdrop of the widened task spectrum, protecting infantry forces in particular is a matter of high importance. It requires providing them powerful and protected means of transport. The Dingo, a light-armoured transport vehicle, is specially tailored for the needs of peacekeeping operations and will enhance the ability to protect own forces in these kinds of missions enormously.
In the armoured transport vehicle GTK, the requirement for protected transport vehicles and conversion kit carriers will be satisfied for a long time by a modern vehicle. It will enhance mobility and survivability immensely.
In addition the new mechanised infantry tank is under development with high priority, and will adapt the fire-power and the mobility of the mechanised infantry according to the challenges of the new century. At the same time, the chassis of the new mechanised infantry tank will be used as a platform for the further development of Armoured vehicles, e. g. the successor of the LEOPARD 2 or the new anti-air-tank.
This new family of armoured vehicles together with the GTK, the DINGO, the Panzerhaubitze 2000 will provide the German Army with armoured and protected highly mobile forces capable of conduction combined arms battle as well as composite force operations.

Helmut Willmann, Generalleutnant
Chief of German Army Staff

Author's preface

With the edition at hand of TANKS OF THE WORLD this standard work appears for the eighth time.
Since the publication of the last edition in 1990 a series of changes have taken place, not so much in the area of completely new developments but more in the field of modernization and upgrading. These are positive effects resulting from the dissolution of the east-west controversies which in former times were the main motive for the rapid succession of acquisition decisions. Also the number of the delivered systems were considerably reduced. On the other hand it was obvious that in many countries comparatively old material remains in service.
All these facts demand putting corresponding priorities as to the presentation of the book, since the total size is limited.
Some chapters on older systems were either totally dropped or were considerably abridged in favour of presenting new vehicles. So the vehicle list offers references to former editions to make finding information easier for the reader.
It was met with general approval that the preceding edition focused on exclusively dealing with systems still in service, and that the inventory numbers were related to the countries concerned. Only in a few isolated cases the data were not available and therefore had to be estimated. The suggestion to submit the whole work as a database on CD-ROM together with an additional index of countries with yearly upgrade was met with a positive echo, too.
As to the supply with photos it has to be mentioned that principally actual photos are used delivered by the producer firms or the Mönch Publishing Group photo archive.
Further I anew have to thank Mr. Steven Zaloga for the permission to make use of his sketches, and thanks also to the renowned Japanese periodical PANZER.

Uelzen, June 2000.

Introduction

1. This book covers all armoured land vehicles in service use. Both unarmoured vehicles and anti-tank weapon systems (unless they are mounted on tank destroyers) are omitted. Major non-serial versions are only summarized at the beginning of each entry but not illustrated by photographs.

2. Armoured vehicles are arranged by country of development and within this by classes and groups according to role. Derivatives locally modified in their user countries are generally shown inside the entry of its country of origin (except totally re-built versions).

3. Following the form of previous editions, the book is divided into Part 1, consisting of type data sheets with drawings and photographs, and Part 2, in which technical data are tabulated.

4. Each entry normally represents a 'main type' which is a basic or parent vehicle, designed to meet the primary role. Successive models and variants involving minor modifications or redesign are grouped under the main type.
The Tables in Part 2 are intended to allow two-way comparison of the characteristics and performance of vehicles with the same role. These tables also enable one to determine the average state of the art by taking the mean of compatible figures.

5. **Type designations.** Whenever known the original vehicle names and classifications are used. Russian names have been transliterated according to the best accepted practice.

6. **Descriptions.** Detailed figures on the main types are given in the Tables of Part 2. The figure *TABLE* in parenthesis refers to it. Only when this is not the case are key figures shown below the heading in Part 1.

7. The **development and production** of main types and their most important derivatives is given in brief, with the aim of bringing out what the development cycle involves.

8. **Distinguishing features** are given to highlight differences from other types. The reference base for these details is simply a predecessor or the parent/basic vehicle, and they are not comprehensive. When the terms 'left' or 'right' are used in describing frontal aspect ('Front'), they refer to the side of the vehicle as seen by a person facing the normal direction of travel.

9. **Other details** are given for vehicle data neither visible from the outside nor integrated in the data tables. In this section major derivatives and sub-variants are described.

10. The vehicle's **role** refers to the different user countries which employ it in the normal mission characteristic. More detailed information are given when the vehicle fulfills a deviate task. The figure in parenthesis indicates the number of vehicles in use.

11. Subjective as they may often be, **assessments** of each main type have been included. These are made from the point of view of an observer who can stand back and look back. These assessments are based on other types of the same class and/or on the general state of the relevant technologies.

12. The basic scale of the **drawings** is 1 : 87. They simply give leading dimensions, so that the relative size can be compared, and highlight characteristic features.

13. **Photographs.** Virtually every model or variant which differs from others in external appearance is included. The quality of prints of types hard to obtain may differ in processing quality.

Armoured Vehicles Today

The previous edition of TANKS OF THE WORLD was published ten years ago and since that time a lot of changes have emerged concerning the policies of security and armament. Only a short time after the unification of Germany the dissolution of the Warsaw Pact followed, and the Soviet union disintegrated into several national resp. federal states with greater sovereignty. Due to the vanishing threat and financial restraints the armament production numbers in Russia drastically sank. The western countries began to reduce the amount of armed forces as well as the equipment numbers. This process has not yet come to an end. So the annual Russian production e.g. of main battle tanks sank from 3000 to a few hundred. Even the Gulf war did not influence this trend.
With regard to the acquisition of material the following has to be stated: the urgency to modernize or upgrade such systems as have already been in service for years has been increasing since the financial means for providing new systems are no longer available in the same amount as they were in former days. The absolutely necessary acquisition of new material ensues in smaller numbers only, often stretched over several years. You also can observe that especially efficient systems are procured, or that minor efficient systems are deliberately put up with. The latter proceeding can, however, soon prove to be rather dubious. Whereas in the first case you would have decided for the technological correct entrance as the precondition for the numbers to be increased if the situation in security politics should alter.
Among the main battle tanks for the time being the measure of all things are: the French LECLERC at the moment under production, the most modern version of the German LEOPARD 2, the Russian T-80, and the American M1 ABRAMS. Also the Israeli MERKAVA demonstrates proof of upgradability, whereas the effectivness of other systems, such as the British CHALLENGER 2, the Italian ARIETE, the Japanese Type 90 or the South-Korean Type 88 have inferior system performance - or they have a deliberately different layout. Besides it is meanwhile recognizable that every potential of growth is limited. Thus consequently innovative considerations for the future development have been initiated, for a longer time already in Russia (T-95), for three years in Germany (NGP), and for a shorter time also in the USA (FCS). There is conformity with regard to significantly increasing the efficiency of the main armament, either by increasing the calibre (135 to 140 mm) or by incorporating hybrid solutions for accelerating projectile speed (e.g. ETC: electric-thermal chemical). Systems of purely electrical propulsion (ETK or EMK) are in the medium term not yet regarded as available, since neither integratable quantities of energy are at disposal, nor are applicable penetrator materials for essentially higher muzzle velocities above 2.400 m/sec. There are differences with regard to assessments which kind of protection should be advisable. Quite certainly a passive all-around protection for the future threats within tactical sensible weight limits is impossible, if you take as a basis the present day protective technologies. It is consistent that there are a series of explorations to neutralize threats by means of „soft“ and „hard kill“ systems; the first include all kinds of deception and deflection systems, the latter include such stand-off measures as blast charges or reactive special armour. Nevertheless - especially with regard to the duel-seeking MBT - the call for a tactically indispensable minimum ballistic protection

might rather lead to vehicles of more than 50 tons, even if the crew should be reduced to three or two.

When conceiving armoured vehicles for the mechanized infantry one cannot but admit that at the moment there is no system which with regard to the armoured protection is capable to closely co-operating with the MBT. Because of the considerably lower protection level such systems must either be echeloned in the rear, or they are mere transport means without major battle value. The experiences of the Israelis in the Lebanon, of the Russians in the Caucasus and of the Western states in the Gulf area clearly show that in this respect re-thinking is indispensable. Hanging tons of weight as additional protection modules on the vehicles affords only limited relief. Till now only the Israelis drew practical conclusions by designing the strongly armoured ACHZARIT, a system which is capable of close joint operation with the MBT. The German reflections in connection with the New Armoured Platforms (NGP) aim into the same direction: to provide the successor of the MARDER with a protection requirement principally comparable to the protection of the MBT. But these considerations are still in an early planning stage.

The greatest spread of armoured systems can be observed with the personnel carriers and reconnaissance vehicles. There is a series of efficient wheeled and tracked vehicles, even offered by the newcomers among the tank producing nations. At present the demand is heated up by the increasing number of peacekeeping missions worldwide.

Anti-tank missile carriers lost their importance, due to the increased range of the tank gun, the improvements of protection and the rivalling abilities of ground-based and airborne target seeking munitions (SMArt a.s.o.).

With the armoured artillery systems the ranges have increased, at the same time the reaction ability and the precision have been improved. The German Panzerhaubitze 2000 that has now gone into production is by far the most efficient system. Whether the US AFAS concept - having to struggle against many setbacks - will justify the high-strung expectations has to be waited for.

Among the armoured anti-aircraft systems a clear gap has been widening between systems widely ineffective against modern aircraft and high-performance systems which can hardly be paid for, and so can be kept in service in small numbers only.

With the combat service support systems as recovery and engineer vehicles it is apparent that for many a country they are eventually regarded under the aspect „nice to have“, but they are beyond price, additionally to the other equipment with armoured vehicles.

This is exemplified by the fact that a great number of aged and really obsolete systems are still kept in service because the financial means for new acquisitions are not avilable. On the other hand there is a trend among the wealthier (e.g. in the Gulf region) to supply themselves with systems of high quality; the imports of the M1A2 ABRAMS resp. of the LECLERC illustrate this.

In future armoured vehicles will play an important part in any kind of armed conflicts. Thus all armies having to replace or renew their equipment will see themselves confronted with the problem whether they are politically and financially able to provide systems really efficient and adequate for the threat. The times when the deficits in combat value could at random be compensated by superior quantity definitely belong to the past, not only since the Gulf War of

1991. This dubiousness is aggravated by the fact that in many areas of technology, e.g. sensors and electronics there are pushes of innovation in rapid sequence with corresponding consequences for the modernity of armoured vehicles. Superiority in range and target acquisition, speed advantage in obtaining and processing information, precision increase and general better performance can soon result in shifts of the tactical level of performance. Modernization can take remedial measures but will run sooner or later against limits which can only be overcome by well-timed new acquisitions which must allow - with regard to the service life - the greatest possible growth potential.

As to the technical layout of future vehicles a trend towards „all electric vehicle" is to be observed which alone can provide the required quantities of electrical energy, increase the mobility performance resulting in a lot of further technical and tactical advantages (range increase, more suitable stowage conceptions, handling advantages a.s.o.).

Greater regard than hitherto must be given to the vehicle protection against top-attack weapons and against mines as well. In order to increase reaction time new concepts of handling and display arrangements must be developed, with a high degree of automation and advanced vehicle management systems. Weapon command and control systems will in future essentially promote the command and control ability, so for instance IFF. In the US Army efforts of this kind are summarized under the term „Digitizing the Battlefield" or „Winning the Information War".

Often reported dead the tank does have a future, which has, of course, to be arranged with regard to the irrevocable frame conditions.

Vehicle classification

The armoured vehicles covered in this book are grouped according to country of origin and role. Classes of vehicle and types within classes are designated as shown below, the abbreviated class title (parentheses) being used at the head of each page.

ARMOURED RECONNAISSANCE VEHICLES (RECONNAISSANCE)	
Armoured vehicle, reconnaissance	AVR
Armoured reconnaissance scout car	ARSC
Armoured reconnaissance fire support vehicle	ARFSV
Armoured reconnaissance infantry vehicle	ARIV
Armoured reconnaissance radar vehicle	ARRV
Armoured NBC reconnaissance vehicle	ANBCRV
MAIN BATTLE TANKS (TANKS)	MBT
TANK DESTROYERS	
Tank destroyer, gun	TDG
Tank destroyer, missile	TDM
ARMOURED INFANTRY VEHICLES (INFANTRY)*	
Armoured infantry fighting vehicle	AIFV
Armoured infantry command vehicle	AICV
Armoured infantry mortar vehicle	AIMV
Armoured infantry radar vehicle	AIRV
Armoured infantry ambulance vehicle	AIAV
Armoured infantry fire support vehicle	AIFSV
ARMOURED CARRIERS (CARRIERS)	
Armoured general purpose carrier	AGPC
Armoured personnel carrier	APC
Armoured command/communication carrier**	ACCC
Armoured ambulance vehicle	AAV
Armoured cargo carrier	ACC
Armoured mortar tractor	AMTC
Armoured mortar carrier	AMC
Armoured police vehicle	APV
ARMOURED AIRBORNE VEHICLES (AIRBORNE)	AABV
ARMOURED ARTILLERY VEHICLES (ARTILLERY)	
Armoured artillery howitzer	AAH
Armoured artillery gun	AAG

* Variants listed under this class are based on an armoured infantry fighting vehicle as opposed to an armoured carrier, and are mainly or solely used within mechanised infantry units.

** This heading covers all armoured command vehicles not classed as armoured *infantry* command vehicles.

Armoured artillery rocket launcher	AARL
Armoured artillery missile launcher	AAML
Armoured artillery observation post	AAOP
Armoured artillery command post vehicle	AACPV
Armoured artillery fire director vehicle	AAFDV
Armoured artillery radar vehicle	AARV

ARMOURED AIR DEFENCE VEHICLES (AIR DEFENCE)

Armoured air defence gun	AADG
Armoured air defence missile launcher	AADML
Armoured air defence fire director vehicle	AADFDV
Armoured air defence radar vehicle	AADRV

ARMOURED ENGINEER VEHICLES (ENGINEER)

Armoured vehicle launched bridge	AVLB
Armoured dozer vehicle	ADV
Armoured engineer vehicle	AEV
Armoured mine clearing vehicle	AMCV
Armoured minelaying vehicle	AMLV

ARMOURED LANDING VEHICLES (LANDING)

Armoured vehicle landing artillery	AVLA
Armoured vehicle landing engineer	AVLE
Armoured vehicle landing infantry	AVLI
Armoured vehicle landing recovery	AVLR

ARMOURED RECOVERY VEHICLES (RECOVERY)

Armoured crane vehicle	ACV
Armoured maintenance vehicle	AMV
Armoured recovery vehicle	ARV

SPECIAL ARMOURED VEHICLES (SPECIAL)

NATO abbreviations armoured vehicle ammunition natures

AP	Armour-Piercing
APC	Armour-Piercing Capped
APC-T	Armour-Piercing Capped - Tracer
APDS	Armour-Piercing Discarding Sabot
APERS	Antipersonnel
APFSDS	Armour-Piercing Fin Stabilized Discarding Sabot
APHE	Armour-Piercing High Explosive
API	Armour-Piercing Incendiary
API-T	Armour-Piercing Incendiary - Tracer
APP	Armour-Piercing Practice
APSVDS	Armour-Piercing Supervelocity Discarding Sabot
AP-T	Armour-Piercing - Tracer
APT-DI	Armour-Piercing Tracer - Dark Ignition
AT	Antitank
DS	Discarding Sabot
BD	Base-Detonating
CANISTER	Canister
HE	High Explosive
HEAT	High Explosive Antitank Tracer
HEATFS	High Explosive Antitank Fin Stabilized
HE/FRAG	High Explosive Fragmentation
HE/FRAG (FS)	High Explosive Fragmentation (Fin Stabilized)
HE/HC	High Explosive Hollow Charge (UK)
HEI	High Explosive Incendiary
HEP	High Explosive Plastic (US)
HEP-T	High Explosive Plastic - Tracer (US)
HE-SD	High Explosive - Self Destruct
HESH	High Explosive Squash Head (UK)
HE-T	High Explosive - Tracer
HET-DI	High Explosive Tracer - Dark Ignition
HVAP	Hypervelocity Armour-Piercing
HVAPDS	Hypervelocity Armour-Piercing Discarding Sabot
HVAPDS-T	Hypervelocity Armour-Piercing Discarding Sabot - Tracer
HVAPDSFS	Hypervelocity Armour-Piercing Discarding Sabot Fin Stabilized
HVTP	Hypervelocity Target Practice
ILL	Illuminating Shell
I-T	Incendiary - Tracer
KE	Kinetic Energy (APFSDS-T)
KE-Üb	Kinetic Energy Practice (DSP)
MZ *or* MP	Multi-purpose, as in HEAT-MP-T = High Explosive Antitank - Multi-purpose - Tracer
P *or* PRAC	Practice
SA	Small Arms
SAPHEI	(Special) Armour-Piercing High Explosive Incendiary
SH	Squash Head (UK)

SH PRAC	Squash Head Practice
SMK	Smoke
T	Tracer
TP	Target Practice
TP-T	Target Practice - Tracer
WP	White Phosphorus
WP-T	White Phosphorus - Tracer

Key to abbreviations

Note. This list excludes – foreign abbreviations, which are expanded where they occur; country abbreviations which refer to the international road traffic identification signs; ammunition and vehicle class abbreviations which are given in the tables immediately preceding this.

AB	airborne, airportable
AD	air defence
ADMG	air defence machine-gun
ammo	ammunition
amph	amphibious
arty	artillery
ATGM(S)	anti-tank guided missile (system)
auto	automatic
BITE	built-in test equipment
cdr	commander
Cl	military load classification (MLC)
CS	coil spring
coax	coaxial (machine-gun)
CW	chemical warfare
ECM	electronic countermeasures
elec	electrical
elec-hyd	eletro-hydraulic
elev	elevation
ESM	electronic support measures
ETC	exhaust turbo-charged
EW	electronic warfare
excl	exclusive
f	front/forward
FCC	fire control computer
FCS	fire control system
fig	figure
GT	gas turbine
H	height
HP	hydro-pneumatic
hyd	hydraulic
IFCS	integrated fire control system
IFF	identification friend/foe
II	image intensification
IL	in-line (engine)
incl	inclusive
IR	infra-red

KE	kinetic energy
km/h	kilometers per hour
kW	kilowatts
L	length
L/	calibre-length (shot travel)
m	metre(s)
maint	maintenance
MG	machine-gun
min	minute
Mk	Mark (i.e. model)
mm	milimetre(s)
MP	multi-purpose
MRS	muzzle reference system
MV	muzzle velocity (V_0)
NBC	nuclear, biological, chemical
NGP	nominal ground pressure
NSWP	non-Soviet Warsaw Pact states
OP	observation post, *or* opposed piston (engine)
OR	operational requirement
r	reverse (gear)
RAP	rocket assisted projectile
RCL	recoilless gun
rds	rounds
recce	reconnaissance
RF	range-finder
rpm	revolutions per minute *or* rounds per minute
SC	supercharged
sec	second(s)
SEL	semi-elliptic leaf spring
SHORAD	short range air defence
SI	spark ignition (petrol engine)
SMS	slaved mounting system
SP	self-propelled
stab	stabilise(-d, -r)
t	(metric) tonne(s)
TB	torsion bar
TC	torque converter
TI	thermal imaging
TMG	turret machine-gun
TOGS	thermal observation and gunnery sight
trav	traverse
UV	ultra-violet
V	V-shaped piston arrangement
W	width
WP	Warsaw Pact
Wt	battle weight

PART 1. LIST OF VEHICLE TYPES

TAM main battle tank
Tanque Argentino Mediano

Models.

TAM MBT
VCA-155 AAH

Non-serial versions.

TH 301 MBT (formerly TAM-4) with Rh 105-30 105 mm gun
DRAGON AADG 2x30 mm turret of Thomson-CSF
VCLC CAL 160 AARL with 36 x 160 mm rockets
VCLC CAM 350 AARL with 4 x 350 mm rockets
VCRT ARV

Development and production.

1974 Light tank development by Thyssen-Henschel under contract from Argentine MoD.
1979 Development trials of 3 prototypes in Argentina. Planned production 512 MBT and AIFV.
1983 Export orders to Peru, Malaysia and Panama did not materialise (neither to Ecuador in 1988). VCA development starts.
1986 Production stops after delivery of 256 tanks.
1994 Delivery of 120 vehicles of a second batch completed.

Distinguishing features. Side. MARDER AIFV hull. 6 roadwheels, top rollers covered by skirting plates, relatively high fighting compartment with long rear mounted turret. Front. High sponsoned hull, turret with sharp edged mantlet.

Other details. Hull modified and suspension redesigned to higher vehicle weight and different moments of inertia. Main armament is the C.360-TAM L51 gun, a further development of the French DEFA gun of AMX 13 (ammunition stowage 50 rounds). One or two Jettison fuel barrels can be carried on rear. (TABLE)
Armoured howitzer with Italian PALMARIA 155 mm howitzer turret on lengthened chassis (7 roadwheels). The hull length is 7.69 m, total height is 2.85 m, the weight is 40 tons. The crew consists of 5 soldiers. 23 rounds are stowed in the turret, 5 in the hull. Fuel capacity is 873 litres. Maximum speed 55 km/h.

Role. In service with Argentina (376 MBT and 19 AAH).

Assessment. Level of protection is only that of an AIFV. Relatively high firepower. Good mobility. Since other versions were adopted, the Argentine Army has a high level of standardization.

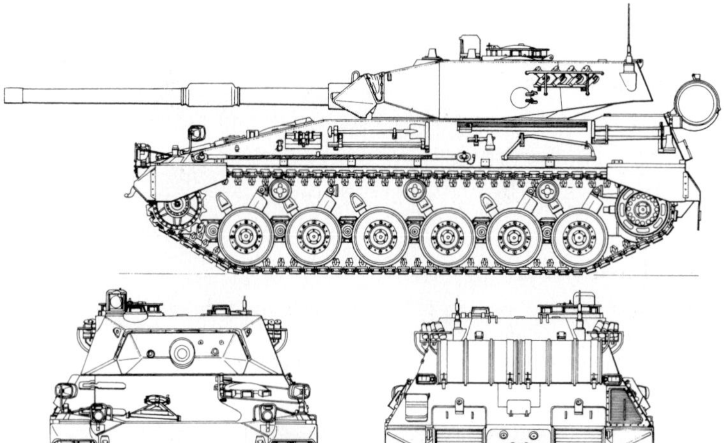

TAM MBT

TAM MBT

PALMARIA AAH

VCTP armoured infantry fighting vehicle
Vehiculo de **C**ombate y **T**ransporte de **P**ersonel

Models.

AIFV with 20 mm RH 202 gun and 1 ADMG 7.62 mm FN60-30
Wt 27.5 t 19.28 kW/t 75 km/h 570 km 530 kW Crew 3+9
VCTM AIMV with 120 mm Brandt AM-50 mortar and 1 MG
H 186 Wt 26 t 20.4 kW/t 61 rounds Crew 5
VCPC AICV with 1 MG in ring-mount
Wt 24,5 t 4 radio-sets 11 m antenna Crew 2+6

Non-serial versions.

VCAmb AIAV
VCLP AVLB in cooperation with MAN

Development and production.

Military requirement for the Argentine Army for a family of armoured combat vehicles in the 25 t class. Development in cooperation with Thyssen-Henschel on basis of MARDER AIFV.
1979 series production starts.
1984 prototypes of VCTM and VCPC for tests. Conceptual studies for AARL with 9xPATT 30/150 rockets and for VCRT.
1986 Production ends
1987 First prototype of VCRT completed with recovery superstructure from MaK of Germany completed.

Distinguishing features. Suspension and hull almost identical with TAM MBT and MARDER AIFV. Two-man turret with rapid fire gun in small drum mantlet. VCTM with two large roof hatches, mortar pointed forward.
VCPC without 20 mm turret.

Other details. MB833 Ka 500 MTU engine, HSWL204 Renk transmission. 8 77 mm smoke grenade launchers. PERI Z11 A1 gunner's sight.

Role. Argentine MechInf units employ 216 VCTP, 50 VCTM and 15 VCPC.

Assessment. Slightly simplified and lower version of MARDER AIFV, nearly its standard. AIMV on MARDER chassis was rejected by German Army.

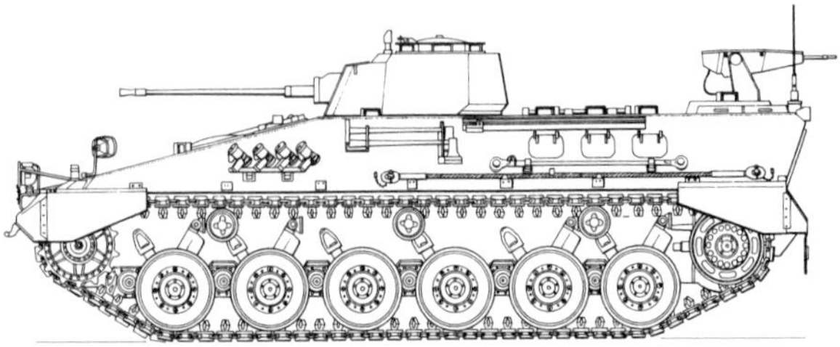

VCTP AIFV

VCTP AIFV

VCTM AIMV

KÜRASSIER SK 105 tank destroyer

Models.

Panzerjäger KÜRASSIER
SK 105 A1 modified to fire the OFL 105 G1 APFSDS ammunition
SK 105 A2
4KH7FA-FA (Fahrschulpanzer) driver training vehicle

Development and production.

1967 first prototype and in 1969 second one by Saurerwerke (later Steyr-Daimler-Puch).
1971 six pre-production vehicles issued.
1985 modifications of the SK 105 to fire APFSDS rounds starts.
1997 Negotiations with Botswana to deliver 32 SK 105.
1998 combat improvment starts.

Distinguishing features. Side. Five medium roadwheels on torsion bars, rear sprocket, three return rollers. Two-part turret, upper part oscillating with long gun and rear bustle. Cupola with seven episcopes on left side. Front. Sponsoned hull with low, sloped upper sides. Boxlike housing on turret roof centre for LASER rangefinder (retrofitted) and IR/white light searchlight on top of it. Double-baffle muzzle break. Export versions with different alterations (sand shields, cyclon filters).

Other details. Tank destroyer with 105 mm gun CN-105-57 and 1 7.62 mm TMG in turret JT 1, a Steyr modification of the FL-12 turret. Chassis 4KH6FA AIFV, diesel engine 6FA rating 236 kW with turbo charger and intercooler mounted at left rear, hydrostatic superimposed steering gear. Later oscillating parts of turret with improved protection and enlarged turret hull. Gun with semi-automatic loader for 12 rounds in two revolving drums. IR-periscope for the commander. 170 vehicles were upgraded with the ZF 6 HP automatic transmission. (TABLE)
Upgrade of 120 vehicles to A2 standard with a computerized fire-control system, eye-safe laser range-finder and the thermal imaging camera LANSA-DOT in place of the former deleted search-light with two monitors (commander and gunner).

Role. In service with the Austrian tank destroyer units (240). Exported to Argentina (106), Bolivia (34), Botswana (20), Morocco (109), Nigeria (25) and Tunisia (54).

Assessment. A lightly protected, mobile TDG with small frontal, but high silhouette. Effective against modern MBT only on short range using APFSDS. Logistic commonality with 4K AIFV. Useful for anti-tank defence, infantry support, reconnaissance missions.

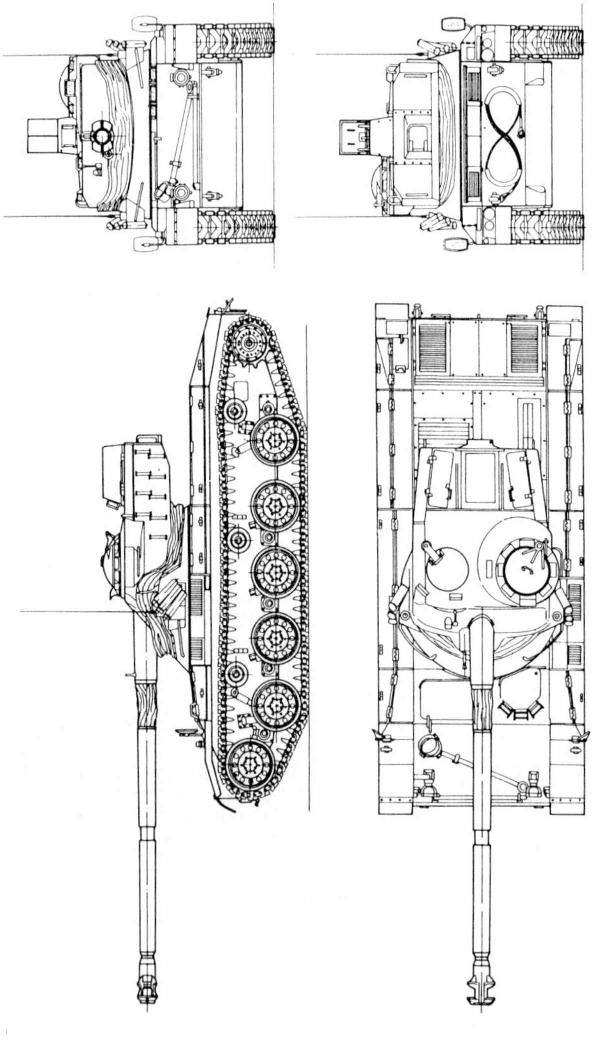

KÜRASSIER TDG

SK 105 TDG

SK 105 A1 TDG

ASCOD armoured personnel carrier
(Austrian Spanish Cooperation Development)

Non-serial version

ASCOD 105 light tank
ASCOD 105 light tank

Development and production.

1990 first prototype completed to meet Austrian and Spanish requirements for a new MICV
1992 second prototype built with 600 HP engine and larger internal space. Both vehicles have 6 roadwheels. Total weight is 22 t.
1994 third prototype issued with seven roadwheels, weight 27.3 t.
1998 initial delivery contract for 123 PIZARRO APC and 21 CP vehicles for Spain
1999 112 vehicles for Austria contracted (delivery 2002 -2004). ASCOD 105 selected by Thailand

Distinguishing features. Side. Protruding front with steep engine deck. Running gear with seven roadwheels and three top rollers covered by skirting plates. Front sprocket. Horizontal roof with vertical hull rear. Box-shaped turret placed centrally. Front. Sponsoned hull with vertical hull sides. Sloped glacis with driver on the left. Turret with vertical sides moved to the right side. Two prominent sight apertures on the turret roof.

Other details. Vehicle is also called ULAN in Austria and PIZARRO in Spain. Main armament comprises a twin-axis stabilized Mauser 30 mm cannon Model F and a 7.62 mm coaxial MG on the left side. Weapon elevation is -10° to +50°. The two-man turret and gun are gyro-stabilized. 200 rounds 30 mm ammunition are stowed in the turret and 700 for the MG. The tanks are equipped with Kollsman DNRS (Direct Night Range Sight). Prototype 3 is powered by the MTU diesel engine type 8V-183-TE 22 with 441 kW.
The ASCOD 105 features the South African LIW LT 105 turret proposed for the upgrade of the ROOIKAT AVR.

Role. In production for the Austrian Army (112) and for Spain (144). Thailand ordered 15 105 mm gun tanks plus one command post and one repair and recovery vehicle for its Marine Corps.

Assessment. Conventional MICV concept with sub-standard protection and average fire-power.

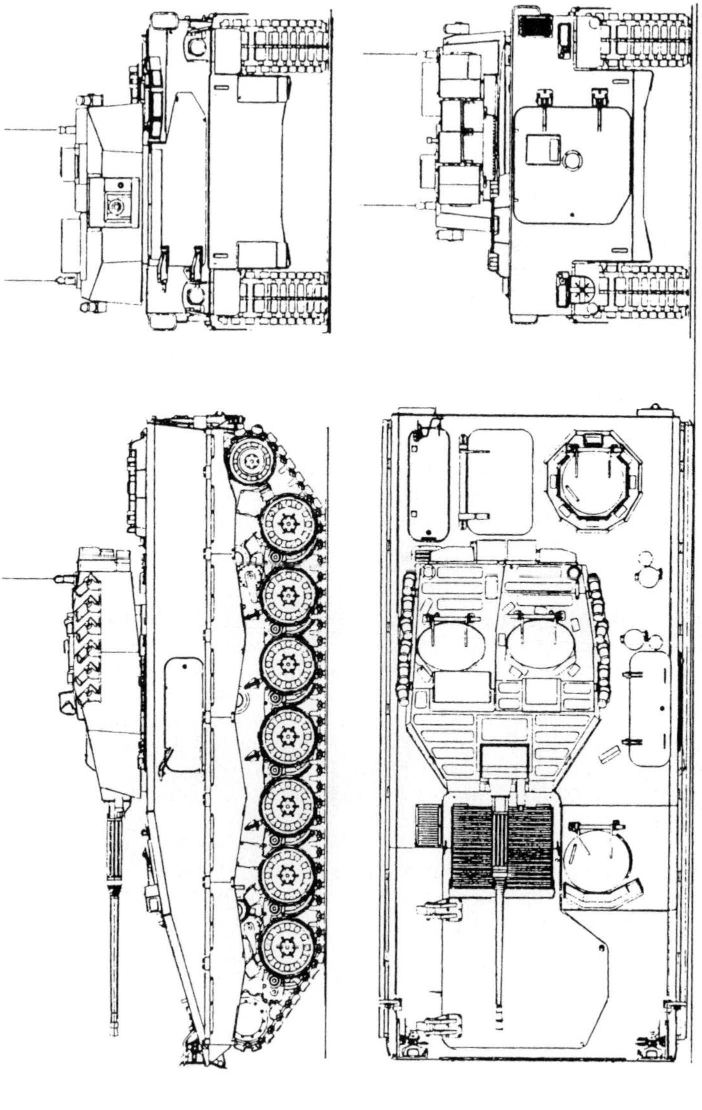

ASCOD AIFV

ASCOD AIFV

4K armoured general purpose carrier

Models.

4K3FA AIFV with 12.7 mm MG
4K3FA-G1 with 20 mm gun
4K3FA FÜ1/FÜ2 ACPV with various radio equipment
4K3FA FS with telex device
4K3FA FÜA 1/A 2/A 3 or A 4 AAOP/AACP
4K3FA FÜFla AADFDV
4K4 F GRW 1 AMC with 81 mm mortar
4K4FA-Pi1/Pi2 engineer versions
4K4FA San AAMB with two stretchers
4K4FA Beob AAOP with Meß-/Beobachtungsgerät 80
4K6FA-B AIFV with 3FA 043 engine

Development and production.

1956 first prototype by Saurer (4K3H).
1960 intermediate version 4K4F introduced with 6-cylinder in-line engine.
1962 production start of 4K3FA with ring-mount for 12.7 mm MG.
1968 delivery of 4K4FA-G1 starts.
1970 end of production.
1975 additional order of 4K6FA-B with deliveries in 1977.

Distinguishing features. Side. Five medium roadwheels on torsion bars, rubber padded tracks, front sprocket, diesel engine right front. Overhang rear with double doors. Low superstructure sloped all-around. Fighting compartment closed by large roof hatches. Front. Slightly sloped sides. Turret for 20 mm cannon or partially protected ring-mount for 12.7 mm MG left front behind driver. At each corner of the fighting compartment hatch is a recess to accept the pivot of an intermediate mounting for the MG 42. Derivatives without turret/ring-mount. FÜFla with armoured casing for auxiliary power-unit on rear plate. FÜ and FÜA versions with traversing base for peribinoculars.

Role. In service with Austria (465, decreasing).

Assessment. Lightly armoured APC with average mobility and limited possibility to fight from under armour. No NBC protection system.

4K3FA APC

4K3FA-G1 APC

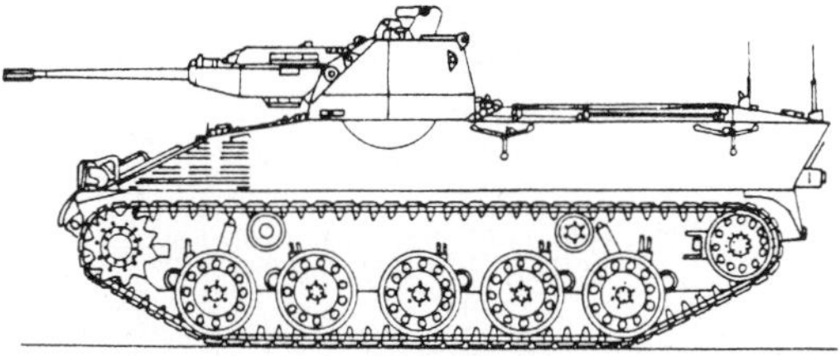

4K4FA-GI AIFV

4K4F GRW2 AMC ▲ 4K4FA San AAV ▼

4K4FA Beob AAOP

4K7FA G 127 armoured general purpose carrier

Models.

4K7FA AIFV with 12.7 mm MG (Greek LEONIDAS 1)
LEONIDAS 2 AIFV (Greece)
4K7FA San AAV with two stretchers
4K7FA GrW 81 AMC 81 mm mortar
4K7FA FÜ ACPV

Non-serial versions.

LEONIDAS 2 with NFT T25 turret
LEONIDAS 2 with DAF/FMC Enclosed Weapon Station with 25 mm KBA-BO2 cannon
LEONIDAS 2 with Santa Barbara TC-13 turret
LEONIDAS with E-90 90 mm gun
4K7FA with Santa Barbara TC-13 turret (M242 25 mm gun) (Spain)
4K7FA with Steyr SP3/300 weapon station (Austrian-Spanish cooparative Development)
4K7FA SPz 1/127, further development of KSPz with SP 1/127 turret

Development and production.

1977 model 4K7FA with stronger engine aimed for export.
1979 serial production starts.
1982 Greek order for LEONIDAS APC.
1987 order for 344 LEONIDAS 2.
1989 production of LEONIDAS 1 stops.
1990 LEONIDAS 2 ordered for the Cypriot National Guard.
1995 production completed.

Distinguishing features. External appearance similar to 4K4FA but longer hull and three return rollers instead of two. One-piece engine grill instead of two-piece of the old series. Three hull bow hatches with horizontal hand grips each (4K4FA has one smaller plate with two vertical grips). Drum-like final drive covers beneath hull bow instead of conical ones. (TABLE)

Other details. LEONIDAS 1 and 2 differ in a number of improvements including firing ports in the rear fighting compartment.

Role. In service with Argentina (10), Austria (26), Bolivia (6), Cyprus (92 LEONIDAS 1, 76 LEONIDAS 2), Greece (175 LEONIDAS 1, 344 LEONIDAS 2), Nigeria (170 including 23 AMC, 2 San, 2 FÅ).

Assessment. A successful AIFV family. Limited possibilities for fighting from under armour, insufficient armour protection.

LEONIDAS 1 APC

LEONIDAS 2 APC

4K7FA San AAV

PANDUR armoured personnel carrier

Models.

PANDUR APC with 12.7 mm machine gun
PANDUR with 25 mm MGTS (Multi-Gun Turret System)
AGMS APC
PANDUR AVR with Steyr SP3/300 turret
PANDUR ACPV
PANDUR AAV
PANDUR ARV

Non-serial versions.

PANDUR with Stear 12.7 mm two-man turret
PANDUR with OTO Melara turret with 25 mm Oerlikon cannon
PANDUR with T25 turret with 25 mm M242 bushmaster cannon
PANDUR with 30 mm Rarden gun turret
PANDUR with 90 mm MECAR KEnerga gun
PANDUR with 90 mm Cockerill gun in CSE turret
PANDUR with UTM-800 HOT launcher
PANDUR with Kvaerner-Eureka TOW launcher
PANDUR 81 mm mortar carrier
PANDUR with 120 mm breech-loaded mortar
PANDUR amphibious version

Development and production.

1979 private venture development by Steyr-Daimler-Puch.
1986 production of 11 pre-series vehicles.
1994 Austrian order for 68 APCs for UN employment.
1995 first deliveries.
1998 delivery contract with Slovenia. First vehicles issued to Belgium.
1999 contract with TACOM for the production of 50 vehicles at AV Technology

Distinguishing features. Side. 6x6 chassis with wedge-shaped front bow, high superstructure with sloped rear hull end with negative angle. Front. Boat-like chassis with inverted upper superstructure.

Other details. Two types of chassis' length differing by 400 mm with different mission specific arrangement. Front engine on the right, driver on the left side. Rear troop compartment with two large rear doors. Two vision blocks on both sides. Central tyre pressure regulation. First and second axle steerable. Reconnaissance version with SP3/300 turret with 30 mm Mauser cannon. Gunner has the Aeritalia P204 sight with x8 magnification, the commander a Philips UA 9126 x6 day resp. x7 night sight.
The AGMS (Armored Ground Mobility System) for the US Army features

appliqué armour, a roof mounted 12.7 mm M2 machine gun, central tyre pressure regulation and automatic drive management.
Fire support version with 90 mm Cockerill Mk. 8 gun with a large range of ammunition but sub-standard penetration capability.
(TABLE ->)
Role. In production for Austria (224 APC 12.7 mm, 45 SP3/300), on order by Belgium (41 APC 12.7 mm, 5 CP, 4 AAV, 4 ARV), Kuwait (70 MGTS, 90 mm Cockerill, CP, ARV, AAV), Slovenia (10 called VALUK) and USA (50 AGMS).

Assessment. Versatile family of wheeled APC with low armour protection.

PANDUR APC

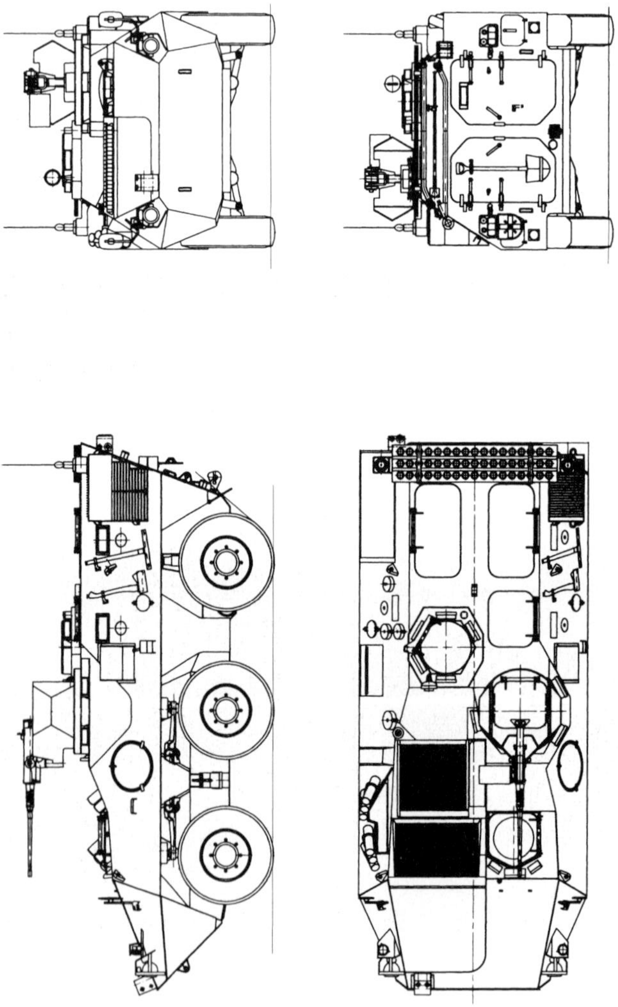

PANDUR APC

PANDUR APC with MGTS

PANDUR AVR

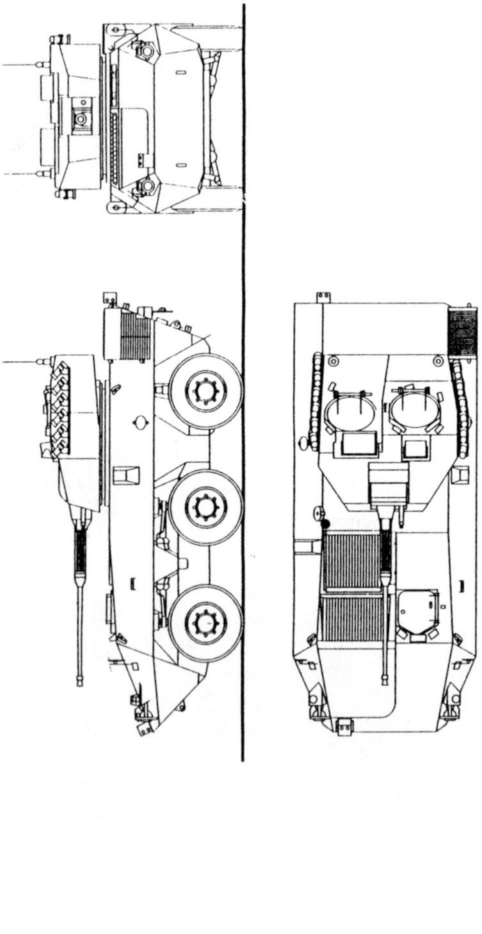

PANDUR AVR

PANDUR AAV

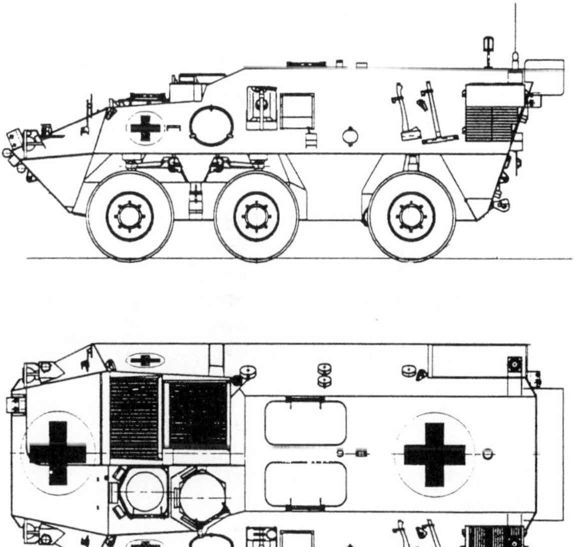

PANDUR AAV

GREIF armoured recovery vehicle

Models.

4KH7FA-SB 20 GREIF ARV
4KH7FA-Pi AEV

Development and production.

Further development based on KÜRASSIER TDG and 4K7FA AIFV.
1974 first prototype completed, introduced in 1976.
1985 ARV production is completed.
19 AEV enter service in Austria.

Distinguishing features. Side. Running gear similar to other members of the light family with five roadwheels, three return rollers. Vertical front with dozer blade. High superstructure with two doors in left side. Crane on raised base right front. Flat rear engine compartment used as load platform. Front. Sides slightly sloped. Driver's seat in centre above front. Crane right side.
AEV with ditcher crane and bigger dozer blade.
(TABLE)

Role. ARV is in service with Argentina (10), Austria (50 ARVs + 18 AEVs), Bolivia (2), Morocco (11) and Nigeria (15).

Assessment. Sound cost effective design. Main components standardized with light family. Suitable equipment and tools.

Greif ARV

4KH7FA-Pi AEV

SIBMAS armoured general purpose carrier

Models.

SIBMAS AIFSV with 90 mm Cockerill Mk III gun
SIBMAS ARV

Development and production.

1976 private venture by BN Constructions Ferrovières et Métalliques.
1978/79 trials in Malaysia.
1983/85 serial production for Malaysia.

Distinguishing features. Side. Large 6x6 wheeled vehicle with wide gap between front and centre axles. Boat-like front. Undercut rear. Turret at front. Engine left rear. Front. High sponsoned hull with slightly sloped sides. Raised driver's hatch central.

Other details. The stressed-skin hull consists of heavily angled armour plates. The vehicle is NBC-proof. The sides have weapon ports and armoured vision blocks. Two side doors, two at rear. ARV with 10.5 t crane and a 20 t winch. (TABLE)

Role. In service with Malaysia (162 AIFSV and 24 ARV).

Assessment. A relatively large vehicle.

SIBMAS AIFSV

JARARACA EE3 armoured scout car

Wt 5.8 t L 4.16 W 2.23 H 1.56 90 kW 100 km/h 700 km Crew 3

Models.

EE3 with 12.7 mm MG on ring mount
EE3 with ET-50 turret and ring-mounted 12.7 mm MG
EE3 with cupola on raised superstructure
EE3 with ATGM MILAN launcher

Development and production.

1979 prototype by ENGESA SA.
1985 model with raised superstructure in production.

Distinguishing features. Side. 4x4 vehicle with pointed bow, flat driver's front with protruding driver's compartment. Side door left, later right side. Engine rear with vertical rear wall. Front. Slightly raised driver's hatch in centre.

Other details. Mercedes-Benz OM 314A turbo-charged diesel engine. 11x20 tyres. 3 driver's periscopes, three roof hatches.

Role. In service with Brazilian reconnaissance units (70), Colombia (50), exported to Cyprus (40), Ecuador (10), Gabon (10), Iraq (300 delivered, status unknown), Uruguay (16).

Assessment. Very simple, versatile scout car.

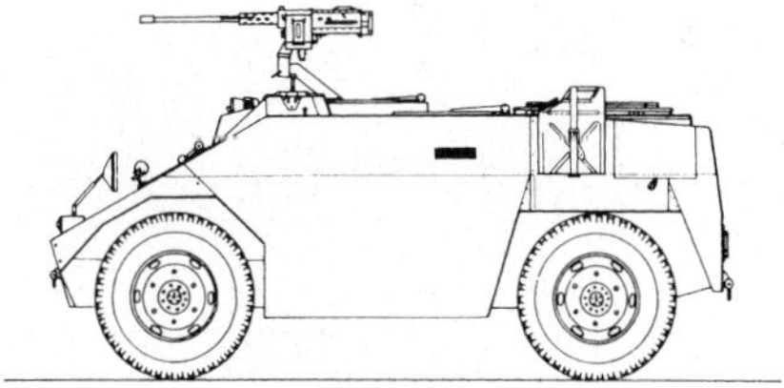

JARARACA ARSC

JARARACA with 7.62 mm MG

JARARACA with 12.7 mm MG

CRR CASCAVEL EE9 armoured reconnaissance vehicle
Carro de Reconhecimanto sobre Rodas

Models.

CASCAVEL 1 with 37 mm gun (ex M3 Stuart) rebuilt to CASCAVEL 2
CASCAVEL 2 with 90 mm gun D921A
CASCAVEL 3 with 90 mm gun EC90/1-36 in turret ET90-1
CASCAVEL 4 with 90 mm gun EC90/1-36 in turret ET90-2
CASCAVEL 5, as 4 with Mercedes-Benz OM 352A diesel engine
CASCAVEL 6
CASCAVEL 7 with Allison Transmission MT-643

Development and production. Developed by Engesa Engenheiros Especializados SA in parallel with CTRA URUTU APC under contract from the Department of Research and Technical Training, starting July 1970, prototype December 1971, production contract 1971.
1972/73 ten pre-production vehicles, slightly shorter.
1974 start of serial production.
1977 production of CASCAVEL 2 for export with French turret.
1979 CASCAVEL 3, manufactured under licence, Cockerill gun.
1980 CASCAVEL 4 with General Motors diesel engine 6V-53N.
1988 order from Venezuela.

Distinguishing features. Box-shaped sponsoned hull in rolled steel plates with sloped front and vertical rear. 6-wheeled vehicle with wide spacing between front and centre axles. Exhaust with large silencer on rear. Long flat turret with large bustle. Near-horizontal glacis. Driver left front with vision flap. Rear engine.

Other details. Bimetal welded armour plate integral unibody structure, thermal and acoustic insulation. Three glass bullet and splinter proof periscopes on driver's roof hatch. Rear towing pintle hook and four lifting eyes, two jerry cans. Rear suspension by leaf springs and double action shock absorbers. ENGESA articulated driven axle "Boomerang" type, with hypoid angular transmission, bevel differential gears, side walking beams with helical gears and differential locking systems. CASCAVEL 4 with central tyre pressure system. Several vehicles were equipped with different types of fire-control systems (OIP of Belgium, Marconi Command & Control Systems, UK). (TABLE)

Role. In service with the Brazilian Army (409 plus 6 Marines), exported to Bolivia (30), Burkina Faso (22), Chile (30), Colombia (127), Cyprus (126), Ecuador (12), Gabon (14), Guayana (6), Iran (40), Iraq (250 delivered), Libya (370), Nigeria (75), Paraguay (30), Qatar (20), Suriname (6), Thailand (56), Togo (36), Tschad (8),Tunisia (18), Uruguay (15), Zimbabwe (90).

Assessment. A notable step forward for Brazil. A well-designed vehicle, but it requires a special flotation kit.

CASCAVEL 2 AVR

CASCAVEL 4 AVR

CASCAVEL 7 AVR

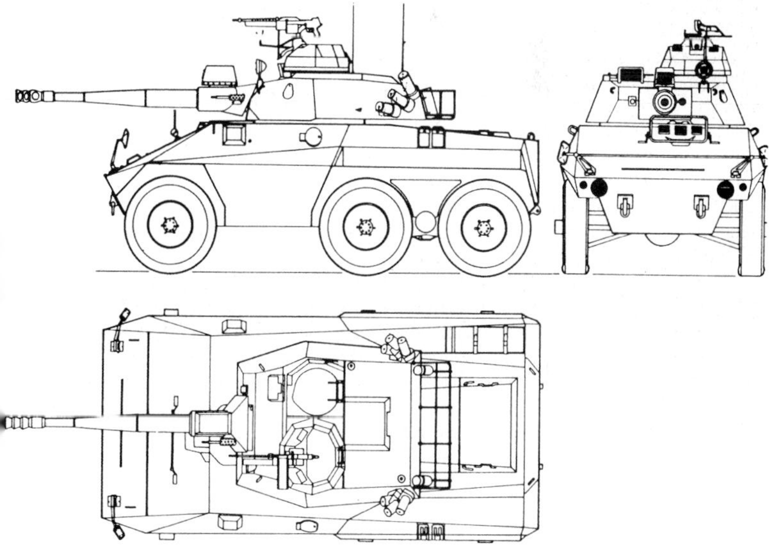

CASCAVEL 5 AVR

CTRA URUTU EE11 armoured general purpose carrier
Carro de Transporte sobre Rodas Anfibio

Models.

URUTU 1 APC (128 kW Mercedes-Benz engine, manual transmission)
URUTU 2 (140 kW Mercedes-Benz engine, manual transmission)
URUTU 3 (156 kW General Motors engine, automatic transmission)
URUTU 4 (URUTU 2 engine, automatic transmission)
URUTU 5 (191 kW Detroit diesel engine)
URUTU with 20 mm turret HS 804
URUTU with ET-20 turret
URUTU with ET-25 turret
URUTU with ET-90 turret
URUTU AMC with 60 mm mortar
URUTU ACCC
URUTU AAMBV
URUTU Cargo Carrier
URUTU ARV

Development and production.

1970 development starts by the Directoria de Pesquisas e Ensino Tecnico with the firm ENGESA, production contract 1972.
1973 URUTU 1 enters production for Brazilian Marines.
1980 URUTU 2 in production for Brazilian Army.
Production for export starts.
1988 order from Venezuela.

Distinguishing features. Six-wheeled vehicle with wide spacing between front and centre axles. Box-shaped rolled steel superstructure with prow front and splashboard. Angled rear with doors, driver front left side. Engine front right. URUTU 5 with larger engine compartment. Two plastic air intake tubes on each side of upper hull swung upwards in swimming mode. Water propulsion by two propellers (not for URUTU 2).

Other details. Rear suspension same as CASCAVEL. Two side doors (on right side omitted on URUTU 5), four hatches for troops, one for the driver and the gunner each. Rear door is pneumatically operated by the driver. Three driver's periscopes, two vision blocks and seven firing ports for troops. Front trim board erected pneumatically. NBC system optional.
ET-20 and ET-25 are turrets developed by ENGESA, armed with 20 resp. 25 mm Oerlikon-Contraves cannons and 7.62 mm coax MG.
(TABLE)

Role. In service with Brazil (175 plus 5 marines), Bolivia (24), Chile (30), Colombia (76), Cyprus (10), Ecuador (18), Gabon (12), Guayana, Iraq (500 delivered, status unknown), Jordan (50+), Libya (118), Morocco, Paraguay (10), Saudi Arabia (100), Suriname (15), Tunisia (18), UAE (30), Uruguay (10), Venezuela (138 + 25 Marines) and others.

Assessment. A significant development for Brazil. Novel features are the vehicle's excellent manoeuvrability in water and its speed. The vehicle can operate in rough sea with waves up to 3 m high. Efficient range, poor firepower.

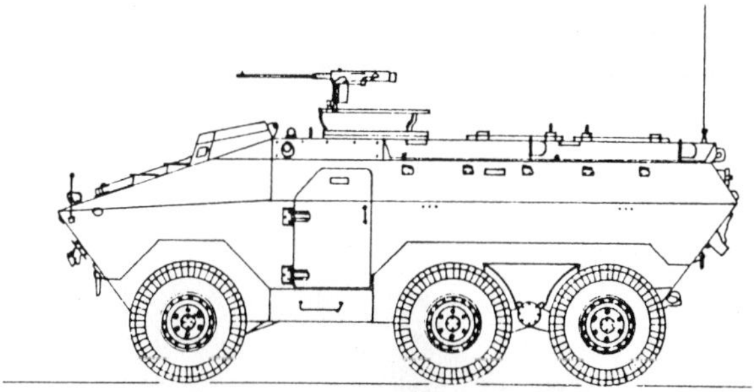

CASCAVEL 2 AVR

URUTU 5 APC

URUTU with ET-25 turret

URUTU with ET-90 turret

URUTU AMC

URUTU AAV

URUTU ARV

VTP-1 ORCA armoured general purpose carrier

Wt 18 t L 7.84 W 2.5 H 2.5 191 kW 120 km/h 1200 km Crew 2+16

Non-serial versions.

VTP-1 with 120 mm mortar
several proposed configurations incl. anti-tank, antiaircraft, artillery ammunition carrier, ambulance, maintenance version

Development and production.

1983 developed under private venture by Industrias Cardoen SA.
1985 production starts for the Chilean Army.

Distinguishing features. Side. 6x6, large distance between first and second axles. Engine front, driver and commander centre. Open top. Side door with hinged shutters, two firing ports. Front. Lower bow with grills. Two rear doors.

Other details. General Motors 6V-53T 6-cylinder water-cooled diesel engine coupled to Allison MT-653-R automatic transmission. 14.00 x 20 tyres. Armour protection 6 - 16 mm.

Role. In service with Chilean Army (100).

Assessment. Largest 6x6 wheeled multi-purpose vehicle, lightly armoured.

ORCA APC

Type 59/69/79 main battle tank

Models.

Type 59 with 100 mm gun (licence version of Soviet T-54 MBT)
Type 59-II with 105 mm gun (P-70 in Pakistan)
Type 69 with 100 mm smooth-bore gun
Type 69-II with original 100 mm gun
Type 79 with 105 mm gun
Various gun/howitzer installations on T-59 chassis (North Korea, status uncertain)

Non-serial versions.

Type 59 with different retrofit kits.
Jaguar MBT (joint venture with US Cadillac Gage Textron).Marksman AADG turret (Marconi CCS) on T-59 chassis.

Development and production.

1956 licence production of USSR T-54 starts.
1982 Type 69 enters production, upgunning of Type 59 with 105 mm gun.
1984 first appearance of Type 69-II and Type 79.

Distinguishing features. All MBT models very similar to USSR resp. Russian T-54 (see there). Minor internal differences. From 1980 all models fitted with IR searchlight, later with laser rangefinder on centre of gun mantlet. 105 mm gun presumeably copy of L7 Royal Ordnance gun. 100 mm gun of Type 69 longer with bore evacuator moved to the gun muzzle. Type 69 all with skirts, refitted to Type 59, too.

Other details. Type 69 equipped with TSFCS (Tank Simplified Fire Control System), gun stabilization, TLR 2 LASER, TGS-C gun sight with spot injection unit and Type 12150L-7BW V-12 diesel engine of 467 kW.
Type 79 with towing eyes under the nose and large sight window in front of the commander's cupola. The tank has a weight of 37.5 t and has an improved day/night main sight.
Several North Korean SP gun mounts on Type 59 chassis have been spotted with 122 mm, 130mm, 152 mm and 175 mm guns/howitzers between 1974 and 1985, also exported to Iran.

Role. In service with the Chinese Army (6000+), Type 59 also with Albania (15), Bangladesh (20), Burma (100 Type 69), Congo (14), Iran (150), Iraq (several hundred), North Korea (175), Pakistan (1200, local upgrading program with 105 mm gun started designated P-70, P-90 as a Type 69 upgraded version in production), Sudan (10 Type 59), Tanzania (26), Thailand (30 Type 69 and 53 Type 69-II), Vietnam (100), Zimbabwe (35 Type 69, 8 Type 59) and Zaire (16).

Assessment. By now obsolete, only useful in third world conflicts.

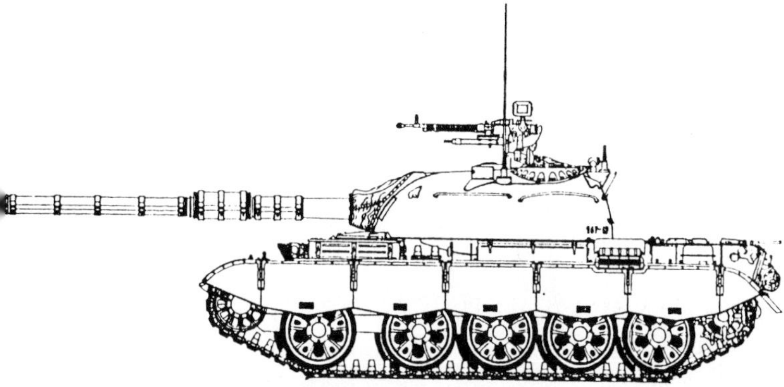

Type 59-II MBT

Type 59-II MBT

Type 69-II MBT

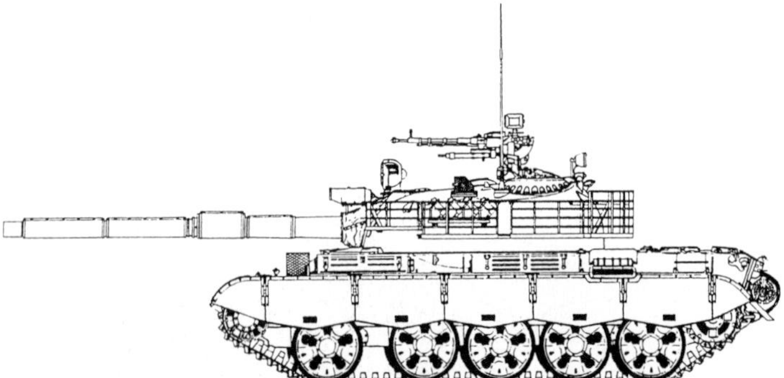

Type 79 MBT

175 mm SP gun on Type 59 chassis (North Korea)

Type 80/85/90 main battle tank

Models

Type 80
Type 80-II
Type 85-II
Type 85-IIA
Type 85-IIM
Type 85-IIAP (Pakistan)
Type 90-II (P-90 in Pakistan)

Non-serial versions

Type 85-III
Type 90-I with 105 mm gun

Development and production.

1978 studies on a follow-on generation tanks starts
1984 first prototype completed called Type 60-III, later Type 80
1988 production start of Type 80 and Type 80-II shortly after
1991 prototype of Type 90 revealed
1992 Type 85-IIM/85-IIAP starts serial production

Distinguishing features. Side. Hull shape similar to Type 79 except of the chassis which has 6 road-wheels and three support rollers (covered by side skirts). Distinctive gaps between the first two wheel stations. Fuel tanks on top of the track guards. Type 80 still has the Type 69/79 cast turret; gun tube with central fume extractor. Type 85/90 tanks with different turret shape, sloped front and rear bustle. Front. Type 80 is identical to Type 69/79, Type 85/90 with dual rubber pad tracks. Turret with steep side walls.

Other details. Type 80/80-II/85-II/85-IIA still armed with 105 mm rifled gun with 46 rounds. Torsion bar suspension with hydraulic shock absorbers on the first two and last two wheel stations. The Type 80 has an individual NBC protection system, the other versions a collectice one. Type 85-IIA is slightly heavier (39.5 t) than Type 85-II and has a combat load of 44 rounds. Type 85-IIM/AP and Type 90 are equipped with a 125 mm smooth-bore gun and automatic loader. Welded turret design with the use of composite armour. Early type still with external laser range finder on the gun mantlet later integrated in the main sight. 125 mm tanks with cross-wind sensors on top of the rear turret roof. Later versions with automatic transmission. Type 90 with increased armour protection including spaced armour track skirts in the front part and a stronger engine (882 kW) in contrast to 537 kW resulting in a higher weight of 48 tons. All versions can ford up to 5 metres depth using a snorkel attached in front of the commander's cupola.

Role. Type 80/80-II/85/90 in service with China, 200 Type 85-IIAP and Type 90-II also delivered to Pakistan.

Assessment. Considerable improvement over Type 69/79 series according to mobility. Type 85/90 comparable to Russian T-90.

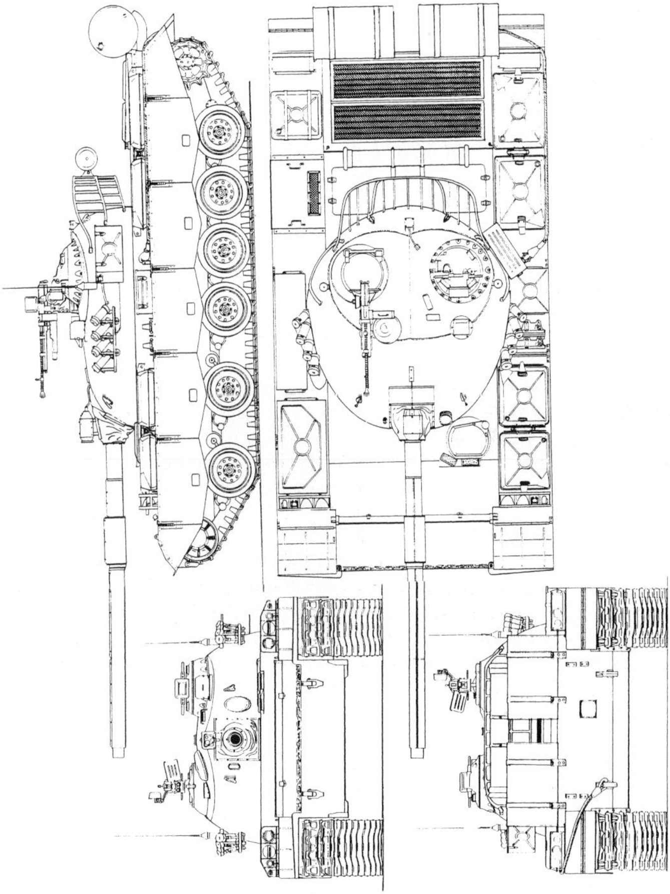

Type 80 MBT

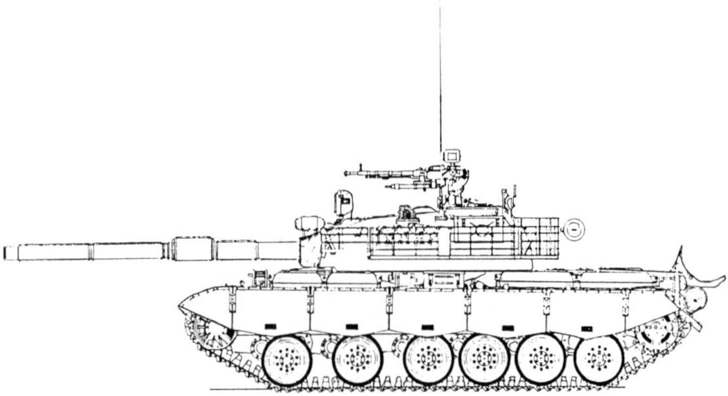

Type 80-II MBT

Type 80-II MBT

Type 85-II M MBT

Type 90-II MBT

YW 531 armoured general purpose carrier

Models.

YW 531 A
YW 531 B
YW 531 C
YW 531 D
YW 531 E
YW 531 H
WZ 701 ACCC
YW 701 A ACCC
YW 702 ACCC
Type 54-1 AAH
Type 70-1 AAH
Type 70 AARL
YW 304 AIMC with 82 mm mortar
Type 82 AIMC with 120 mm mortar
YW 750 AAV
YW 751 AAV

Non-serial version.

YW 531 C TDM with 4-tube Red Arrow 8 AT missile launcher

Development and production.

1967 production start using Soviet assemblies manufactured under licence. Early designation K-63. Production was terminated in the late 80s.

Distinguishing features. Side. Four large 'wafer' roadwheels each side (similar to Soviet light armoured vehicle family). Engine in front third of vehicle. Nose slopes back under front hull bow. Vertical rear. The main lower front plate is vertical; behind it is a near-horizontal glacis plate or engine deck with a hatch in the middle. Driver front left, commander front right. Additional all-round vision cupola behind driver. Over the middle of the rear crew compartment is a circular opening with a MG ring mounting. Two other hatches over the rear part of the crew compartment. Front. Hull sponsoned to almost full width. Folded splashboard on engine deck.

Other details. Initial version (and WZ 701) have Type 6150L water-cooled engine with 192 kW, later models powered by BF8L 413F water-cooled diesel engine with 235 kW. YW 531 A and B have one firing port on left side of rear compartment, YW 531 C and H two ports. YW 531 C has two roof hatches behind the driver's hatch, older models only one. The YW 531 H has the oblong roof hatches above the rear compartment elongated. This model is equipped with a circular gun shield with open top, probably retrofitted to older models, too.

Command post and ambulance vehicles have a box-like superstructure (WZ 701 is 2.2 m high and has a weight of 13 t, YW 750 has a weight of 12.8 t, the YW 751 has a new cdr's cupola with a 12.7 mm ADMG (500 rounds) and provides space for a two-men medical crew and four stretchers or eight seated patients. Its weight is 13.8 t).

Mortar carriers with big roof hatch.

Type 54 AAH with simple mounting with frontal gun shield but without armoured cabinet. Combat weight is 15.3 t, its height is 2.69 m. The 122 mm (copy of the Russian M-30) howitzer has an elevation of -2.5 to +63 degrees. Its traverse is 22.5 degrees to both sides. Ammunition stowage is 40 rounds and 1000 for the MG. The howitzer is operated by a crew of seven.

The Type 70-1 is has an almost identical arrangement but uses a chassis which has five roadwheels and three return rollers.

Type 70 MRL with 130 mm 19-tube multiple rocket launcher has a weight of 14 t, a height of 2.63 m and a crew of 6.

All models amphibious, propelled by the tracks, later models with air-ventilation.

(TABLE)

Role. In service with Albania (103), China (5000), Iraq (approx. 500), North Korea (500, also modified license version NKN M-1973), Pakistan, Sudan (40), Tanzania (20), Thailand (410 YW 531H), Vietnam (400), Zaire (12) and Zimbabwe (30).

Assessment. Relatively simple design more to be classified as an APC rather than an AIFV, since the crew has only limited facilities for fighting mounted from under armour. Good cross-country mobility.

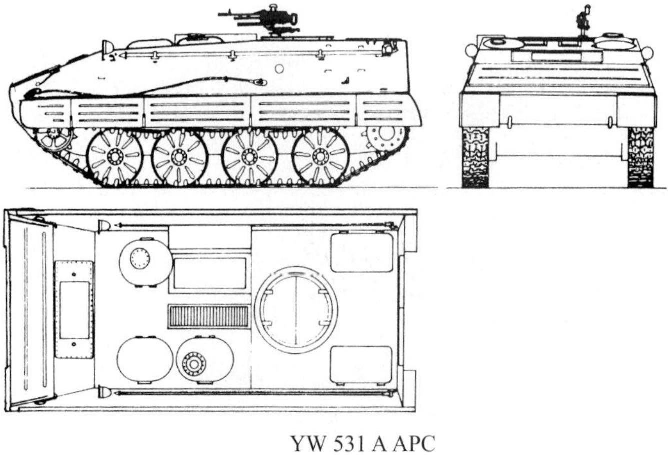

YW 531 A APC

YW 531 A APC

YW 531 C APC

YW 531 H APC

YW 701 ACCC

YW 750 AAV

Type 70 AARL

YW 534 armoured general purpose carrier

Wt 14.3 t L 6.15 W 3.13 H 2.56 235 kW 65 km/h 500 km Crew 15

Models.

YW 534
YW 307 AIFV
Type 85 AIFV
Type 85 AICV
Type 85 ARV
WZ 751 AAV
Type 85 AAH
Type 85 AARL

Non-serial versions.

YW 309 with copy of BMP-1 turret
NVH-1 AIFV
NVH-4 AIFV
YW 534 with quadruple Arrow 8 ATGM launcher on raised hull
Type 85 AMC with 120 mm breech-loaded mortar
Type 90 AFV family (improved Type 85).

Development and production.

1985 enlarged version of YW 531 issued by NORINCO.

Distinguishing features. Side. Five roadwheels with different gaps, three return rollers (the rear ones closer together), sprocket front, side skirts. Layout similar to YW 531. Firing ports in sides and rear. Front. Sloped upper sides. High bow. Driver left, engine right front. AICV and AAV with box-like superstructure.

Other details. 12.7 mm MG with allround shield resp. 25 mm gun turret (YW 307) rear of centre. YW 307 with gun in oscillating mounting, height 2.93 m. Weapon elevation is from -12 to +52 degrees. The YW 307 has a weight of 15.4 t and has a crew of 10 soldiers. YW 534 has an ammunition stowage of 1100, the YW 307 of 400 rounds 25 mm and 1000 7.62 mm MG rounds. Amphibious, driven by tracks at maximum speed of 6 km/h. This family has a NBC system and driver's IR night driving equipment. Engine is the Deutz BF8L 413F air-cooled diesel.
The AAH is equipped with the Chinese copy of the Russian 122 mm D-30 howitzer mounted in an open superstructure protected with a front shield. The combat weight is 16.5 t, its overall height 2.85 m. The ammunition stowage is 40 rounds, the howitzer is operated by a crew of six.
The multiple rocket launcher is equipped with 30 130 mm Type 85 rockets. The combat weight is 14.5 t, its height in travelling position is 3.02 m.

Role. In production for the Chinese Army. Type 85 AIFV in service with Sri Lanka (18 + 1 AICV + 2 ARV), Myanmar (150) and Thailand (1400) Thailand also operates 20 130 mm multiple rocket launchers.

Assessment. Slightly improved standard compared with the YW 531.

YW 534 APC

YW 307 APC

Type 54-1 AAH

Type 85 AARL

WZ 523 (Type 86) armoured personnel carrier

Wt 11.2 t L 6.02 W 2.55 H 2.73 121 kW 80 km/h 600 km Crew 12

Development and production.

1984 first appearance. Entire Chinese product.

Distinguishing features. Side. 6x6 vehicle with short, high bow and long box-shaped hull and vertical rear. Second and third wheel closer together. Three firing ports covered by shields in each side. Small side door behind forward compartment. Partially armoured ring mount top centre. Front. High bow with trim vane mounted on glacis plate. Three front windows with shutters hinged at top. Driver seat in centre.

Other details. Main armament is a 12.7 mm MG with 600 rounds ammunition stowage. Two forward oval hatches, two large hatches over troop compartment, one rear door opening to right. Ten infantry soldiers are carried in the rear compartment. EQ6105 petrol engine centre right with exhaust pipe right hull edge. Fuel capacity is 255 litres. In water the amphibious vehicle is propulsed by two propellors at rear. Speed in water is 7 km/h

Role. In service with the Chinese Army and Sri Lanka (12).

Assessment. Simple design based on standard truck elements. Good road and water mobility, firepower marginal. Vehicle could provide basis for many variants.

WZ 523 APC

WZ 551 armoured personnel carrier

Wt 15.3 t L 6.65 W 2.8 H 1.95 188 kW 85 km/h 600 km Crew 3+9

Models

WZ 551 APC
WZ 551 APC with 23 mm cannon
WZ 551 (4x4) with 4 Red Arrow 8 ATGM'
WZ 551 AAV

Non-serial versions.

WZ 551B with 14.5 mm turret
NGV-1 with 25 mm Giat Dragar turret
WZ 551 with a 25 mm cannon in the oscillating turret
WZ 551 with WZ 501 73 mm turret
WZ 551 with one-man turret with 4 PL-9 ATGM
WZ 551 (8x8) AAH with 122 mm howitzer

Development and production.

1986 first prototype with roof-mounted 12.7 mm MG issued.

Distinguishing features. Side. 6x6 vehicle with high front part, almost vertical glacis, splash-board folded on almost horizontal front plate. Flaps in rear fighting compartment. High one-man turret with externally mounted gun in a drum-like assembly or pintle-mounted machine-gun. Exhaust pipe on upper hull end on left side. Front. Slightly tapered front part. Horizontal angle bow plate. Two large vision flaps in front plate. Rear. One large door with vision block and firing port flap. Two propellers in drum-shaped housing on either side under the vehicle.

Other details. 23 mm cannon with quick barrel-change handle. Flash eliminator on muzzle. Two roof hatches in the front for the commander and the driver, four oblong ones on top of the crew compartment. Air-cooled Deutz BF8L413F V-8 diesel engine. First and second axle steerable. Ambulance version with a raised super-structure.

Role. In service with the Chinese Army and the army of Bosnia-Herzegovina.

Assessment. Performance comparable to the French VAB APC. Only minor armour protection against small-arms fire.

WZ 551 APC

Type 83 armoured artillery howitzer

Wt 30 t L 7.0 W 3.24 H 3.5 382 kW 55 km/h 450 km Crew 5

Models.

Type 83 AAH
Type 89 TDG
Type 89 AARL
Mine-clearing rocket launcher
Trench Digger

Non-serial versions.

Type 83 AAH with 130 mm field gun 51-1
Type 83 AAH with 155 mm howitzer
Type 83 AARL with 2x152 mm GSL 111 rocket launcher

Development and production.

1984 first appearance.
1989 production of 120 mm TDG starts.

Distinguishing features. Side. Six roadwheels, two centre pairs close together, front and rear wheels with destinct gap. Drive sprocket front, three return rollers. Sloped engine front cover. Large turret with slightly sloping front and rear walls and sloping rear part of roof. Howitzer in drum mantlet with bore evacuator and large double baffle muzzle brake. ADMG on front turret roof in open ring mount. Front. Grill at right front plate, sponsoned hull, driver on the left. ADMG right, periscope on left turret roof. Large hull rear door.

Other details. The AAH is armed with the 152 mm Type 66 howitzer (30 rounds stowage). A 12.7 mm ADMG, a 7.62 mm MG and a Type 40 rocket launcher are fitted. The howitzer elevates from -5 to +65 degrees and fires projectile up to 17 km. The engine is of the 12150L diesel type. Fuel capacity is 885 litres.
120 mm TDG mounts an MBT-like turret with steep front and side plates. The smooth-bore gun has about L-44 calibre length and has a fume extractor near the tube middle.
The Type 89 multiple rocket launcher has a 4x10 rounds box-like reload magazine in the front vehicle section and the 4x10 traversable launcher behind of it. The 122 mm rockets are of different types (HE fragmentation, HE incendiary, bomblet or minelaying). The range of the latter is 7 km, the other have a range of 20 km maximum. Recently extended-range versions were offered with a range of up to 30 km. The vehicle's combat weight is 29.9 t. The height in travelling configuration is 3.18 m. Late production vehicles feature an armoured cab on the left hand side of the launcher for the layer.

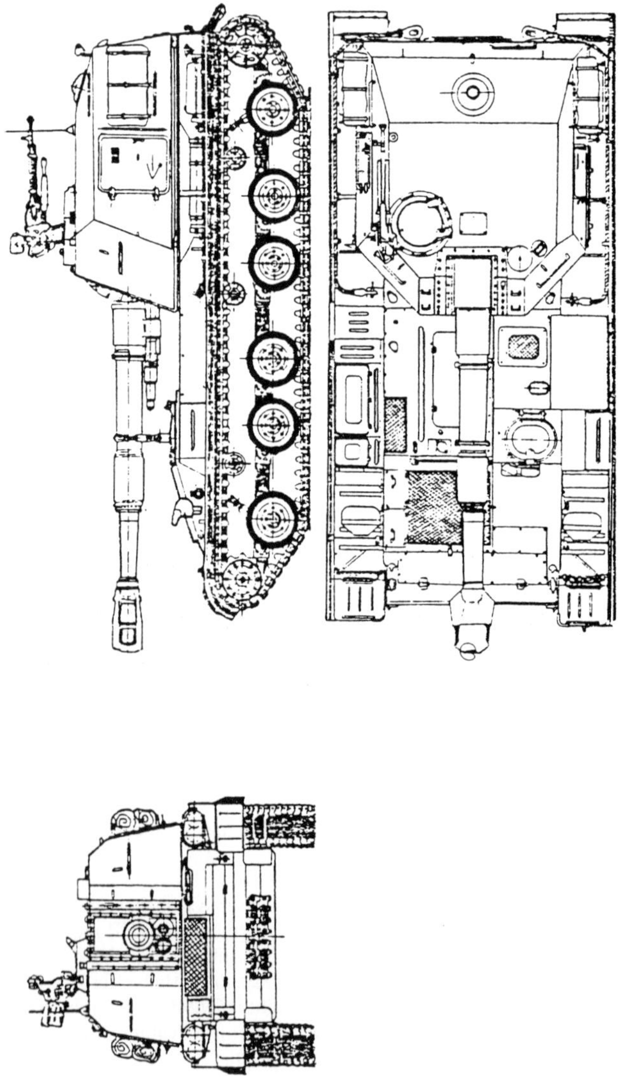

Type 83 AAH

The rocket launcher has a low turret carrying launching rails for two 425 mm Type 762 mine clearing rockets with a weight of 600 kg and a range of 800/1000 m. Combat weight is 27 t. Vehicle's length is 7.18 m.
The trench digger has a front-mounted cab and a rear-mounted trench digging equipment similar to the Soviet MDK-2. Its combat weight is 30 t, length is 10 (operating) and 8.07 m (travelling). Its width is 3.6 m, operating height is 4.2 m.

Role. In service with the Chinese Army.

Assessment. AAH and TDG of average standard. First large armoured vehicle family of indigeneous Chinese construction.

Type 83 AAH

Type 83 TDG

Trench Digger

Type 80/88 armoured air defence guns

Wt 31 t L 8.42 W 3.31 H 2.75 426 kW 50 km/h 420 km Crew 6

Models.

Type 80 AADG
Type 88 AADG

Development and production.

1980 Type 80 issued by NORINCO based on licence produced Soviet assemblies.
1986 issue of Type 88 AADG.

Distinguishing features. MBT chassis of Type 59 or Type 69. Type 88 has a box-shaped turret with angled upper sides. Type 80 very similar to Soviet outdated ZSU-57-2 but with five roadwheels and track skirts. Large open-topped turret with flat sides curved at top. Two long cannons with perforated muzzle brake.

Other details. Type 88 has a two man fully enclosed power operated turret armed with two 37 mm cannons of a cyclic rate of 380 rpm each at a MV of 1000 m/s, range is 4 km. Maximum traverse speed is 60ˉ/sec, maximum elevation speed is 50ˉ/sec, with elevation from -7ˉ to +87ˉ. Combat weight is 34.2 t. Type 80. Main armament consists of two Type 59 57 mm cannons used either in a semi-automatic electro-hydraulic mode or manually with -1 to +81 resp. -5 to +85 degrees elevation. The guns have a cyclic rate of fire of 105 to 120 rounds/min each. Effective anti-aircraft range is 5.5 km. A total of 300 rounds is carried.

Role. In service with the Chinese Army (Type 80) and Thailand (25 Type 88).

Assessment. Fair-weather weapon of very limited capability against high performance aircraft.

Type 80 AADG

Type 88 AADG

Type 84 armoured vehicle launched bridge

Development and production.

1984 beginning of development by NORINCO.
1986 first prototype issued.

Distinguishing features. Chassis of T-69 MBT with a bridge launching mechanism similar to the German BIBER AVLB. The 18 m bridge extends horizontally in two parts spanning a gap of about 16 m. Vehicles up to a weight of 40 tons may cross the bridge.

Role. Service use in the Chinese Army, several delivered to Thailand.

Assessment. Modern system reducing the bridgelayer's risk being spotted when in operation as in the case of vertically folding scissor-type model.

Type 84 AVLB

Type 653 armoured recovery vehicle

Wt 38 t L 7.3 W 3.3 H 3.0 426 kW 45 km/h 420 km Crew 3

Development and production.

1985 issued and produced by NORINCO.

Distinguishing features. T-69 MBT chassis with box-like crew compartment and telescopic jib crane on right side which lays back alongside hull when not in use. Front mounted dozer blade.

Other details. The crane has a lifting capacity of 10 t maximum to be traversed through 360°. Maximum lifting height is 6 m. It has a hydraulic front mounted winch with a nominal capacity of 35 t on a single cable and a maximum recovery capacity of 70 t. The dozer can excavate 100 m3 of earth per hour.

Role. In service with China and Thailand.

Assessment. Not up to standard of modern ARV.

Type 653 ARV

LOV armoured general-purpose carrier

Wt 9.2 t L 5.89 W 2.39 H 2.1 96 kW 100 km/h 800 km

Models.

LOV-IZV AVR
LOV-ABK ANBCVR
LOV-OP APC
LOV-Z ACCC
LOV-UP1 artillery forward observer vehicle
LOV-UP2 gun fire command post
LOV-ED electronic warfare carrier
LOV-RAK AARL

Development and production.

1992 development contract for the Torpedo company.
1993 first production vehicles issued by RH ALAN.
1997 modernized hull shape with bullet-proof windshields for both driver and codriver.

Distinguishing features. Side. Two small wheels on each side with coffin-like superstructure. Pointed front with horizontal roof, straight rear with negative angle. Small door in the front section above front wheel. Front. Slightly sloped glacis. Upper hull sides slope inwards. Raised driver's cabin on left side.

Other details. The LOV is based on the chassis of the 4x4 7 ton truck Torpedo HV. The vehicle has a hull of all-welded steel armour. The APC version has a rear compartment for 10 soldiers. They may fight from on-board either using two roof hatches or a firing port on both hull sides resp. in the left aft door. Armament consists of a roof-mounted 12.7 mm M2 MG in an open mount only providing lateral and rear protection.
The recce version is operated by a crew of four and has additional wide-range radio equipment.
The command post vehicle has a six-man crew and also additional radio systems.
The AARL consists of a 8.5 t vehicle with its hull rear cut down and fitted with a 24 round 128 mm multiple rocket launcher. It can be traversed a full 360° arc. The rockets have a range of 8.5 km, in an extended range version 13.5 km. An additional 24 rockets are stored in the rear section.
All vehicles have a 50 kN winch in the front and a central tyre pressure adjustment system. The engine is a Deutz BT6L 912S turbo-charged 6 cylinder diesel coupled to a ZF Z5-35S manual transmission with 5 forward and 1 reverse gears.
A slightly enlarged version LOV-2 has been reported as a follow-up model.

Role. In production for the Croatian Army.

Assessment. Low-cost carrier concept; APC role only in low-intensity scenario due to low level of protection.

LOV-IZV AVR

LOV-OP APC

LOV-RAK AARL

LOV 2-Z ACCC

OT-64 (Obrnemy Transporter) armoured personnel carrier
Polish: SKOT (Stredni Kolovy Obojzvelny Transporter)

Models.

OT-64 (SKOT)
OT-64A (SKOT-2 A)
SKOT-2 AP (OT-64C(2)
OT-64C(1) with OT 62 TOPAS 2 AP turret (Morocco)
OT-64R-2M/SKOT-WD R-2M batallion command carrier
OT-64R-3MT/SKOT-WD R3M regiment command carrier
OT-64R-4MT/SKOT-WD R4 division command carrier
SKOT-WD R-3Z ACCC
SKOT-WD R6 ACCC
SKOT-WD R-2AM AACPV
SKOT-WD S-260-ART artillery resupply vehicle
OT 64ZDR AAV
SKOT-WI S-260-INZ AMLV
WPT ARV
DTP-64 ARRC

Development and production.

1964 production start of indigenous Czech development based on TATRA T813 8 t truck.
1980 end of series production.

Distinguishing features. Side. Eight-wheeled vehicle with large centre distance between two middle axles. Large coffin-like superstructure with high front and vertical rear. Driver's hatch proud of hull. Front. Upper part of sides slightly sloped. AICV with heightened fighting compartment and four antenna bases. ACCC with rail antenna.

Other details. Roof hatches over crew compartment at rear. Two front axles steerable. Water propulsion by two propellers. NBC collective protection/ventilation system can be fitted. Semi-automatic Wilson gearbox. Tyre pressure regulation system. SKOT-2 A and SKOT-2 AP with turrets on small octogonal fixed base, the latter one with rounded turret front.
The command versions differ in radio equipment.DTP (Dilna Technické Pomoci) ARV with 0.8 t crane.
(TABLE)
Command and control versions with different radio equipment according to command level.

Role. In service with Cambodia (30), Czech Rep. (100), Iraq, Libya (50), Poland (734), Slovakia (33), Sudan (35), Syria (100) and Uruguay (60).

Assessment. Large AIFV for carriage of two sections. No facilities for crew to fight under full armour protection. Awkward arrangement of hatches. Retrospective fitting of turret has improved firepower.

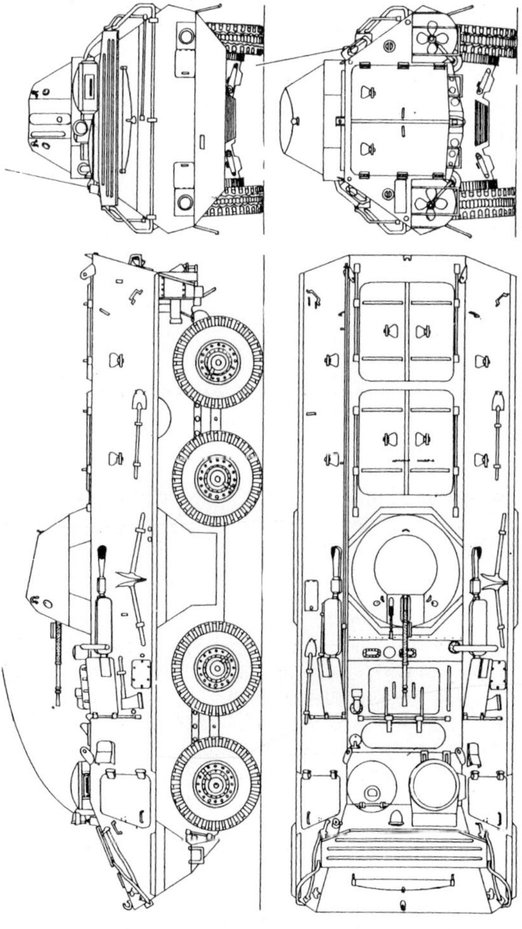

OT 64 SKOT-2AP APC

OT-64 APC

OT-64A APC

OT-64 R-2M ACCC

OT-64 R-3MT ACCC

OT-64R-4MT ACCC

SKOT-WD R-2AM ACCC

VZOR 77 152 mm DANA self-propelled artillery howitzer
Delo Automobilny Nabijene Automatkly

Models

DANA AAH
ZUZANA 155 mm AAH (Slovakia)

Non-serial versions.

DANA with L45 or 47 barrels
STROP AADG with twin 30 mm cannons 2A38
Polish Radwar N-22 radar on DANA chassis

Development and production.

1979 first appearance. Modified TATRA-815 chassis of 10 t truck with partially armoured superstructure and turret.
1982 first prototype of a 155 mm version.
1989 development start of STROP AADG.
1997 troop-trials of STROP.

Distinguishing features. Side. 8-wheeled vehicle with stabilisers which swing out between front and rear pairs of axles and one in the rear. Low slightly sloped armoured frontˇdriver's cab with vertical rear. Armoured flaps with vision slits over windscreens. 2 hatches in roof. Welded turret with plates angled outwards (double trapezium configuration). Upper part of turret cleft for gun. ADMG and hatch for commander and detachment on top of turret and large square hatch on front of right half. Gun barrel projecting over driver's cab with rear third stepped and rectangular muzzle brake. Long rod-like linkage above breech extends forward half-way along barrel. Box-shaped rear with engine cooling air louvres at top and side. Spade centre rear. Front. Armoured turret with upper part protruding over driver's cab and extending slightly beyond chassis width. Trapezoidal cupola with vision slit on left half.

Other details. Steering is power-assisted on front four wheels with central tyre pressure regulation system for driver to adjust whilst on the move. Ordnance elevates from -30 to +60o to both sides. Turret traverse and gun elevation hydraulic with manual emergency controls.
(TABLE)
ZUZANA is equpped with a 155 mm gun firing NATO standard ammunition.

Role. In Czech (273) and Slovakian (135 ZUZANA converted from DANA) divisional artillery units, exported to Libya (80) and Poland (111).

Assessment. First AAH on wheeled chassis. Highly mobile, but ligthly armoured. 155 mm version can fire NATO standard ammunition.

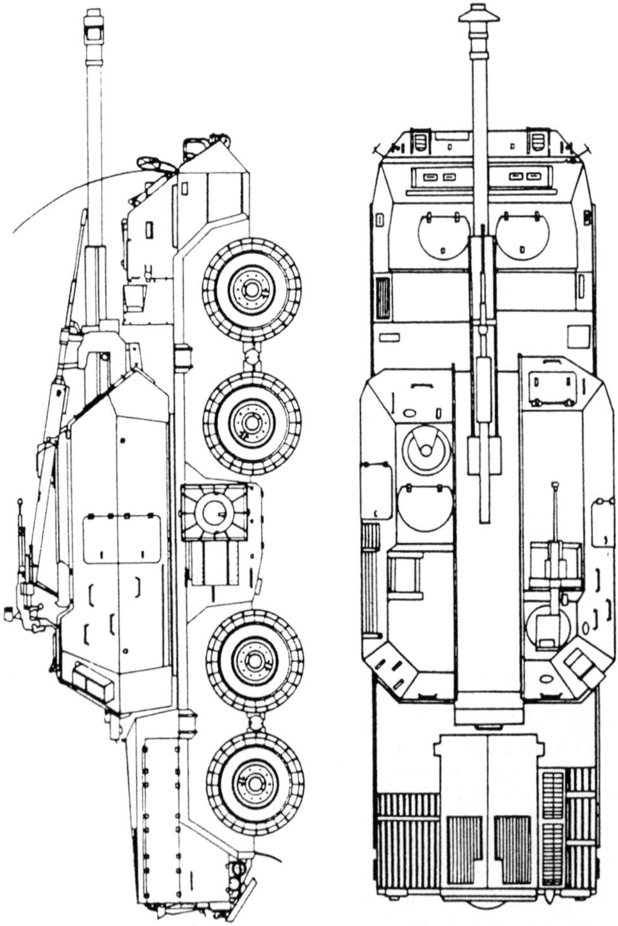

VZOR 77 AAH

DANA AAH

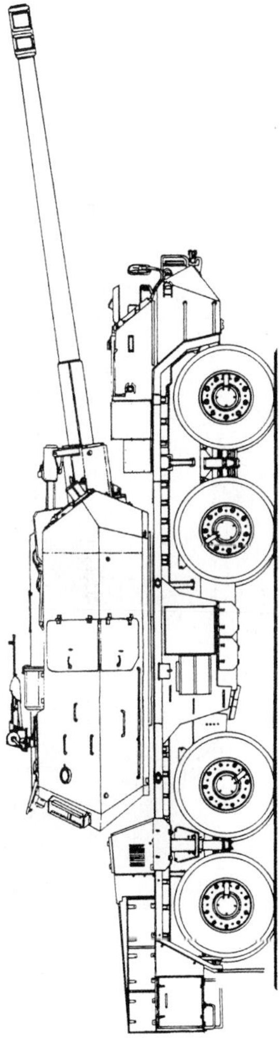

ZUZANA AAH

ZUZANA AAH

FAHD armoured general purpose carrier

Models.

FAHD APC
FAHD with SAMM BTM 208 K turret
FAHD with Creusot-Loire TLi turret
FAHD 30 with BMP-2 turret
FAHD anti-riot vehicle
FAHD ACCC
FAHD AAV
FAHD ARRV

Non serial-version

FAHD AMLV

Development and production.

1983 development of a WALID APC successor using the Thyssen-Henschel model TH 390 on Mercedes-Benz LAP 1117/32 truck chassis.
1985 start of series production for Egypt, export orders followed.
1993 upgrade program to FAHD 240 (HP of new engine).

Distinguishing features. Side. Two wheels on each side. Very large, box-shaped superstructure with slightly protruding upper bow. Side doors in front of forward axle. Horizontal roof and slightly sloped rear. Four firing ports and vision blocks in upper part of hull between the two wheel bases. Front. Grill low down in nose of vehicle. Cdr's and driver's front compartment with large windscreen which can be covered by two armoured shutters hinged at top. Hull sides sloped inwards starting from wheel line.

Other details. Front mounted DB OM-352 turbocharged 6-cylinder diesel engine with automatic transmission. Central tyre pressure regulation. NBC system and front-mounted 5 t winch optional. Wheelbase is 3.2 m, the turning radius 7 m; tyre size is 13 R20.
FAHD 30 mounts the BMP-2 turret with 30 mm cannon and has a weight of 12.5 t, a height of 2.71 m and a crew of 10.
The BTM 208 K turret is armed with a 12.7 mm and one 7.62 mm MG.
An upgrade package is offered resp. used by the Egytian Army including a new HP 750 Nm MD turbocharged engine with 181 kW.
The ambulance version is able to transport 4 sitting and 4 stretcher patients.
The repair version has a rear mounted 2.5 tons lifting boom and repair equipment.
The anti-riot version has a steel grader beneath the front bow and a small turret with changeable armament resp. smoke, tear gas launchers.

Role. In service with the Egyptian Army (158), in service also with Algeria (100), Kuwait (110 delivered, most of them lost), Oman (17) and Zaire (20)

Assessment. A typical example of an easy to manufacture, light carrier of limited cross-country ability but good road mobility. Useful especially for internal security purposes.

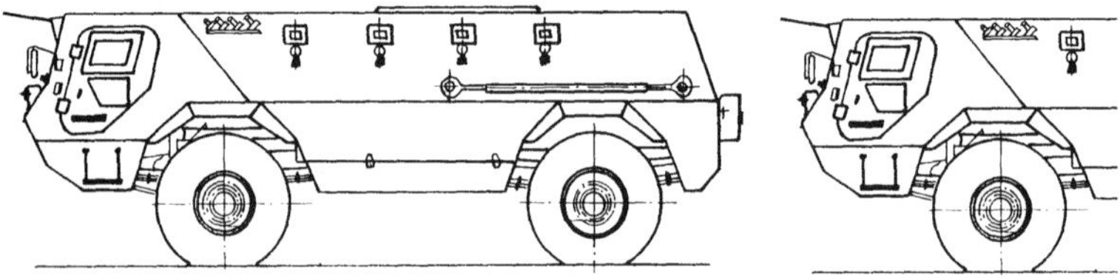

FAHD APC

FAHD APC

FAHD with Creusot Loire TLi turret

FAHD 30 APC

FAHD anti-riot vehicle

FAHD AARV

FAHD AAV

PASI armoured personnel carrier

Models.

XA-180 APC
XA-180 with 12.7 mm DShKM MG turret
PASI-83 with 12.7 mm NSVT MG turret
XA-185
XA-186 with Helio Mirror 7.62 mm MG turret
XA-188
XA-181 with CROTALE NG ADMS
XA-180 with Ericsson air defence radar
XA-180 ACPV
XA-180 AAV
XA-200 APC

Non-serial version.

XA-185 with LAV-25 turret

Development and production.

1982 prototypes for replacement of BTR-60 PB built by SISU and Valmet (Type 1912-6).
1983 SISU version (using many components of the SISU SA-150 VK (4x4) truck is selected and production starts (Finnish designation PASI-83).
1988 selection of CROTALE NG ADMS by Finland.
1995 delivery of XA-185 to Norway.
1996 competition won for the delivery of XA-188 to the Netherlands.
1999 Finnish order of 48 XA-200 vehicles to be delivered 2000/01

Distinguishing features. Side. 6x6 wheeled vehicle with equal distance between axles. Amphibious bow front, trim vane on nose top. Hull rear almost vertical with flat roof. Commander and driver at front, each with side door. Three vision blocks and firing ports on each hull side. Shrouded propeller under each hull side at rear (except XA-181, XA-186 and XA-188). Front. Hull sides almost vertical. Two windscreens covered by shutters hinged at top. Rear. Two doors, one with vision block and firing port.

Other details. Two forward axles steerable. Engine (Valmet 6-cylinder water-cooled turbocharged diesel) rear of driver at left side, air in- and outlet on roof. Over the top of the crew compartment are two roof hatches hinged in the centre to be locked vertical if required.
XA-185 version with improvements in the power train including an increased engine performance and new axles.
XA-186 (not amphibious) and XA-188 with modified interior and larger payload resulting in a weight increase of 700 kg. These two versions are provided with appliqué armour.
(TABLE)

The XA-2000 is the latest version in production. The Finnish vehicles are equipped with the Helio FVT800 one-man turret with 12.7 mm and 7.62 mm MG. The Valmet 612 engine has 202 kW. Total weight in 22 t. Vehicle height is 2.6 m, the range is 600 km.

Role. In service with the Finnish Army (485 + 20 with CROTALE NG), Denmark (20 XA-188), Ireland (2 in UN role), the Netherlands (92 XA-188). Norway (48 XA-186), Sweden (15) and Multinational UN detachments.

Assessment. Versatile, mobile APC useful for different applications.

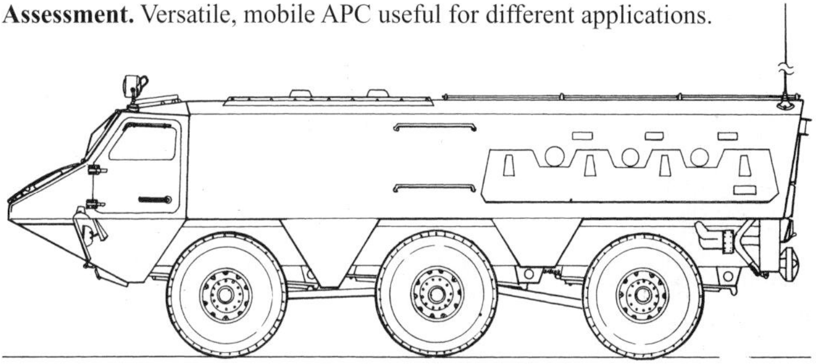

XA-180 APC

XA-180 with 12.7 mm DShKM MG turret

XA-185 APC

XA-186 APC

XA-181 with CROTALE NG ADMS

XA-180 AAV

VBL armoured reconnaissance scout car (ULTRAV-M11)
Véhicule Blindé Léger

Models.

VBL reconnaissance
VBL anti-char moyenne portee
VBL with PL 127 turret
VBL long (wheelbase)
VBL with 12.7 mm MG mount (Gabon)
VBL with TOW ATGM mount (Oman)

Non-serial versions.

VBL with different weapon stations and radar systems offered

Development and production.

1978 competitive development by Panhard and (later rejected) Renault for French Army requirement of an ultra-light multi-purpose reconnaissance vehicle below 3.5 t.
1984 prototypes available, studies for about 20 derivatives. First export order by Mexiko.
1985 VAB adopted by the French Army.

Distinguishing features. Side. Small 4x4 vehicle with flat forward engine compartment, slightly sloped driver's front and bulky crew compartment. Single forward-opening side door with window in upper part. Front. Flat bow, hull above wheel arches slopes inwards. Two large angled front windows.

Other details. Upper sides of the rear hull slope inwards. Large rear door with a window opens to the left. In the rear hull roof there is a single-piece circular hatch that opens to the right used by the operator of the MILAN ATGM launcher. Both commander and driver (left side) may use roof hatches. 9x16 combat tyres with pressure regulation system. NBC system standard for French Army vehicles. Fully amphibious within two minutes, propelled in the water by a propeller attached to the hull rear.
Standard armament application is a circular rail DARD II mounting with one 7.62 mm MG.
(TABLE)
Anti-tank version with MILAN ATGM launcher is slightly heavier (3.7 t) and has 6 missiles on board.
PL (Panhard Levasseur) 127 turret version introduced in the French Army; the Kuwaiti version retain the 7.62 m MG, the Omani turret has a rear extension housing MG boxes.
The long version is 21 cm longer (3.92 m length) with a wheelbase of 2.75 m. Gabon operates the VBL with a circular rail CTM 105 mount with 12.7 mm MG 600 rounds).

Role. In recce service in the French Army, 922 and 120 long wheel-base vehicles currently on order, most of them delivered.(Exported to Benin (10), Cameroon (5), Djibouti (7), Gabon (12), Greece (34), Indonesia (18), Kenia (20), Kuwait (20), Mexico (40 incl. 8 with MILAN), Niger (7), Nigeria (72), Oman (42 + 8 long wheelbase versions), Portugal (14 plus 4 with MILAN), Qatar (16), Rwanda (16) and Togo (2).

Assessment. Low price as a result of mass-produced components in standardization with commercial components. Low running costs. Adaptable to many applications. High mobility.

VBL AVR

VBL AC

VBL with PL 127 turret

VBL with PL 127 turret (Kuwait version)

VBL with PL 127 turret (Oman version)

VBL long

VBL with 12.7 mm MG mount

VBL with TOW ATGMS

ERC armoured vehicle reconnaissance
Engin de Reconnaissance Cannon

Models.

ERC 90 F4 Sagaie
ERC 90 F1 Lynx
ERC 20 Kriss (Gabon)
ERC 60-20 Serval
ERC Sagaie 2 with TTB 190 turret (Gabon)

Non-serial versions.

Offered with numerous weapon applications.

Development and production.

1975 private venture of Panhard aimed for export.
1978-1980 evaluation by the French Army and acceptance.
1984 delivery of the first vehicles to the French Army, presentation of Sagaie 2.

Distinguishing features. Side. 6x6 wheeled vehicle with large gap between first and second wheel, centre axle can be raised. Low, sloped glacis. Rear-opening door on the left side behind front wheel. Box-shaped wheel arches over centre and rear axles, sheet metal sides ribbed horizontally. Different turret and weapon arrangements. Sagaie 2 slightly longer. Front. Well sloped prow-front with driver's hatch in upper part. Large sheet metal coverings over front wheels.

Other details. GIAT 90 mm F4 gun in the TS 90 turret is able to fire APFSDS rounds with 1275 m/s, not possible with the 90 mm F1 gun of the Lynx.
ERC 90 F1 Lynx has the TJN 2.90 combined day/night periscope from SOPELEM and the TCV 107 LRF (CILAS-SOPELEM). A total of 21 rounds for the main gun is carried. Vehicle's weight is 7.6 t.
Hispano-Suiza Serval 60/20 turret is armed with a 60 mm breech/muzzle-loaded mortar, a 20 mm automatic cannon mounted externally at the rear of the turret and a coaxial 7.62 mm MG. The mortar is either of the HB 60 short or long range type with a range of 2600 m or 500 m in the direct fire mode resp. 5000 m or 1000 m. Ammunition stowage is 50 mortar bombs.
Sagaie 2 with two XD3T Peugeot turbo-diesel of 73 kW each and larger 13.00 x 16 tyres. The TTB 190 turret is offered with different options of armour thickness up to a total weight of 4.2 t, it has two different ammunition stowage arrangements in the turret bustle (35 rounds/13 for ready use or 32 rounds/8 in a drum magazine) and uses the P204 day/night sight and the LRS5 NG sight for the gunner.
(TABLE)

Role. In service with Argentina (12 with Lynx turret), Chad (4 Lynx), France (192 TS-90), Gabon (6 Sagaie 2 TTB 190 and 4 ERC 20 Kriss), Ivory Coast (7 with TS 90 turret), Mexiko (120 with Lynx turret), Morocco (20 LYNX), Niger (ERC 60/20 Serval), Nigeria (46 TS-90).

Assessment. Sound family of reconnaissance vehicles with anti-tank and supporting fire capability.

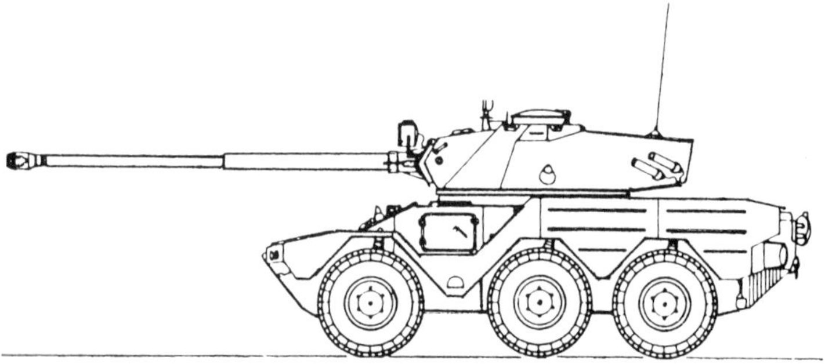

ERC 90 F4 SAGAIE AVR

ERC 90 F4 SAGAIE AVR

ERC 90 F1 LYNX AVR

ERC 20 KRISS AVR

ERC 60-20 SERVAL AVR

ERC SAGAIE 2

VBC 90 armoured vehicle reconnaissance
Véhicule Blindé de Combat

Wt 13.5 t L 8.135 W 2.5 H 2.55 173 kW 92 km/h 1000 km Crew 3

Development and production.

1978 prototype by Renault aimed for export.
1981 first production vehicles completed.
1985 production ends.

Distinguishing features. Side. 6x6 wheeled chassis with equal spacing based von VAB with very high hull and sloped glacis plate with protruding driver's cabin roof. Horizontal hull top with raised engine deck and vertical hull rear. Louvres on left side of hull at rear. Flat turret has a long-barreled gun with single baffle muzzle brake. Front. Prow-shaped front. Driver's bay with three windows. Turret with sloped sides, raised portion on the left. Commander's cupola on the left turret roof.

Other details. The GIAT TS 90 turret uses the SOPTAC 11 fire-control system with CILAS TCV 107 LRF. Twenty rounds of ready-use ammuntion are stowed in the turret, further 25 are carried in the hull. The gun fires different types of ammunition including APFSDS. On the left side of the gun is a coax 7.62 mm MG. Manual turret traverse only. Unlike the VAB the VBC is not amphibious.

Role. In service with the French Gendarmerie (28) and with Oman (6).

Assessment. Logistic standardization with VAB. Firepower inferior to AMX 10 RC.

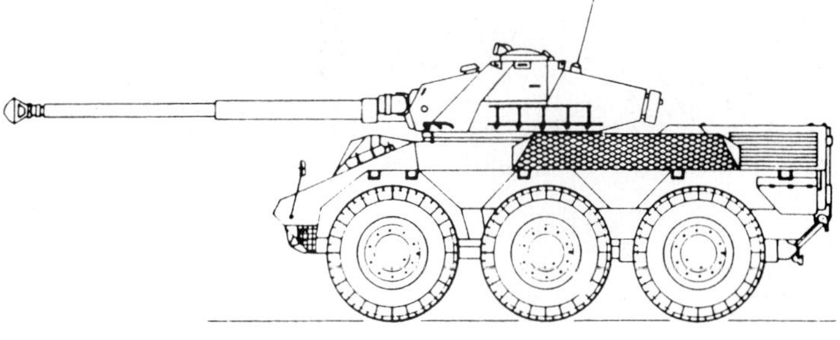

VBC 90 AVR

VBC 90 AVR

AMX 10RC armoured reconnaissance vehicle
Roues Cannon

Non-serial version.

AMX 10RC with GIAT TML 105 turret

Development and production

1970 development of a successor of the Panhard EBR commences at the Atelier de Construction d'Issy-les-Moulineaux. The highest possible degree of standardization of assemblies with the AMX 10 AIFV is called for.
1976 serial production starts.
1985 replacement of the older Hispano-Suiza HS-115 engine by the Baudoin GF11SRX diesel of 221 kW.
1987 production ceased for the French Army after 284 vehicles.

Distinguishing features. Side. 6x6 wheeled vehicle with evenly spaced axles. Low front with prominent overhang and raised driver's hatch. Sponsoned hull. Long, low turret with tapered mantlet and sloped rear. Cupola with periscope. High engine compartment aft. Vertical hull rear. Long smooth-bore gun with bottle-shaped muzzle brake. Front. Vertical hull sides. Heavily sloped turret sides.

Other details. Wheeled vehicle with skid steering, the wheels not being steerable in the normal sense. The wheels have independent hydropneumatic suspension and can be raised and lowered by driver. Fully amphibious, water propulsion is by two hydrojets on hull rear. Trim vane on hull bow has to be erected. NBC collective protection system.
Most of the French vehicles were retrofitted with applique armour plates on the turret, hull bow and sides, a Thomson-CSF DIVT 16 thermal camera and a modified gun to fire APFSDS rounds.
(TABLE)

Role. In service with the French Reconnaissance units (337). Exported to Morocco (108 non-amphibious) and Qatar (12).

Assessment. Good firepower (integrated fire control system), even against MBT at medium ranges. Good mobility with long road range. Low level of protection.

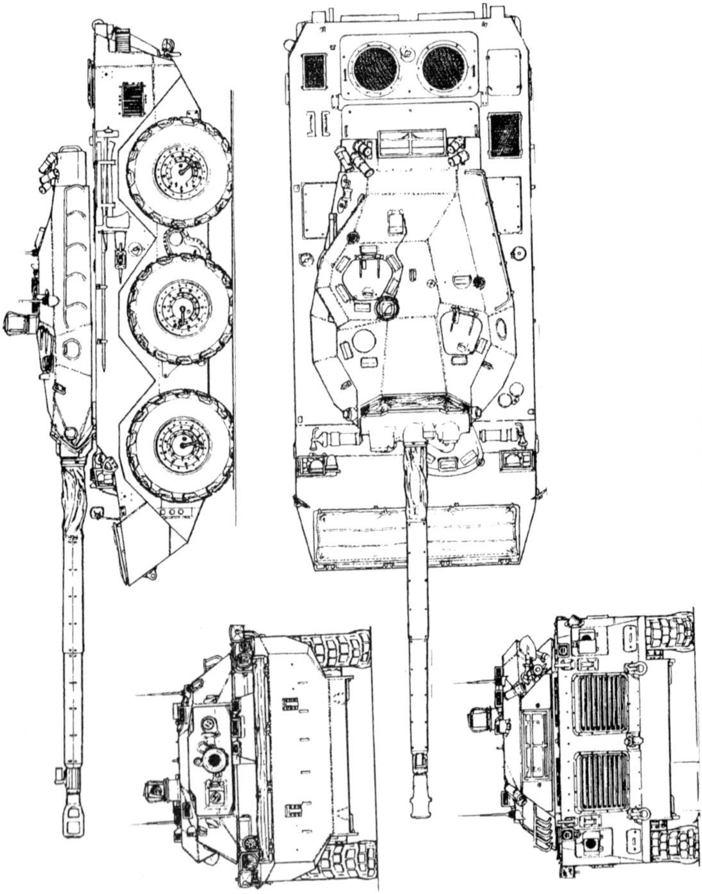

AMX 10RC AVR

AMX 10RC AVR

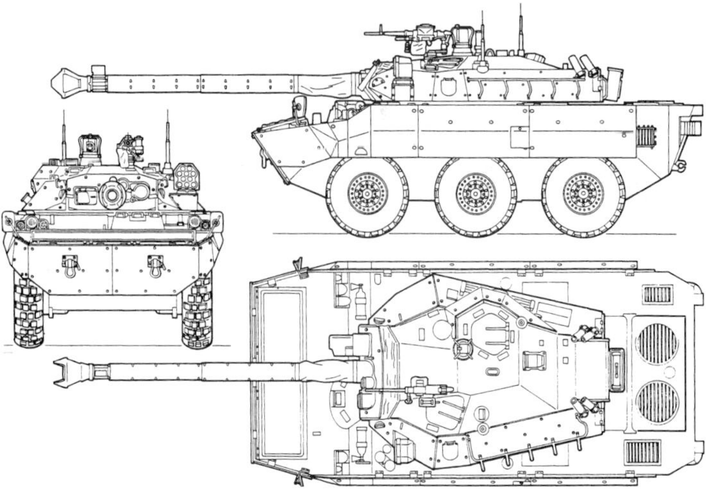

AMX 10RC with add-on armour

AMX 10RC with add-on armour

AMX 30 main battle tank

Models.

AMX 30 (AMX 30 E = Spanish licence version)
AMX 30 B2
AMX 30 S
AMX 30 EM2 (Spain)
AMX-30V (Venezuela upgrade)
AMX 30 DCA AADG (Saudi Arabia)
AMX 30 B2 DT AMCV

Development and production.

1957 development initiated by Atelier de Constructions d'Issy-less-Moulineaux against a joint Franco-German-Italian OR.
1967 series production starts.
1982 first delivery of new AMX 30 B2, in addition conversion of old AMX 30 to the AMX 30 B2 configuration starts.
1983 Spanish licence production stops.

Distinguishing features. Side. five roadwheels and five return rollers. Rear sprocket. Shallow upper hull, engine rear, hull rear vertical, silencer on track guard above drive sprocket. Very long egg-shaped turret with turret bustle extending over engine deck, high cupola. Gun barrel with several steps on thermal sleeve. Front. Sponsoned hull with sloped sides above track guards. Well sloped glacis plate with driver's hatch in upper left part. Range-finder exits protruding on each side of turret, external mantlet. High cupola to right with numerous vision blocks and side-mounted ADMG.
AMX 30 B2 without range-finder exits, TV camera front right.

Other details. Main armament is the CN-105-F1 105 mm gun (L56) with an ammunition stowage of 47 rounds. A model F2 20 mm cannon is mounted coaxially on the left side of the main gun (1050 rounds). Exported tanks and the first issued for the French Army were armed with a 12.7 mm coax MG (600 rounds). A 7.62 mm ADMG model F1 is mounted to the right cupola side (2050 rounds). NBC collective protection system. Optical coincidence range-finder M208 operated by the commander, some AMX 30 received the M496 LRF. AMX 30 E with different radio system and 7.62 mm coax MG and ADMG.
AMX 30 B2 with LRF and computerized fire-control system but still without gun stabilisation system. Partly retrofitted with CASTOR thermal sight. New ENC200 transmission (4-ratio and reverser powershift), torque converter with automatic disconnection, hydrostatic steering system. Four regiments were equipped with GIAT G2 explosive reactive armour modules.
(TABLE)

AMX 30 S with reduced power (456 kW) optimized for the use in hot climate. Use of the CILAS/SOPELEM M409 day/IR sight.
AMX 30 E modernisation programm in two steps. Replacement of the engine by the German MTU 833 KA-501 (625 kW) pack with ZF LSG 3000 automatic transmission, installation of a fire-suppression system and a crew bay system (AMX 30 EM1). Second phase (AMX 30 EM2) with the installation of the computerized fire-control system Mk 9 developed by ENOSA.
AMX-30V with Israeli MATADOR fire-control system with LRF and light intensification night channel, weapon stabilization system and AVDS-1790-2C air-cooled engine (551 kW) with semi- automatic Allison CD-250-GA 3-speed transmission. Fuel capacity increased up to 1298 l.
The air-defence version AMX 30 DCA carries a TG 230A turret with twin 30 mm cannons of a cyclic rate of fire of 600 rpm. Both, the gunner and the commander are equipped with the SAGEM weapon sight with a magnification of x1 and x6 and an APX M250 periscope for ground targets. The RD515 OEIL NOIR radar acts as a range-finder. Elevation of the guns is -5 up to +85 degrees, the effective range is 3 km.
AMX 30 B2 DT (Démineur) with gun removed and equipped either with the RAMTA or the PEARSON plough systems, a GIAT magnetic field generator and a marking launcher. Vehicle's weight is 35.5 t, total length is 10.4 m, height 2.86 m. Remote-control mode possible. The vehicle can be equipped with URDAN roller system in place of the plough (total weight 41 t).

Role. In service with France (166 B2), Bosnia-H. (42 ex-UAE), Chile (32), Cyprus (102 B2), Greece (154 reserve), Qatar (24 AMX 30 S), Saudi Arabia (290 stored AMX 30 S, 53 DCA), Spain (160 EM2), Venezuela (81).

Assessment. Average fire-power of AMX 30 B2 when APFSDS is used. Good mobility. Armour protection not up-to-date.

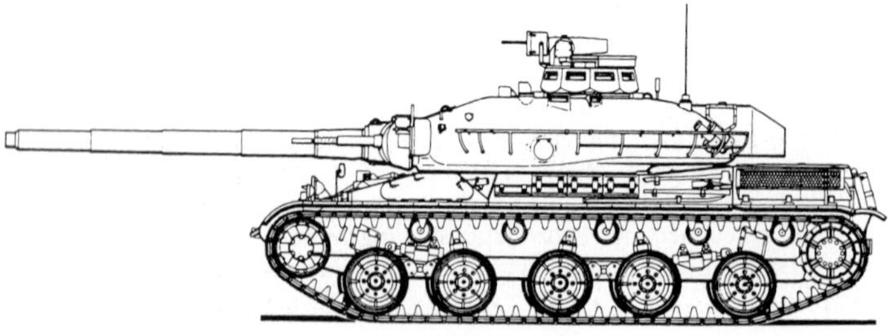

AMX 30 B2 MBT

AMX 30 B2 MBT

AMX 30 B2 with add-on armour

AMX 30 EM2 MBT

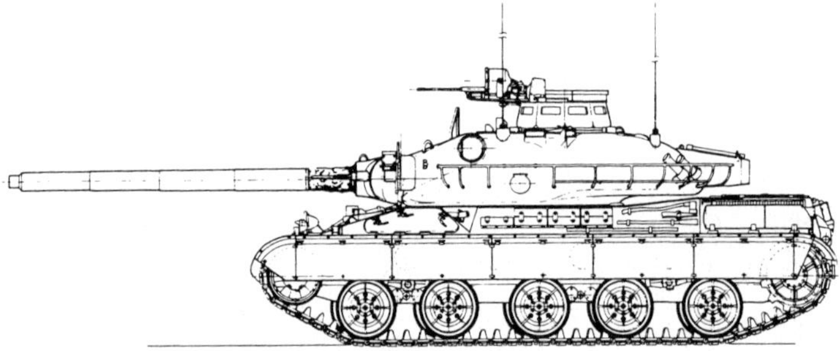

AMX 30 S MBT

AMX 30 V

AMX 30 DCA AADG

LECLERC main battle tank

Development and production.

1983 end of Franco-German MBT project, new French design definition.
1986 end of definition phase.
1989 first of six prototypes issued.
1991 start of series production.
1993 export order from the UAE.

Distinguishing features. Compact vehicle design due to small hyperbar engine concept with six roadwheels. Long flat two-man turret with main sight on right side of the mantlet and rotating periscope on left front turret roof. Long gun with thermal sleeve without fume extractor.

Other details. Main armament is the CN 120-26 120 mm gun with a calibre length of 52. APFSDS rounds can be fired at muzzle velocities exceeding 1750 m/s. A 12.7 mm coaxial MG and a 7.62 mm ADMG are used. In the rear of the two-man turret an autoloader with 22 rounds is integrated. Another 18 rounds are stowed in a drum located to the right side of the driver. Other features are an advanced fire-control system, electric turret drive, thermal imaging, independent stabilised cdr's daysight periscope HL-70, a battlefield managing system, an auxiliary power unit, special armour. The commander is also provided with a 15 cm TV monitor displaying the gunner's vision.
A hydro-pneumatic damper system is used. Explosive suppression system and a NBC system will be integrated. On either rear side of the turret there are nine launcher tubes for various types of grenades (smoke, infra-red decoy and anti-personnel). The tank is amphibious using a snorkel.
(TABLE)
For hot climate employment the tank is also offered with an MTU 883 V-12 diesel engine with 1103 kW coupled to a Renk HSWL 295 TM five gear automatic transmission, improved cooling and filter system. Other modifications for the UAE version include the installation of the GIAT FINDERS (Fast Information Navigation Decision and Reporting System), additional side amour protection, externally mounted auxiliary power unit and an airconditioning system. The commander's SFIM HL-80 stabilized sight has a laser range finder and a thermal imager. The roof mounted 7.62 mm MG can be aimed and fired by remote control.

Role. Under production for the French and the UAE Armies. Current French requirement is 420, 310 are contracted so far. The UAE ordered 388 of the tropicalised version.

Assessment. Due to digital vehicle electronics this tank represents a high standard of modern MBT technology partly ahead of Leopard 2 A4 or M1A1 ABRAMS.

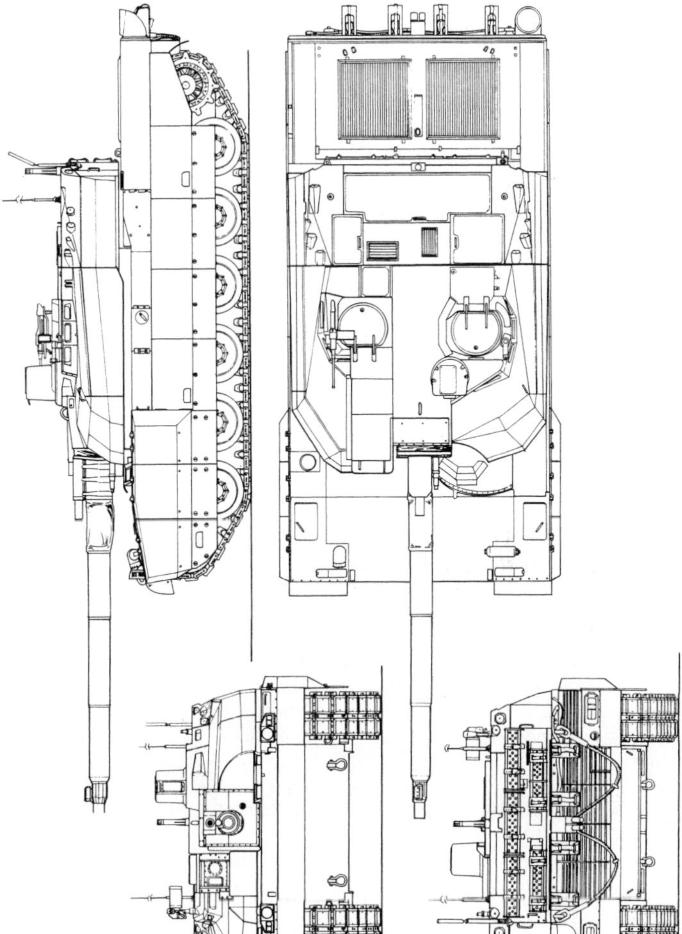

LECLERC MBT

LECLERC MBT

LECLERC tropicalisé MBT

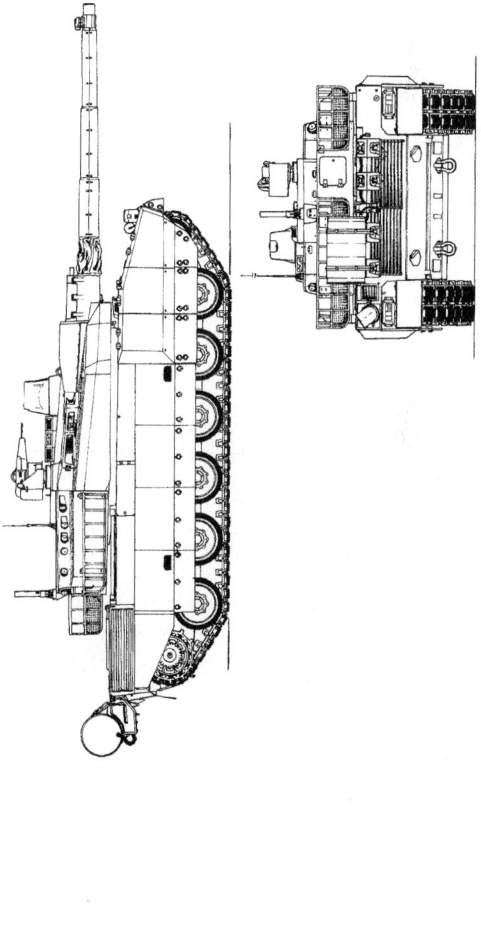

LECLERC tropicalisé MBT

AMX 13 light tank

Models.

AMX 13 Model 51 with 75 mm gun in FL-10 turret
AMX 13/90 with 90 mm gun in FL-10 turret
AMX 13/105 with 105 mm gun in FL-12 turret
AMX-13 SM 1 (upgraded Singapore version)
AMX 13 155 mm F3 AAH
LAR-160 AARL
AMX 13 D ARV
AMX 13 VCG AEV
AMX 13 bridgelayer

Development and production.

1946 developed by Atelier de Construction d'Issy-les-Moulineaux; development authority Direction des Etudes et Fabrications d`Armement (DEFA).
1948 first prototype.
1962 105 mm version, large numbers manufactured for export.
1967 re-gunned with 90 mm (bored-out) tube. FL-12 turret exported to Austria (KÜRASSIER) and Brazil (SUCURI).
1987 more than 3000 vehicles produced, many conversions and retro-fitting kits.

Distinguishing features. Side. Five roadwheels, two top rollers, front sprocket. Low hull with sloped front. Large oscillating turret at rear with bustle extending to rear; long gun with muzzle brake in upper part of turret pivots on lower part with canvas cover joining them. Commander's domed hatch cover. Front. Driver front left with engine to his right. Cupola left.

Other details. Upper part of turret is rigid with the gun mount and elevates with it. Turret bustle with autoloader, twelve ready rounds in two magazines, rest stowed seperately. Numerous conversion kits are offered for modernisation (fire-control systems, diesel engines).
Artillery system with 33 calibre manually loaded 155 mm F3 gun with maximum range of 21.5 km. Length 6.22 m, weight 17.4 t.
LAR-160 version with a 2.8 t elevatable launch pod with two magazines of 18 Israel Military Industries rockets each. The calibre is 160 mm, missile length is 3311 mm with a span of 350 mm. The weight is 110 kg with a 50 kg warhead. The range is 12 - 30 km according to type of ammunition.
Armoured recovery vehicle with front-mounted A-frame, a 15 t winch with 50 m cable, secondary winch with 120 m cable. Wt 15.3 t.
Bridgelayer with 25 t bridge of 14 m length. Wt 19.7 t.
(TABLE)

Role. In service with Algeria (50), Argentina (55 and 24 AAH, retrofitted with KHD V-8 diesel of 191 kW), Chile (47 and 12 AAH), Cyprus (12 AAH), Dominican Republic (2), Ecuador (104 105 mm gun tanks, 81 retrofitted with Soptac 18 fire-control system and 10 AAH), Indonesia (125 AMX 13/90, 6 ARV and 4 AVLB), Ivory Coast (5), Lebanon (35 AMX 13/75, 15 AMX 13/105), Mexico (136), Morocco (80 AAH), Peru (80 AMX 13/105, 30 AMX 13/75, 12 AAH), Qatar (22 AAH), Singapore (350), Venezuela (31 AMX 13/90, 36 AMX 13/75, modernized with 132 kW 6V33T General Motors diesel engine with Borg Warner automatic transmission and Soptac 18 fire-control system; additionally delivered from France 31 refitted AMX 13/90, 36 LAR-160).

Assessment. At time of introduction very advanced standard. Highly versatile, simple and proven hull. Firepower gradually improved by up-gunning also APFSDS ammunition offered. Useful only for low-intensity conflicts.

AMX 13 SM 1 TDG

AMX 13 155 mm F3 AAH

LAR-160 AARL

AMX 13 VCG ARV

AMX 10P armoured infantry fighting vehicle

Models.

AMX 10P
AMX 10PC ACCC
AMX 10P HOT TDM (Saudi Arabia)
AMX 10 P25 (Singapore)
AMX 10 PAC 90 TDG (Indonesia)
AMX 10 AMB AAV
AMX 10 TM AMTC
AMX 10 VOA AAOP
AMX 10 SAO AAOP
AMX 10 VFA AACPV with ATILA control system
AMX 10 VLA AACPV with ATILA
AMX 10 SAF AACPV
AMX 10 RATAC AARV
AMX 10 SAT AACPV
AMX 10 RAV ACC
AMX 10 ECH AMV

Development and production.

1965 developed by Atelier de Construction d'Issy-les-Moulineaux.
1968 first prototype completed with one-man turret.
1973 pre-production starts with 80 vehicles, troop trials.
1978 development of AMX 10 PAC 90.

Distinguishing features. Side. Five roadwheels, front sprocket, three top rollers. Upper part of track covered with skirt. Boat-like hull front, vertical rear with large power-operated ramp. No firing ports in the sides. 2-man turret with unarmoured external mounting, allowing cannon to be elevated in AD role. Front. Pointed hull front with nose inverted back under hull, sloped glacis plate with driver on upper left side, horizontal roof. Turret mounted centre of roof, two outward opening roof hatches above troop compartment. Vertical hull sides.

Other details. The basic AIFV version is armed with a 20 mm cannon and a 7.62 mm coaxial MG in the Toucan II turret NBC protection system is standard. Different weapon and turret installations were tested.
AMX 10P HOT with turret replaced by two-man LANCELOT turret with four ready-to-launch HOT missiles and 14 reload missiles carried in the hull rear. Wt 14.1 t.
AMX 10P Marines specially modified for naval operations with more powerful water jets, transparent trim vane, four bilge pumps, new air intakes, streamlined track links. Baudouin 6 F11 SRX diesel engine of 203 kW. Two-man turret removed, GIAT CIBI 50 with externally-mounted 12.7 mm MG. Landing party of 13 soldiers.

AMX 10P 25 Marines AICV with GIAT one-man DRAGAR turret with GIAT 25 mm cannon.
AMX 10 PAC 90 Marines with GIAT TS 90 turret armed with 90 mm gun. 20 rounds (12 HEAT and 8 HE) are carried in the turret, another ten in the hull. Four infantry men are carried in the rear compartment. Wt 14.8 t, H 2.73 m.
AMX 10 AMB carries three stretchers or one stretcher and four seated patients.
AMX 10 TM tows a 120 mm Thomson-Brandt mortar MO-120-RT-61 with 60 rounds of ammunition. Toucan I turret, crew of 6.
AMX 10 RATAC with roof-mounted puls Doppler radar, total height is 2.84 m. Crew 5. Fitted with navigation system.
AMX 10 ECH with a crane of 6 t lifting capacity. Crew of five.
ACCC and AACPV have different additional radio and communication equipment.
(TABLE)

Role. In service with France (335), Indonesia (24 Marines + 10 PAC 90), Morocco (10), Saudi Arabia (500 incl. 24 RATAC), Singapore (22 P25 + 22 PAC 90), UAE (23).

Assessment. A highly mobile vehicle with average fire-power. The very cramped dimensions make crew comfort minimal. Protection for the external mounting is inadequate. The basic vehicle is in a highly suitable weight class for all kinds of derivatives.

AMX 10P AIFV

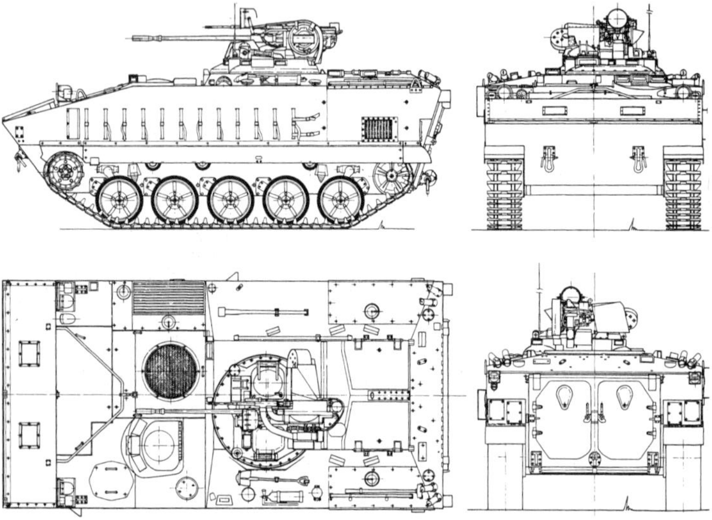

AMX 10 P 25 AIFV

AMX 10 PAC 90 TDG

AMX 10 AMB AAV

AMX 10 VOA AAOP

AMX 10 SAO AAOP

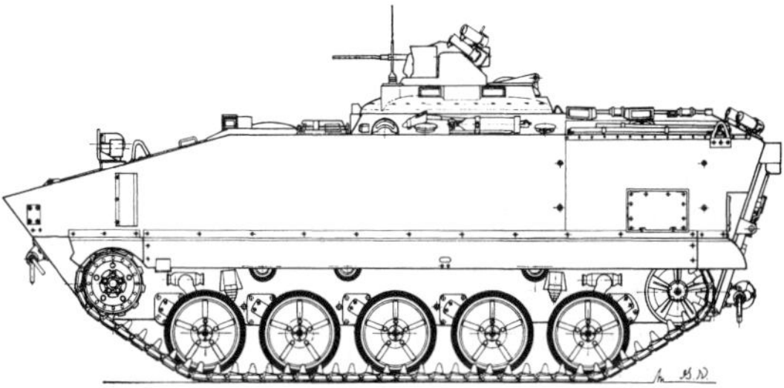

AMX 10 RATAC AARV

AMX 10 SAT AACPV

AMX 10 ECH AMV

VAB armoured general purpose personnel carrier
Véhicule de l'Avant Blindé

Models.

VAB VTT (Véhicule transport de troupe) with CB.52 7.62 mm MG ring mount
VAB VTT with CB 127 gun ring shield with 12.7 mm MG
VAB VTT (6x6) with CB 127
VAB VCI (Véhicule de combat de l'infanterie) (4x4 or 6x6) Toucan
VAB VCI T 20
VCAC HOT Mephisto TDM
VCAC HOT UTM 800 (6x6) TDM
VMO Véhicule maintien de l'ordre (6x6) with 7.62 mm MG in CB 80 mount
VAB PC (Poste commandement) ACCC
VAB RASIT with ground surveillance radar
VAB Transmission ACCC with RITA Automatic Integrated Transmission Network
VAB RATAC with artillery radar
VAB FDC fire direction centre
VAB FOO forward observation officer vehicle linked to ATILA system
VAB FOA (Véhicule d'observation d'artillerie) AAOP
VAB DANAOS AAOP
VAB VIT (Véhicule d'implantation topographique) AACPV
VAB Bromure electronic warfare vehicle
VAB Sanitaire AAV
VTM 120 (Véhicule tracteur de mortier) towing a 120 mm mortar
VPM 81 (Véhicule porte mortier) with a Royal Ordnance 81 mm mortar
VAB VDAA (Véhicule défense anti-aérienne) with ESD TA 20/RA 20S twin 20 mm turret AADG
VAB Cargo ACC
VAB Echelon AMV
VAB ARV
VAB Genie AEV
VAB Reco NBC

Non-serial versions.

VAB NG (new generation with higher mobility and protection)
X8A 8x8 trials vehicle
VAB has been offered with numerous (turret) applications including the GIAT DRAGAR 25 mm turret and for the anti-aircraft role with Matra SATCP (Santal).

Development and production.

1967 development starts of a multi-purpose armoured vehicle for a wide diversity of roles.
1973 tests of 4x4 and 6x6 vehicles starts in competition between Panhard and Saviem/Renault, latter one to be adopted.
1976 first vehicles (4x4) handed over to the French Army.
1985 procurement of 135 VCAC Mephisto for the French Army.
1993 production stops.
1996 VAB NBC version enters service.

Distinguishing features. Side. Four or six-wheeled vehicle with high front and almost vertical glacis. Square bin between wheels, 6x6 vehicle equally spaced road wheels. Two resp. three flaps in side of rear fighting compartment. Exhaust pipe on upper right side of hull. Different turret/armament arrangement. Front. Two large vision flaps in glacis. Slightly tapered front with sharp concave horizontal angle. Splashboard. Horizontal roof and almost vertical hull rear with two large doors.

Other details. Amphibious vehicle equipped with NBC protection system.
VCAC HOT Mephisto with four ready-to-launch and eight reserve missiles, UTM 800 version has 16 reserve missiles.
VAB command and communication versions carrying different sets of radio and communication systems.
VAB DANAOS (day and night artillery observation system) are rebuilt vehicles with the LOA 20 cupola on the centre turret roof with day and thermal sight including a laser rangefinder. The vehicle is also provided with the NSM 20 land navigation system.
VAB Sanitaire can carry four stretcher or five seated or two stretcher and ten seated patients.
VTM 120/VPM 81 have a crew of six, a TLi 52 A turret and carry 70 resp. 120 mortar bombs.
Cargo space is 6.5 m3, loading height is 740 mm.
VAB Echelon has a crew of three and is equipped with tools. Armed with the TLi 52 A turret.
ARV version with a 14 t crane, a 70 t winch of 60 m cable length, crew of four either with 12.7 mm or 7.62 mm MG in the rotating CB shield or with TLi A cupola.
VAB Genie with engineer equipment, crew of nine.
VAB NBC recce version with BrÅcker MM1 mass spectrometer, detection and alerting systems, crew of four.
(TABLE)

Role. In service with Argentina (2 6x6), Brunei (26 4x4), Central African Republic (8 4x4), Cyprus (18 with UTM 800, 27 with T 20, 100 with CB 127 all 4x4), France (3840 plus 33 Gendarmerie and 135 HOT), Ivory Coast (9 VMO, 4 PC 6x6, 12 VDAA), Lebanon (95/85), Mauritius (11 VMO), Morocco (245)

VAB/VMO, 12 VTM, 68 VAB PC, 20 ECH, 49 VCI Toucan, all 6x6), Norway (3 6x6), Oman (21 6x6, 4 MEPHISTO, 9 VDAA (6 without radar), 2 San), Qatar (133 VCI, 24 UTM 800, 5 VPM81, 15 VMO, 5 Cargo, 5 San), UAE (20 VMO), Zaire (12).

Assessment. Largely manufactured from commercial assemblies. Therefore running costs relatively low. Useful APC without combined arms fighting capability, only lightly armoured.

VAB VTT APC

VAB VTT with CB 127 mount

VAB VCL (4x4) TOUCAN

VAB VCI (6x6) TOUCAN

VAB VCI T20

VCAC HOT Mephisto TDM

VCAC HOT UTM 800 TDM

VMO APC

VAB PC ACCC

VAB RITA ACCC

VAB RATAC AARV

VAB VOA AAOP

VAB DANOS AAOP

VAB VIT AACPV

VAB Bromure

VAR Sanitaire

VTM 120 AMTC

VAB VDAA AADG

VAB Genie AEV

VAB Reco ANBCRV

VBCI/GTK/MRAV armoured general purpose carrier family

Provisional data: W 2.99 H 2.376 (roof) 103 km/h 1100 km

Development and production.

1990 tactical requirement issue in Germany for a replacement vehicle for M113 and FUCHS carriers.
1993 Joint military requirements between France and Germany.
1994 French-German MoU for joint development for VBM (Véhicules Blindés Modulaires) and GTK (Gepanzertes Transportkraftfahrzeug). Decision in favour of a wheeled solution. A consortia formed by GIAT, Panhard, Mercedes and Krauss-Maffei expected development order.
1996 British Army enters the project requiring a Multi Role Armoured Vehicle. Two bidding consortia gather ARGE (GIAT, Krauss-Maffei/Wegmann/Mak-Rheinmetall, GKN) and TEAM (Panhard, Henschel/KUKA, Alvis V./VDS). The French designation changes to VBCI (Véhicule Blindé de Combat d'Infanterie).
1997 The Netherlands join the project.
1998 ARGE concept is selected.
1999 France does not sign contract.
2004 scheduled production start.

Distinguishing features. Side. 6x6 or 8x8 vehicle with box-shaped superstructure, pointed bow and vertical rear. Front. Sponsoned hull with steep side walls. Boat-shaped sloped front plate with engine grill on the left and driver's compartment on the right side. Different weapon stations on proposal.

Other details. This vehicle family forms the platform for numerous mission-specific layouts and configurations including APC, CP, ambulance, mortar carriers etc.
The vehicles attain superb mobility due to the most advance axle layout in the armoured wheeled vehicle market (the two front wheel sections are steerable). A modular protection concept including appliqué plates and mine protection is realized.
The 8x8 variant is 7.88 m long and will be powered by 460 kW engines, the 6x6 vehicle is 650 mm shorter and features a 400 kW engine, all EURO-III Deutz versions coupled to ZF 6HP-900 fully automatic transmissions. The 6x6 vehicle's weight will be 26.5 t the 8x8 can exceed 33 t. The wheelbase of 2.582 m is identical, as well as the ground clearance of 504 mm. A ditch of 2 metres can be crossed, the vehicle climbs up to 60 % slopes. The turning radius is 15 m. Maximum crew is 10 resp 11 for the 8x8. The vehicle can wade up to a depth of 1.5 m without any preparation.

Role. At the moment a first lot of 600 AGPCs will be produced for Germany, Great Britain and the Netherlands.

Assessment. Most advance armoured wheeled vehicle concept with outstanding mobility performance data. Different hull modules can be connected to the base structure. High ballistic protection up to 20 mm penetrators and against blast mines.

VBCI/GTK/MRAV AGPC

VCR armoured general personnel carrier
Vehicule de Combat à Roues

Models.

VCR/TT APC (Transport de Troupe)
VCR/TT (4 x 4) APC (Argentina)
VCR/PC ACC (Poste de Commandement)
VCR/TH TDM (Tourelle Hot)
VCR/IS AAV (Intervention Sanitaire)
VCR/AT AMV (Atelier Véhicule)

Non-serial version.

VCR/AA with RBS 70 anti-aircraft missile system

Development and production.

1977 developed as a private venture by Panhard as a further development of the AML series using many of the same assemblies.
1979 series production starts for export only.
1980 4x4-version in production.

Distinguishing features. Side. 6 x 6 vehicle. Sloped glacis, step up to hull roof which extends to vertical rear. Front. Tapered glacis, narrow front. Vertical hull sides except for rear troop compartment which slopes inwards mid-way up. Commander's cupola rear of driver on left side, engine right. Main armament on forward part of roof. Rear troop compartment has two upward-opening flaps in each side.

Other details. Centre axle can be raised. Horizontal mudguard over wheels. UAE version has double rear doors.
VCR (4x4) has a weight of 7.8 t and a length of 4.87 m. In place of the middle set of wheels there are two water propulsion jets.
VCR/PC with additional radio equipment. Wt 6.17 t.
VCR/TH with UTM 800 HOT turret with four ready-to-launch missiles.
Wt 7.4 t, crew of four. 14 additional rounds stored inside.
VCR/IS with heightened hull carries four stretchers or six sittingand two stretcher patients; total weight is 6.53 t.
VCR/AT has only towing capabilities, equiped with repair facilities.
(TABLE)

Role. In service with Argentina (24 4x4), Iraq (100 VCR/TH delivered, status unknown), Mexico (40 VCR/PC), UAE (82 VCR/TT,AT,IS,PC).

Assessment. Cost-effective vehicle because of use of proven assemblies. Useful for low-intensity conflicts.

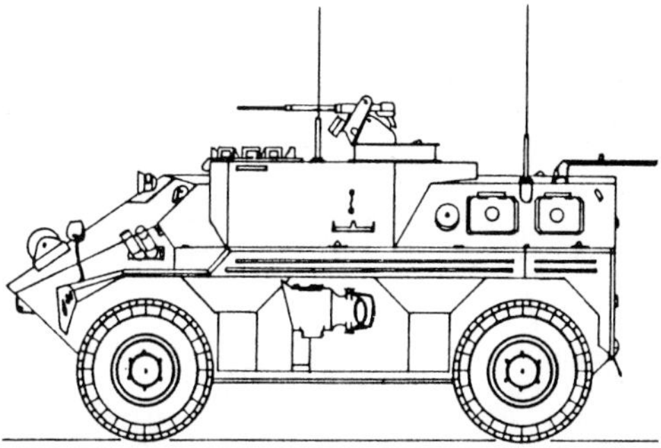

VCR/TT (4x4) APC

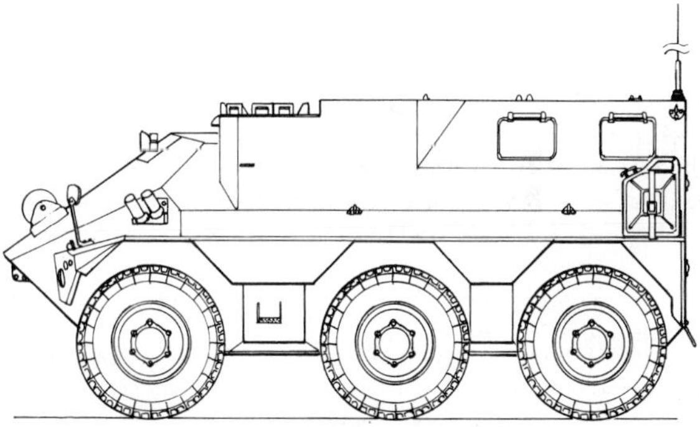

VCR/PC ACCC

VCR/TH TDM

VCR/IS AAV

M3 armoured general personnel carrier

Models.

M3 with Creusot-Loire STB rotary shield with 7.62 mm MG
M3 with Creusot-Loire CB gun ring with 7.62 mm MG
M3 with Creusot-Loire CAFL38 turret with 12.7 mm MG
M3 with Creusot-Loire TLi 38 turret with 7.62 mm MG
M3 with Creusot-Loire TL 2i turret with twin 7.62 mm MG
M3 with Creusot-Loire CB 127 ring shield with 12.7 mm MG
M3 with Creusot-Loire CB 20 ring shield with 20 mm cannon
M3 with GIAT Toucan I turret with 20 mm and 7.62 mm MG
M3/VPC ACCC
M3/VDA AADG
M3/VSB with RASIT battle field surveillance radar
M3/VSB with ESD RA 20S airspace surveillance radar
M3/VAT AMV
M3/VLA AEV
M3/VTS AAV
Buffalo APC
Buffalo ACCC

Development and production.

1969 first prototype as a private venture of Panhard.
1971 series production started for export only.
1985 Buffalo version issued.

Distinguishing features. Side. Two large road wheels with forward opening door, box-shaped armoured hull with compound angled front. Chassis with protruding mud-guards. One spare wheel left side. Three hatches each side for mounted fighting. Vertical hull rear. Front. Pointed front with sloped glacis, driver in centre position, horizontal roof, armament in the middle. Vertical hull sides with upper part sloping inwards. Two large rear doors.

Other details. The APC is amphibious without preparation, being propelled by its wheels. Different configurations of main armament. Optional equipment like NBC protection system, air-conditioning system, front mounted winch and/or shield, smoke dischargers.
M3/VPC with additional communications equipment.
M3/VDA with Hispano-Suiza H 20 R/H 20 S turret with two HS 820 SL guns
M3/VAT with maintenance equipment, crew of five.
M3/VLA with a removable hydraulically operated dozer blade, crew 6
M3/VTS can carry four stretcher or six sitting (resp. 2/3), crew 3
Buffalo is an improved version of the M3 with a Peugeot PRV petrol or a XD 3T diesel engine of 107 resp. 70 kW and additional stowage room. Wt 6.2 t, L 4.6, W 2.4, H 2.0, 100 km/h, range 900 km. (TABLE)

Role. In service with Algeria (33 APC, 5 VPC, 2 VAT, 2 VLA, 2 VTS), Angola (ex-Portugese), Bahrain (100), Benin (Buffalo), Burkina Faso (13), Burundi (9), Chad (10, some with Toucan I), Columbia (Buffalo), Gabon (10), Iraq (approx. 200), Ireland (60), Ivory Coast (14), Kenya (8), Lebanon (15), Malaysia (37), Mauritania (25), Mexico (6 VTS), Morocco (40), Niger (14, 10 VDA), Nigeria (4 RASIT, 4 VTS, 4 VAT, 6 VPC), Portugal, Rwanda (12 M3/Buffalo), Saudi Arabia (150), Senegal (14), Sudan (100, 2 with RASIT, 1 work shop, 2 with RA-20 radar and 13 with Toucan I 20 mm cannon), Togo (3), UAE (300) and Zaire (70).

Assessment. A practical and relatively cheap vehicle, useful for Police operations.

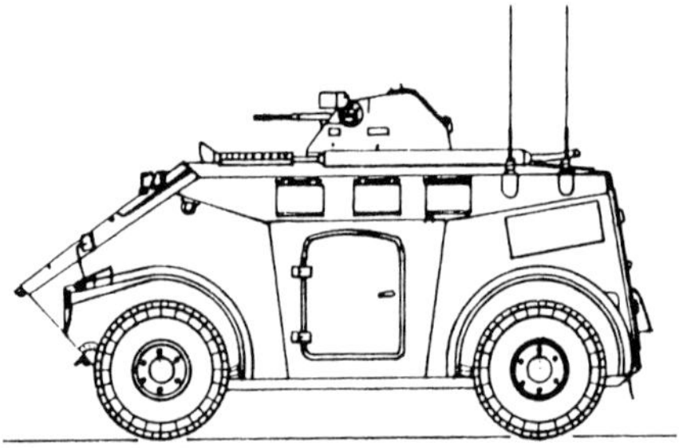

M3 with TL 2i turret

M3 APC

M3/VAT AMV

BUFFALO APC

VXB-170 armoured personnel carrier

Wt 12.7 t L 5.99 m W 2.5 m H 2.05 m 125 kW 750 km Crew 12

Models.

VXB-170 Group Transport Vehicle
VXB-170 with BTM 103 turret with 7.62 mm MG and 40 mm grenade launcher
VXB-170 Platoon Command Post
VXB-170 Section Command Post

Development and production.

1967 development starts by Berliet in the VAB competition (BL 12 prototype).
1971 troop trials of improved version with Gendarmerie Nationale.
1973 serial production starts, several prototype versions.

Distinguishing features. Side. Amphibious 4-wheeled vehicle with squat sides. Hull with prow front and sides sloped all-around. Raised driver's hatch extends from glacis to roof. Three roof hatches for crew, one gull-wing door each side, four flaps each side for fighting from under armour. Engine compartment left rear with grills in roof. Front. Lower part of hull vertical, upper half sloping slightly inwards. High boat-like front. Large wind-screen to front and sides.

Other details. The chassis is in pressed steel with armour plate at the front. The suspension is separately variable for each axle by means of an electrical adjustment system allowing the vehicle to be tilted to front or rear to improve obstacle crossing. Differential lock and winch. Allison TX200 transmission. 14.00 x 20 runflat tyres. Most of the VXB-170 are equipped with a hydraulically operated obstacle clearing blade at hull front.

Role. In service with French Gendarmerie (155), Gabon (10), Gambia and Senegal (12).

Assessment. Average armoured carrier for police roles.

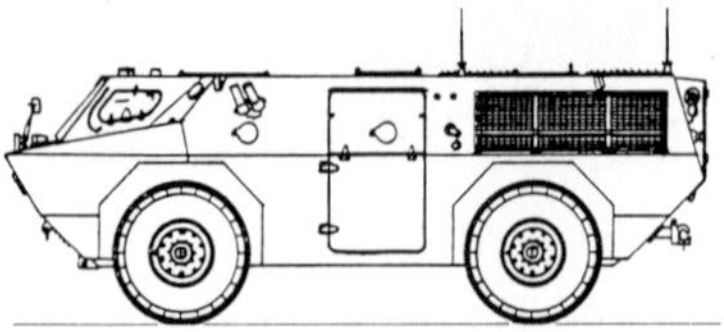

VXB-170 APC

VXB-170 APC

Canon de 155 mm automoteur modéle F1
(155 Grande Cadence de Tir (GCT))

Models

GCT 155 AUF1
GCT 155 AUF1 T
GCT 155 AUF1 TA
GCT 155 AUF2

Non-serial version.

155 GCT on T-72 hull.

Development and production.

1969 development starts to meet a French Army requirement for a SP gun with a rapid fire capability.
1972 first prototype completed.
1977 series production starts for Saudi Arabia.
1979 introduction into the French forces begins.
1982 orders placed by Iraq.
1991 combat improvement starts.
1996 last deliveries.
1999 contract for the modernization of a first batch of 10 vehicles to AUF1 TA standard

Distinguishing features. Side. AMX 30 MBT hull, lightened by 2 t. Large turret with sloped front and vertical rear. Long gun in mantlet with protruding buffers and recuperator. In each side of the turret there is a large entrance hatch for the crew, with two large rear hatches for loading with ammunition. Front. Vertical-sided turret wider than hull. The commander's cupola is on the right and the loader's hatch on the left of the turret roof. A 12.7 mm or a 7.62 mm ADMG can be mounted on the turret.

Other details. There is an automatic loader and combustible cartridge cases are used. Six charges being used. The vehicle has an adjustable suspension and is provided with an auxiliary engine for NBC protection. Two smoke dischargers are fitted on both turret front sides each.
Combat improvement includes an automatic loading system (4 rounds in 45 sec) and and fire control data link with forward observer resp. fire control centre. The 155 AUF1 T has a number of improvements including a new APU, an improved automatic loading system and an inertial goniometer.
(TABLE)
An upgrade program has just started involving 174 vehicles. 104 are planned to be converted to the AUF1 TA standart with the Atlas cannon fire control system that succeeds the Atila, as well as replacing the existing powerback by a Renault VI E9 engine coupled to an automatic gearbox. The remaining 70 howitzers, designated AUF2, will receive a new 52 calibre ordonance and an entirely automatic system for handling, loading and firing the propellant charges and shells.

Role. In service with France (179 + 74 155 AUF1 T), Iraq (85 delivered, status unknown), Saudi Arabia (51).

Assessment. The use of an automatic loader and combustible cartridge cases offers a high rate of fire.

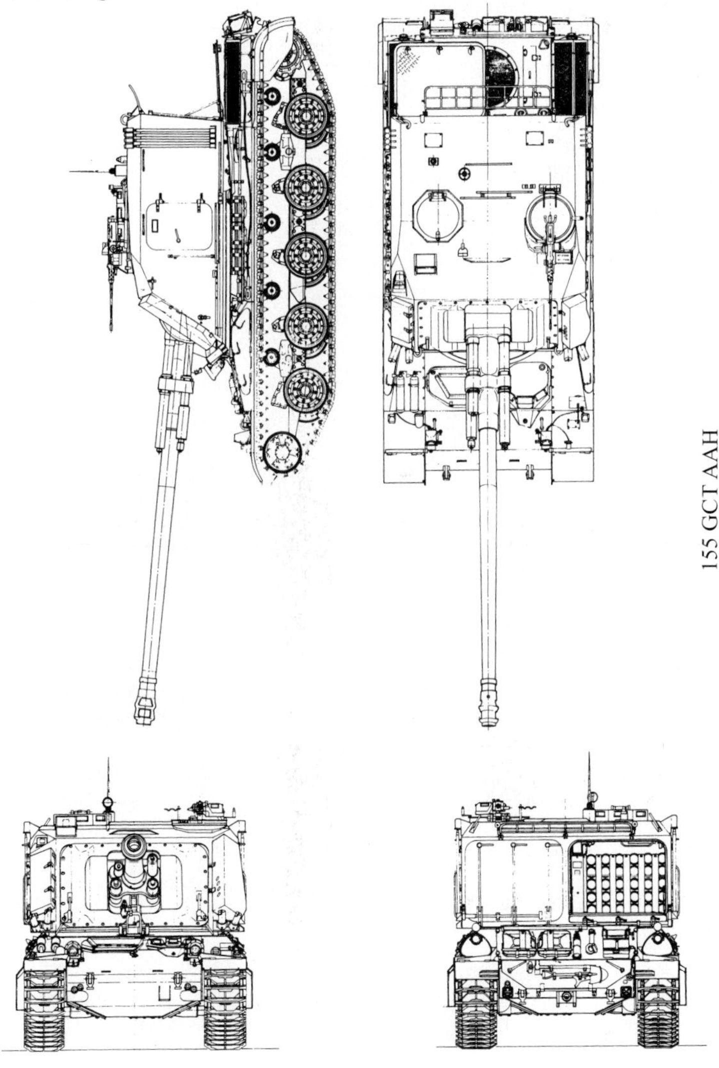

GCT 155 AUF 1 AAH

CROTALE/SHAHINE armoured air defence missile launcher

Wt 14.8 t L 6.22 W 2.65 H 3.43 70 km/h 500 km crew 2

Models.

CROTALE 1000 (South African CACTUS)
CROTALE 2000
CROTALE 3000
CROTALE 4000
CROTALE acquisition unit
CROTALE 4000 acquisition unit
SHAHINE on AMX 30 hull
SHAHINE acquisition unit on AMX 30 chassis
SHAHINE 2
SHAHINE SICA on P-6R vehicle

Development and production.

1964 South African development contract with Thomson-CSF for a mobile airportable, all-weather, low-altitude SAM capable of engaging aircraft up to speeds of Mach 1.2 (85 % of costs paid by ZA).
1971 delivery of seven platoons to South Africa, two firing and one acquisition units ordered by France, later adopted for airfield defence.
1973 final deliveries to South Africa, introduction into Libya. (series 2000) built.
1975 SHAHINE version ordered by Saudi Arabia.
1975 series 3000 in production.
1978 Saudi Arabian order for airfield defence.
1983 series 4000 under production.
1984 SHAHINE 2 ordered by Saudi Arabia (older units retro-fitted).

Distinguishing features. CROTALE launcher vehicle, side. 4x4-wheeled vehicle with vertical lower front, shallow glacis and driver in raised hatch. Front. Sponsoned hull with vertical sides. Surveillance vehicle with oval radar antenna. Launcher vehicle with circular tracking and guidance radar, with 2x2 launch containers for missiles beside it. When in action the vehicle is supported on three hydraulic jacks. Series 4000 units with extendable mast for radio link system. SHAHINE launcher with 3x3 missile containers.

Other details. Firing unit. Monopulse tracking/guidance radar (K-band), command transmitter (X-band), IR error measuring system for lock-on phase, SN1050 digital computer and TV system for tracking targets at low angles of elevation. Units not operable, when on the move, linked together via cable. Crotale 4000 have LIVH radio link. Missile. Single-stage solid fuel motor, which accelerates the missile to an initial velocity of 800 m/sec, maximum speed is 2.5 Mach, range 8.5 km, 15 kg warhead. Acquisition unit. Pulsed Doppler target

acquisition radar (S-band), IFF system, SN1050 digital computer, command post station capable of acquiring and tracking 12 targets simultaneously, and engaging 6 at once. Acquisition range approx. 18 km.
SHAHINE launcher unit comprises a monopuls J-band fire-control radar with a range of 17 km, one or two missiles can be guided. The R460 missile has an improved propulsion rocket motor, maximum range is 14 km. The acquisition unit has a pulse-Doppler E-band surveillance radar with a range of 18.5 km. Up to 40 targets can be registered, 18 to be handled simultaneously. SHAHINE 2 with acquisition radar's range increased (19.6 km) and the altitude capability is 6 km instead of 3 km before.
(TABLE)

Role. In service with Egypt (12), France (30), Libya (9), Pakistan (12), Saudi Arabia (32 acquisition/28 SHAHINE-1 and 36 SHAHINE-2 fire units for the army, 16 CROTALE acquisition and 48 CROTALE fire units for the air force), South Africa (7), UAE (3).

Assessment. Effective system for medium heights. Splitting the system over two vehicles gives rise to problems in mobile roles.

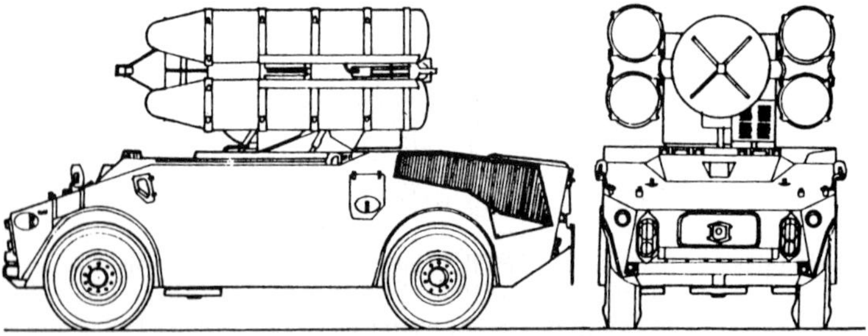

CROTALE AADML

CROTALE 3000 AADML

CROTALE 4000 AADML

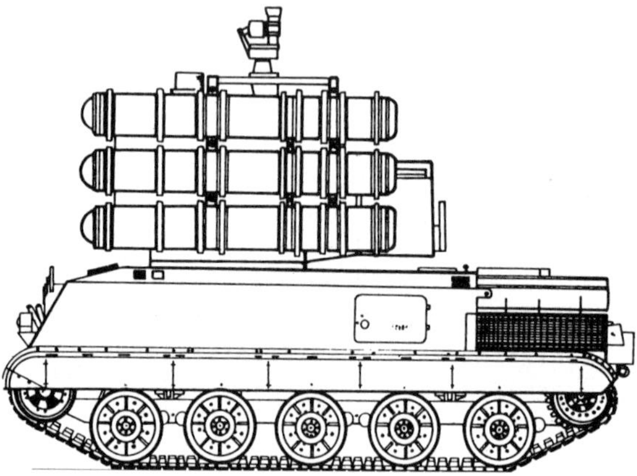

SHAHINE AADML

SHAHINE AADML

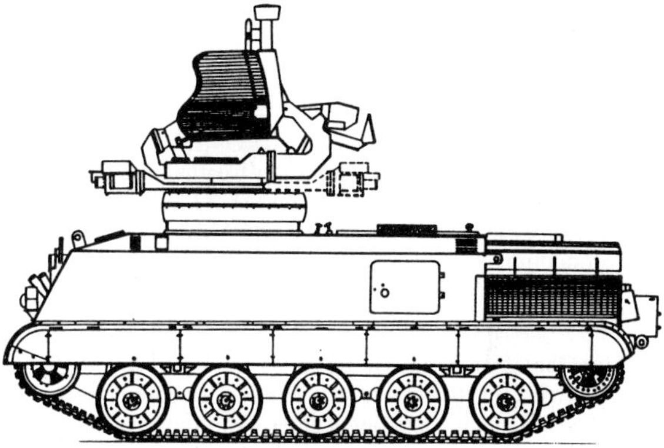

SHAHINE acquisition unit

SHAHINE acquisition unit

AMX 30 R armoured air defence missile launcher
with SAM ROLAND

Development and production.

The ROLAND is a joint Franco-German (EUROMISSILE) development. The original French OR called only for a fair weather version.
1971 development trials with prototype on AMX 13 hull with trailer.
1973 trials on prototype of complete system on AMX 30 hull.
1977 series production starts.
1987 production of ROLAND 3 commences.
1988 first deliveries to Spain.

Distinguishing features. Side. Hull similar to AMX 30 D ARV with high armoured superstructure. Turret with traversable launching ramps to accept the missile in its transit/launch container at sides. Fold-down combined surveillance/target acquisition antenna on turret rear. Front. Sponsoned hull with vertical sides. Narrow turret. Oval radar antenna.

Other details. For weapon system see entry under Germany. Initial version ROLAND 1 only with optical tracking.
(TABLE)

Role. In service with France (53 ROLAND 1, 98 ROLAND 2), Iraq (60 delivered, status uncertain), Mexiko (30), Nigeria (16), Quatar (3), Spain (18).

Assessment. A relatively heavy hull which does however ensure high mobility for this air defence system. Value of ROLAND 1 limited due to the lack of all-weather capability.

Top: AMX 30R with Roland 1
Down: AMX 30R with Roland 2

36

36
I 854 0005

AMX 30 PP armoured vehicle launched bridge
(Char Poseur de Pont)

Development and production.

1969 development starts by Direction Technique des Armaments Terrestres (DTAT).
1971 pre-production commences, troop trials up to 1975.

Distinguishing features. An armoured superstructure together with various systems to support and allign the scissor type bridge, mounted on the AMX 30 MBT hull. The MLC-50 bridge consists of two parts which are folded together for transport and extend to a length of 22 m. The bridge can be layed across a 20 m gap within 5 minutes. The bridge is 3.1 m wide, 3.92 m with widening panels. Like the AMX 30 MBT the bridgelayer too has an NBC protection system. The design of the bridgelaying mechanism allows the bridge to be laid on slopes up to 30 %, the maximum cant being 20 %.
(TABLE)
A MLC-65 bridge can be used, too.

Role. In service with France (28) and Saudi Arabia (12).

Assessment. This vehicle has almost the same mobility as other tanks, the only limitations being the heavier weight, its length and its higher centre of gravity.

AMX 30PP AVLB

AMX 30 D armoured recovery vehicle
(Char de Dépannage)

Models.

AMX 30 D ARV
AMX 30 D (S) ARV
AMX 30 DI ARV
AMX 30 EBG (Engin Blindé Génie) AEV

Development and production.

1967 development starts at AMX, prototype in 1970.
1975 production starts.
1981 prototype of AMX 30 EBG.
1988 issue of first AMX 30 EBG.

Distinguishing features. AMX 30 MBT hull with box-shaped superstructure with turret-like raised centre portion and cutaway right-hand side for crane jib with 270° traverse. MBT commander's cupola with ADMG and spotlight left front. Driver's hatch with three periscopes forward of this. Carrier for replacement power-pack at rear. Front dozer blade, which acts as support for lifting operations.

Other details. Behind the blade is an exposed auxiliary winch. A snorkel can be fitted for fording depth up to 4 m.
AMX 30 D (S) is export model for Middle East with sand shields over upper parts of tracks, engine of reduced power (456 kW) so optimized for desert operations.
AMX 30 DI has increased crane lift of 15 t and crane traverse of 240°.
AMX 30 EBG similar to ARV with small turret on centre with a 7.62 mm MG, two smoke dischargers on each side, a single 142 mm demolition gun firing a 17 kg charge up to 300 m. Each side of this are two mine-launching tubes for five 1.8 kg mines each up to 250 m. Large dozer blade (3.5 x 1.1 m) with six scarifying teeth to rib up surface when tank is driven in reverse. Crane on right side with 7.5 m telescopic jib with 15 t lift and 3600 traverse, can be fitted with a powered earth auger of 2.2 m diameter. Winch with 80 m cable, 15 - 20 t pull. 50 kW auxiliary engine for hydraulics. The tank is provided to launch the trailer-borne MLC 65 TCP assault bridge using the standard scissors action. The bridge weighs 4.8 t and has a width of 4 m. The length is 14 m. (TABLE)

Role. ARV in service with Cyprus (2), France (134), Qatar (1), Saudi Arabia (57), Spain (10), UAE (4), Venezuela (4). AEV only in service with the French Army (71) and Saudi Arabia (12).

Assessment. Both vehicles almost comparable to ARV/AEV apart from LEOPARD 1 versions. Highly versatile, good mobility. Well devised equipment.

AMX 30 DI ARV

AMX 30 EBG AEV

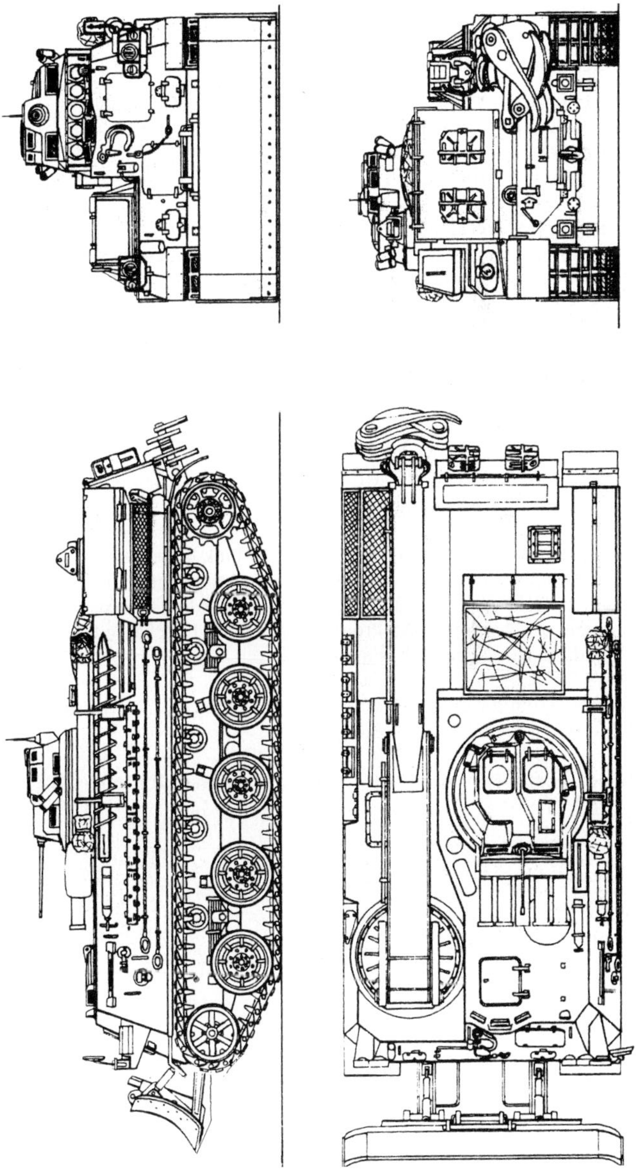

AMX 30 EBG AEV

DNG char de dépannage (nouvelle generation) ARV

Development and production.

1990 studies starts on a new ARV with GIAT and MaK (Germany).
1993 order by United Arab Emirates.
1994 first prototype with six roadwheels issued.
1997 roll-out of first production vehicle.

Distinguishing features. Side. LECLERC MBT elongated chassis (seven roadwheels) with high superstructure in the front with distinctive step down to engine compartment. Sloped hull front with dozer blade under forward part. Two large side doors on left side, crane jib in stowed position runs whole length of right hand side. Front. Left part of the tank with steep glacis with driver's position and flat cdr's cupola with AA machine gun. On the right hand side turntable with crane. Full width dozer blade raised when not in use.

Other details. The DNG uses the complete recovery equipment identical to the German Büffel supplied by MaK.
Corresponding to the LECLERC MBT the DNG is also offered in the tropicalised version with MTU 883 engine.
(TABLE)
The DNG can easily be converted to an engineer tank by replacing the regular dozer blade by a Pearson full-width mine plough with an integral GIAT Demeter 2 magnetic field generator. The hydraulic arm can be fitted with a bucket, auger, grapple or a hammer. Two Pronit mineclearing linecharge boxes can be installed on both rear sides of the engine deck.

Role. In production for the UAE (46 ordered in the tropicalised version). Initial French order amounts to 15 vehicles.

Assessment. High-performance ARV capable to recover MLC 60 MBTs.

LECLERC DNG ARV

FENNEK armoured vehicle reconnaissance

Wt 9.0 t L 5.3 W 2.5 H 2.2 177 kW 110 km/h 850 km Crew 3 (provisional data)

Development and production.

1985 German requirement for a light AVR.
1988 winning concept was the ZOBEL from GST.
1989 comparative trials of the ZOBEL with the French VBL.
1994 political decision in favour of a German/NL project.
1996 the industrial team DAF/Wegmann starts delivery of four trials vehicles.
1999 revised version delivered for trials.
2000 procurement contract expected.
2003 start of serial production planned.

Distinguishing features. Side. Flat bulky design with two large wheels. Operator's cabin with a large side window. One entrance side door with a long window. Bow-shaped hull front. Front. Wedge-shaped front bow. Operator's cabin covering whole width with large front window and two roof hatches with periscopes.

Other details. The vehicle is equipped with a retractable 1.5 m mast fitted with a sensor pod containing a thermal camera, a day sight and laser rangefinder. It is operated from a swivel-mounted console. The vehicle is also provided with an onboard hybrid GPS navigation system and a battlefield management system. Different weapon installations are possible either 7.62/12.7 mm MG or a 40 mm grenade MG (on demand by Germany). The German Army plans a twin-vehicle role with one vehicle equipped with the full sensor suite, the companion vehicle provided with a small aerial mini-aircraft LUNA with a range of approx. 25 km and a ground sensor equipment.
The engineer version (Pionierkundungsfahrzeug) will be equipped with special terrain data recce and navigation systems.

Role. Light recce vehicles for division/brigade level. Initial requirement for Germany is 178, 24 (total 636), 50 as a terrain recce vehicle for armoured engineers, and 220 for the Dutch forces.

Assessment. Top mobility vehicle with a wide range sensor equipment.

FENNEK AVR

LUCHS armoured vehicle reconnaissance

Development and production.

1965 start of development of a complete range of vehicles.
1968 testing of nine prototypes of DB and the Joint Project Office (JPO) formed by several manufacturers.
1971 the Daimler-Benz design is chosen and Thyssen Henschel is awarded a contract for series production.
1975 production starts until 1978.
1985 retrofitting (LUCHS A2) until 1987.

Distinguishing features. Side. 8x8 chassis with large gap occupied by a bin with a flap between second and third axles. Superstructure with sloped and undercut front and rear. Small turret at front. Front. Narrow hull between wide wheels with upper part sponsoned. No turret overhang. Periscope ring left.

Other details. Turret armament consists of the Rheinmetall 20 mm cannon and one external ADMG on the cdr's ring mount. All-wheel drive and steering with coil spring suspension and rigid axles. Natural swimmer (trim vane erected), propelled in water by two rear-mounted steerable propellors. Signaller/reverse driver left behind turret. Engine right rear, NBC collective system.
The A2 modernisation program consists of integration of a tank thermal sight, a double-feed system for the cannon, new combat tyres, and a new long-range radio set (XK 405). (TABLE)

Role. In service with the German Army (409). Most of the vehicles will be phased out after the introduction of the FENNEK. Approximately 100 will be kept in service.

Assessment. A measure of standardization has been achieved by the use of commercial assemblies. Highly mobile vehicle with poor firepower.

LUCHS AVR

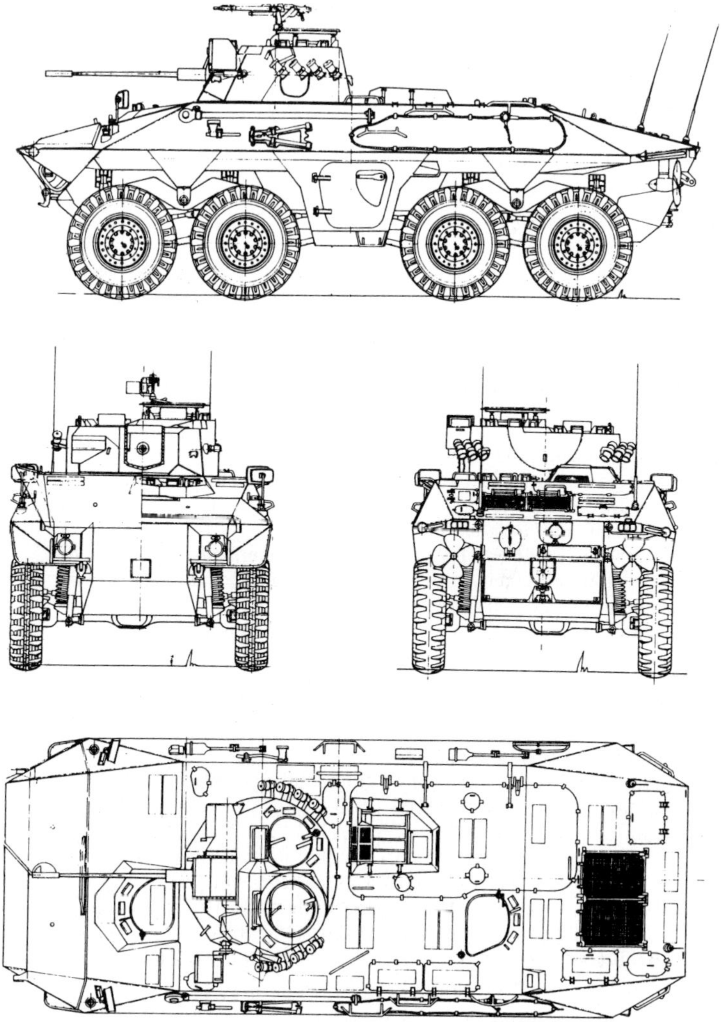

LUCHS AVR

LEOPARD 1 main battle tank

Models.

LEOPARD 1 A1
LEOPARD 1 A1A1
LEOPARD 1 A2
LEOPARD 1 A3
LEOPARD 1 A5
LEOPARD 1 A5 (BE)
LEOPARD 1 A5 DK
LEOPARD AS 1
LEOPARD C 2
LEOPARD TU 1
LEOPARD GR 1
LEOPARD 1-V
LEOPARD driver training tanks

Development and production.

1956 issue of a military requirement for a 30 t-tank.
1957 development starts jointly with France, Italy joined 1958.
1965 first production vehicle built by Krauss-Maffei handed over, 400 LEOPARD (1) of the first batch in production.
1968 production for Belgium (324) starts.
1969 first vehicles (468 A1) produced for the Netherlands.
1970 retrofit improvement starts with modified optics and weapon stabilisation system (A1 standard).
1971 export to Norway (78 A1 without weapon stabilisation) and to Italy (200 A1).
1974 sixth lot issued (A4 with integrated fire-control system), MaK joins in the production, licence production in Italy (720 A2).
1975 retrofitting of A1 with add-on armour and skirts (A1A1).
1976 production of A3 DK and AS 1.
1978 export of C 1, production of 60 driver training tanks.
1982 export of 77 TU 1 with EMES 12A3.
1983 export of 106 GR 1 with EMES 12A3.
1986 upgrading of 1276 A1A1 to A5 standard starts, delivery of first modernized Dutch Leop 1-V (verbeterd) with EMES 12A3.
1990 Danish A3 brought to A5 standard together with 100 A3 and 10 A4 additionally procured from Germany.
1993 Belgium starts upgrade program with SABCA fire-control-system.
1999 upgrade to C2 standard starts in Canada.

Distinguishing features. Side. Seven roadwheels, four return rollers, sprocket rear. Well sloped glacis plate, horizontal hull top, raised engine deck, vertical rear. Either rounded or box-like turret with bustle. Either round or pointed external mantlet with long gun, fume extractor nearly centered. Most tanks with wavy side skirts. Front. Sloped front with driver on the right side. Cast turret with sides sloping inwards or plated turret with wedge-shaped mantlet and sloped sides. No commander's cupola, external ADMG mounting.

Other details. The tank is equipped for submerged driving up to 4 m with a full-width tube on the cdr's hatch. NBC collective protection system. All tanks are equipped with a weapon stabilisation system, those not being retrofitted with a thermal sight are provided with a light-intensification camera PZB 200 on the gun mantlet. The LEOPARD 1 A2 has thicker cast turret armour.
The LEOPARD 1 A3/A4 has spaced turret armour.
The AS 1 has the A3 turret and is equipped with the SABCA computerized fire control system with laser rangefinder.
The C2 retains the C1 chassis but is retrofitted with ex-German A5 turrets and additionally protected with MEXAS add-on armour.
The LEOPARD 1 A5 has the KAE EMES 18 fire-control system with thermal sight and laser rangefinder.
(TABLE)

Role. In service with Australia (90 AS 1), Belgium (132 A5 (BE) upgraded with SABCA fire-control system with thermal sight, 33 A1 not modernized), Brazil (173 ex-Belgium A1 plus 2 driver training tanks), Canada (114 C2, Denmark (230 A5 with A3 turrets), Germany (580 A5, 54 A5 are earmarked for the use as forward artillery observer tanks, 60 driver training tanks), Greece (106 GR 1, 3 ex-GE A3, 75 ex-GE A5 turrets on A2 chassis, 170 ex-NL 1V, 2 ex-NL 1V2/A5), Italy 120 ex-German A5 turrets on A2 chassis, 267 A2 used as spa-reparts), Netherlands (30 tanks converted to driver training tanks), Norway (111 A5 NO with Geadrive 90 BL electric turret drive, 59 A1 NO) and Turkey (77 TU 1, 150 converted German A3 with EMES 12A3, 5 ex-GE A3, 165 ex-GE A1A1, 2 driver training tanks).

Assessment. Robust and reliable design with above average mobility. NATO-standardised 105 mm ammunition, only latest KE design is able to cope with modern tank threat. Modernized versions (A5 etc) offer employment of this tank in a tank destroyer role.

LEOPARD 1 A1 MBT (Norway)

LEOPARD 1 A5 MBT

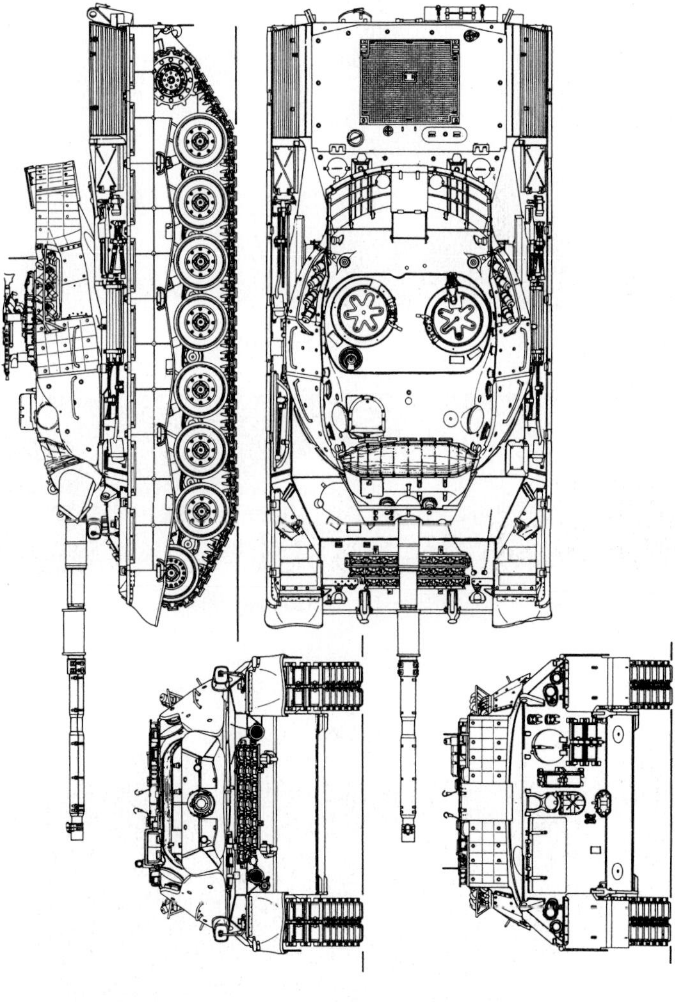

LEOPARD 1 A5 MBT

LEOPARD 1 A5 (BE)

LEOPARD 1 A5 DK

LEOPARD AS 1 MBT

LEOPARD TU 1 MBT

LEOPARD 1-V MBT

LEOPARD 2 main battle tank

Models.

LEOPARD 2 A4
LEOPARD 2 A5
LEOPARD 2 (NL)
PANZER 87, LEOPARD (CH)
Stridsvagn 122
LEOPARD 2 driver training tank

Non-serial version.

LEOPARD 2 with 140 mm smoothbore gun (CH)

Development and production.

1970 development of new MBT starts in place of cancelled MBT 70.
1973 manufacture of 17 different prototypes.
1975 further development of LEOPARD 2 AV (4 PTs).
1979 production start of the first batch (380).
1981 export to the Netherlands.
1985 fifth batch (370) in production (LEOPARD 2 A4) with digital fire-control computer and crew-bay system. All older tanks are gradually being brought to this standard. Export of first two Swiss tanks.
1987 sixth batch (150) in production, half of them with improved special armour.
1989 procurement of seventh (100) and eighth (75) batch.
1992 decision to upgrade 699 tanks from 1994 onwards (increased armour protection, thermal sight cdr's periscope, electric turret drive), number later reduced to 225.
1994 Sweden decides to procure LEOPARD 2.
1995 first LEOPARD A5 handed over to the German Army. Development starts of further upgrade step with L55 gun tube and new KE ammunition (LEOPARD 2A6). Step III upgrade program (new 140 mm turret) is abandoned.
1996 Spanish plans to co-produce 390 starting in 1998 are postponed.
1997 German MoD decides to upgrade a second batch of 125 tanks.
1999 Spanish contract for 219 Leopard 2E.
2001 Scheduled start of conversion for Germany and the Netherlands.

Distinguishing features. Side. Seven roadwheels, four return rollers, sprocket rear. Top run of track covered by armour and rubber skirts. Sloped glacis, horizontal hull, upward-sloping engine deck, almost vertical rear. Very large angled turret with vertical front and long bustle. Large rotating head periscope in front of commander's hatch. Relatively short-barreled gun with several steps and fume extractor offset half-way point. Front. Shallow sponsoned hull with verti-

cal sides, large overhanging turret with vertical sides. Small rectangular external mantlet. Large window sight with armour covers in right turret front. Rotating head periscope on right in front of periscope ring. External ADMG on left (loader's) turret ring.
A5 and Strv 122 with distinctive wedge-shaped turret front armour.

Other details. Main armament is the 120 mm Rheinmetall smooth-bore L44 gun. Full primary stabilisation of main-sight (combined day/thermal sight) and cdr's periscope in azimuth and elevation. Fire-control computer with laser range-finder. Collective NBC system. Submerged driving up to four metres using a snorkel on top of cdr's hatch.
A5 with new generation special armour, add-on armour on turret front, seperate thermal sight integrated in commander's periscope, electric turret drive.
A6 will be fitted with the new L55 gun tube to fire the new KE ammunition DM53.
Strv 122 like A5 plus add-on hull front and turret roof armour, tank command and control system (TCCS) with seperate data transmission radio.
(TABLE)

Role. In service with Austria (115 ex-NL A4), Denmark (51) Germany (1614 A4, 350 A5 in production with conversion to A6 as follow-on program, 31 driver training tanks), Netherlands (180 A5, 136 A4 on sale, 14 earmarked for conversion as mine-clearing tanks, 20 driver training tanks), Spain (108 German A4 on lease until 2001; government agreement to co-produce 219 A5 from 2000 - 2007), Sweden (35 delivered from Germany, 85 more are license-produced 160 Strv 121 (German A4) on permanent lease), Switzerland (380, 3 driver training tanks).

Assessment. Together with the M1A1/A2 ABRAMS and the LECLERC the LEOPARD 2 marks the highest current standard in MBT technology. Good firepower, excellent armour protection and high mobility.

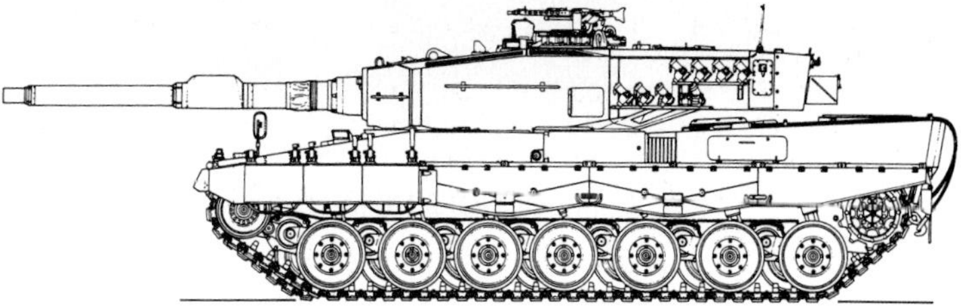

LEOPARD 2 MBT

LEOPARD 2 A4 MBT (Spain)

LEOPARD 2 NL MBT

PANZER 87 LEOPARD MBT

LEOPARD 2 A5 MBT

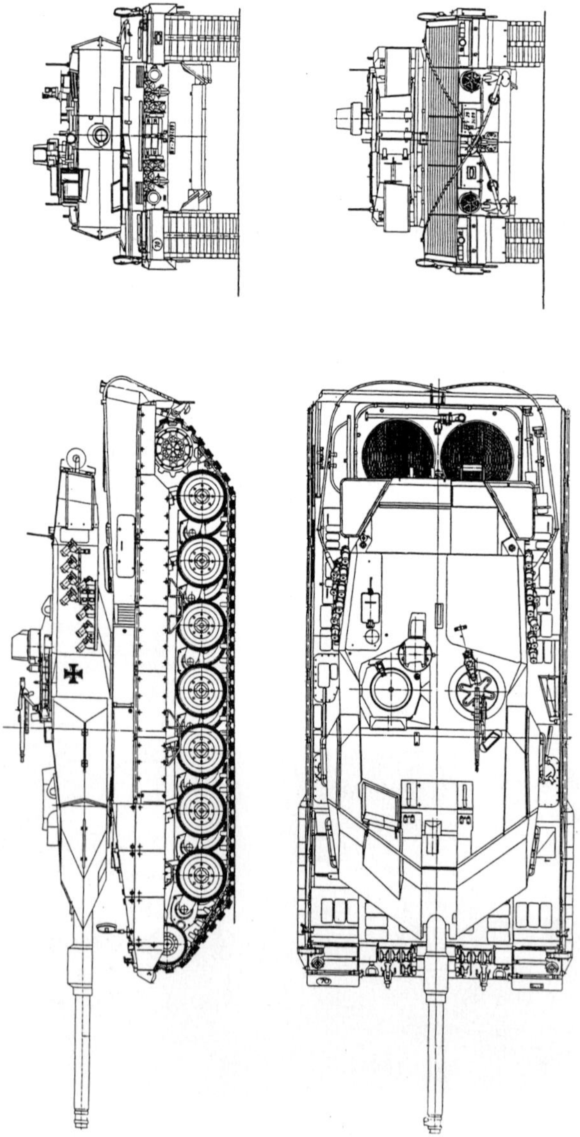

LEOPARD 2 A5 MBT

Stridsvagn 122 MBT

LEOPARD 2 A6 MBT

Raketenjagdpanzer JAGUAR tank destroyer, missile

Models.

JAGUAR 1
Beobachtungs- und Führungspanzer ACCC

Development and production.

1965 production start of 750 Jagdpanzer Kanone (JPz 4-5).
1967 series production of RJPz 2 (SS11 ATGM) by Henschel and Hanomag (370 built) starts.
1972 Belgian order for improved version of JPz.
1974 concept studies for integration of HOT ATGM carried out by Thyssen Henschel and MBB. Selection of K3S concept of MBB.
1978 first deliveries of JAGUAR 1. Prototypes of modified JPz 4-5 with TOW ATGM.
1983 conversion of 162 JPz 4-5 into JAGUAR 2 starts until 1985, phased out in 1995.
1984 first trials with a JAGUAR 1 equipped with an Eltro RZ 1201 thermal sight.
1985 HOT-2 missile enters production with a 150 mm calibre warhead.
1992 HOT 2T with tandem warhead starts production.
Upgrade of 181 JAGUAR 1 with NZBG thermal sight.
1996 delivery of 90 surplus JAGUAR 1 to Austria (18 used for spareparts).

Distinguishing features. Side. Five roadwheels, three return rollers, sprocket rear. Upper part of suspension covered by wavy skirt. Sloped glacis with raised superstructure behind, drop-down engine compartment rear. Sloped rear. Front. Sloped glacis with driver on the left upper side. Sloping sides.

Other details. JAGUAR 1 is armed with the HOT-2 of 4000 m range and an armour piercing capability of 1300 mm RHS. The gunner tracks the target through a periscopic sight which is mounted in the forward part of the roof and has a two-stage magnification (x4 or x12). The rotating ready-to-launch magazine contains eight wire-guided missiles. Total stowage is 20. Cdr's cupola is behind the periscope right of the missile magazine with a cdr's observation periscope behind. The tank is provided with a 7.62 mm bow MG on the right side. (TABLE)
The BeoFüPz is a re-built JPz 4-5 with its 90 mm gun removed. It is used as an forward artillery/mortar observer vehicle.

Role. In service with Austria (72 JAGUAR 1, 23 BeoFüPz) and Germany (181 Jaguar 1, 170 re-built TDG as BeoFüPz).

Assessment. HOT 2T is capable to deal with modern MBT. However only the upgraded JAGUAR 1 has night-fighting capability.

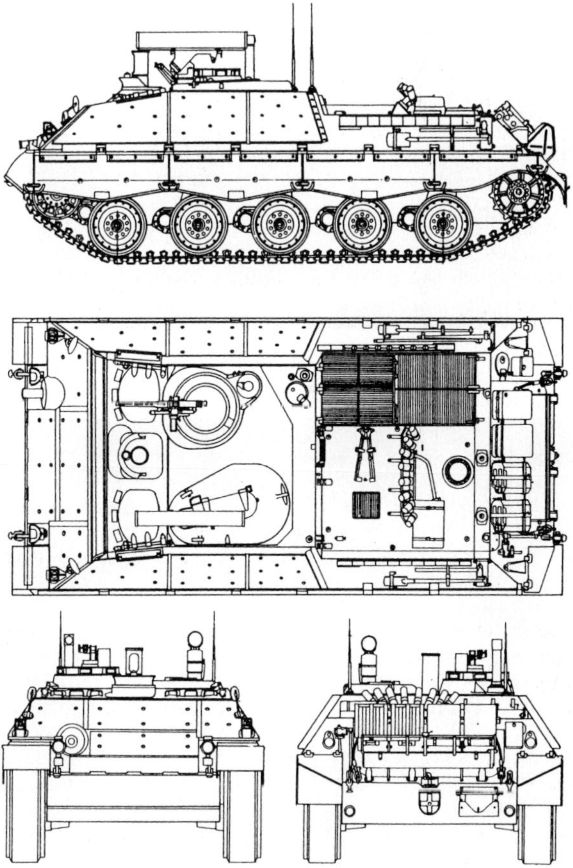

JAGUAR 1 TDM

JAGUAR 1 TDM

JAGUAR 1 TDM with NZBG

BeoFüPz ACCC

MARDER 1 armoured infantry fighting vehicle

Models.

MARDER 1 A3
Fahrschulpanzer (driver's training tank)

Non-serial version.

MARDER 1A5 with KUKA E4 30 mm turret

Development and production.

1960 order for a light armoured vehicle family centering around the section AIFV. Meeting this requirement threw up exceptional difficulties with the result that the TDG and TDM variants were given priority; their development was successfully concluded in 1967.
1966 new requirement for 2-man turret with external mounting.
1970 production run of 1929 vehicles (1045 Rheinstahl, 871 MaK).
1974 follow-up order for second batch totalling 2136.
1979 modification to A1 (+) (674), A1 (-) (350), A1A (1112) standard improving night sight capabilities.
1983 modification to A1A2 and A2 (thermal sight) standard starts.
1989 upgrading to A3 standard starts. The MICV is now called MARDER 1 because a successor development is named MARDER 2 (later abandoned).
1998 upgrading of 2100 MARDER 1 A3 completed.
1999 requirement for additional mine protection issued.

Distinguishing features. Side. Protruding front with high lower front. Running gear with skirting plates, six roadwheels and three top rollers. Front sprocket. Horizontal roof with vertical hull rear. Two circular firing ports on each side in rear troop compartment. Large turret with sloping front, sides and rear with externally mounted cannon and smoke dischargers on left side. Front. Sponsoned hull. Sloped glacis plate with driver on the left. Hull sides slope inwards above suspension. Turret with high external mounting, commander's cupola on the right with MILAN ATGMS adaptable outside on turret ring.

Other details. All MARDER retrofitted with dual ammunition feeding system also for firing APDS rounds. MARDER 1 A3 with external additional armour plates, external box-like mounting of the coaxial MG on the left turret side. Additional stowage boxes on hull sides, deleted firing ports. Modified final drives due to weight increase by 5 t to 34 t. (TABLE)

Role. Only in German service (2100 AIFV and 36 driver's training tanks). In 2002 63 vehicles will be provided with additional mine protection.

Assessment. Fast and agile vehicle but very large and complex. External cannon mounting makes the silhouette very high. Insufficient firepower against fairly armoured new generation AIFV. Only deep-fording capability. A3 version with increased armour protection; insufficient mine protection.

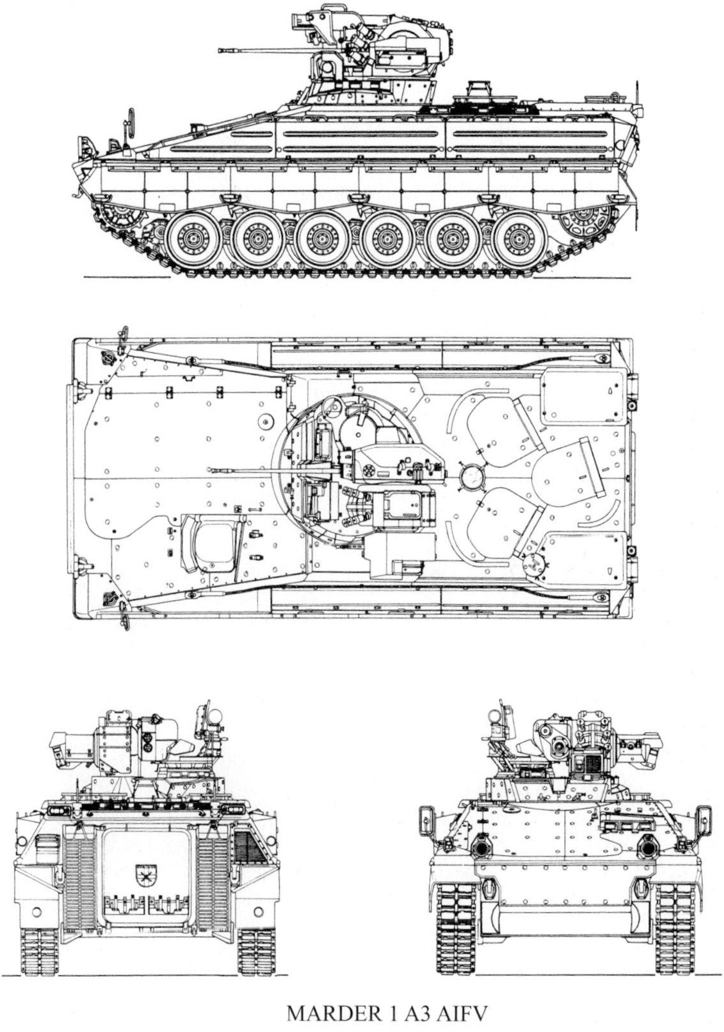

MARDER 1 A3 AIFV

MARDER 1 A3 AIFV

TPz 1 FUCHS armoured general purpose carrier
Transportpanzer

Models.

TPz 1 APC
TPz 1 engineer version
FUCHS with 12.7 mm MG externally in KUKA turret
TPz 1 A1 EloKa ELINT version
TPz 1 A2 ACCC
TPz 1 AAV
TPz 1 A3 ANBCRV (US M93 FOX)
M93A1 (USA) ANBCRV
TPz 1 RASIT
TPz 1 HELAS ELINT version

Non-serial versions.

SUPERFUCHS with 313 kW engine and 606A2 KUKA turret
TPz 1 EOD (Explosive Ordnance Disposal)

Development and production.

1964 start of conceptual work on a family of lightly armoured multi-purpose carriers.
1970 trials of GB consortium and Daimler-Benz.
1971 comparative development trials.
1974 contracts placed on DB for prototypes and pre-production vehicles.
1979 start of series production.
1987 issue of TPz 1 RASIT.
1988 first deliveries of NBC version.
1997 German requirement for additional 56 vehicles in the APC version with a 20 mm weapon station issued.

Distinguishing features. Side. 6x6 vehicle with long boat-shaped hull with pointed front, horizontal roof and vertical rear. Forward-opening door each side of front driver's compartment. Exhaust pipe on left side of hull. Front. Sponsoned hull sloping inwards above large road-wheels. Wide driver's windscreen with armoured shutter hinged at top. Twin doors at rear.

Other details. TPz 1 APC for the use as an armoured infantry section vehicle (crew 1+8) has a pivot mount for ATGM MILAN on the rear hull roof.
The engineer version exists in two sub-variants, one for an engineer recce team (1+3) or as a command vehicle, the second as an armoured engineer section carrier recognizable by two drum-like containers for barbed-wire.
The ambulance version can either carry 4 stretcher or 8 sitting casualties.
TPz 1 A1 EloKa is fitted with a EK 33 jammer and has a 15 kW generator in the extended left rear door. It has only one roof hatch instead of three. When in

action a line of 13 antennas will be placed on the roof top. This type is non-amphibious.
TPz1 A2 has additional radio sets (up to four) and a 5 kW generator in the left wing of the rear door.
TPz 1 A3 has an extended rear hull with upper openings for the use of chemical detection devices and markers. On the vehicle's sides are a nuclear radiation detection probe and dose rate counting tubes. A computer-controlled mass spectronometer identifies agents and levels of contamination then formats data for transmission. The tank is provided with a land navigation system.
37 are upgraded with data communication interfaces with the central computer unit, new precision Gamma ray detector.
The US version M93 is currently being upgraded to M93A1 with an integrated sensor suite with the mast-mounted M21 stand-off detector (range up to 5 km) and a GPS system.
TPz 1 RASIT with elevatable French RASIT battlefield surveillance radar system (range 20 km) and land navigation system.
TPz 1 HELAS is provided with a passive electronic radar detection system. Signal processing includes identification, classification and threat assessment. In action a mast antenna is hydraulically erected.
For peace-keeping missions 124 vehicles were provided with IBD MEXAS add-on armour.
(TABLE)
For the employment with crisis-reaction forces different weapon stations are currently under trials.

Role. In service with the German Army (1038 incl. 220 TPz 1, 220 TPz 1 engineer v., 172 EloKa, 134 TPz 1 A2, 114 TPz 1 A3, 110 RASIT). 14 vehicles were exported to The Netherlands in the EloKa and 4 in the HELAS version, Venezuela (10 with 12.7 mm MG) and the USA (123 of which 60 vehicles were donated during 1991 Gulf War). Great Britain received 11 and Israel 8 ex-German TPz 1 A3 during 1991 Gulf War.

Assessment. Extremely agile and versatile carrier with numerous applications. Immune to small arms fire and shell splinters. Easy servicing thanks to good accessibility and quick assembly changes.

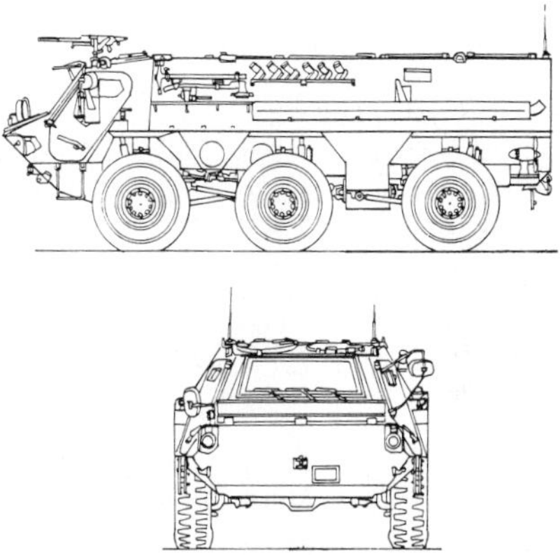

TPz 1 FUCHS APC

TPz 1 FUCHS APC

TPz 1 engineer version

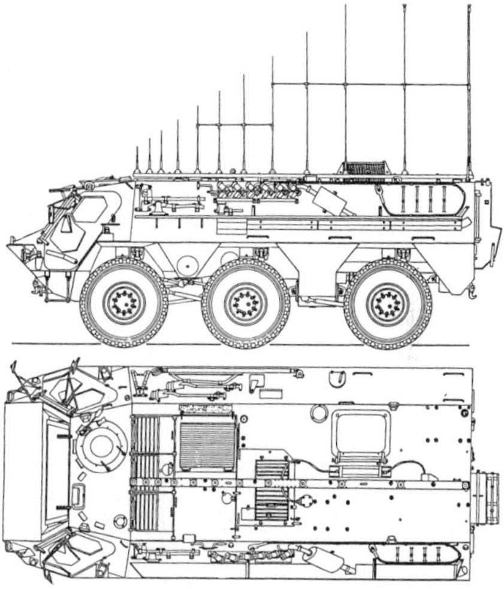

TPz 1 A1 ELINT version

TPz 1 A1 ELINT version

TPz 1 A2 ACCC

TPz 1 AAV

TPz 1 A3 ANBCRV

M93A1 ANBCRV

TPz 1 RASIT ARRV

TPz 1 HELAS ELINT version

WIESEL airborne weapon carrier

Models.

WIESEL with 20 mm cannon
WIESEL with ATGWS TOW 2

Non-serial versions.

WIESEL with twin MILAN ATGWS launcher
WIESEL with twin HOT ATGWS launcher
WIESEL with SAMM BTM 208 turret with 12.7 and 7.62 mm MG
WIESEL AARV with RATAC-S radar system
WIESEL 2 Extended Base Vehicle with an additional road-wheel
WIESEL 2 with electric drive
WIESEL 2 with 120 mm mortar

Development and production.

1969 order for a weapon carrier for use in airborne and infantry formations and complying with the internal loading requirements of the CH53G helicopter. The vehicle must be operational immediately after landing.
1970 first prototype TERRIER by Faun (4x4 vehicle).
1971 studies carried out by Rheinstahl, Food Machinery Corp. San José, Daimler Benz, Hotchkiss and Faun led to the OR being restricted to a carrier for airborne use only, if possible light enough for 2 to be transported in one CH53G.
1973 development proposal were submitted by Faun for a wheeled and by GST, Hopp, Porsche and Rheinstahl for tracked vehicles.
1975 Porsche prototype is selected for the project definition phase and 6 prototypes with 20 mm or TOW 1 armament underwent trials.
1978 development is cancelled owing to financial stringency. Proposals of a whole family of WIESEL vehicles were made on the basis of a private venture.
1984 resumption of program, troop trials with modified prototypes.
1989 start of production at Krupp MaK.
1992 last deliveries.
1997 prototype of WIESEL 2 ambulance version delivered.
2005 start of delivery of 66 WIESEL 2 mortar carriers planned.

Distinguishing features. Side. Tracked vehicle with four roadwheels, low idler, front sprocket and one top roller. Sloped front, vertical rear. Horizontal louvres with exhaust pipe to immediate rear with gauze-type cover. Front. Hull sides slope inwards to above track guards with air outlet on left and stowage boxes on right side. Driver on the right front in a raised hatch, engine on the left. TOW on pedestal mount without armour protection.

Other details. The vehicle uses a commercial Audi engine coupled to an automatic ZF transmission. The track is of an endless rubber type.
WIESEL with 20 mm cannon armed with KUKA E6-II-A1 turret with dual-feeding system (60 rounds on the left and 100 rounds on the right side with different type of ammunition. Additional 240 rounds are stored internally. Elevation is from -10 to +45 degrees and traverse 110 degrees to both sides, both operated manually. A Zeiss PERI Z-16 sight is used with x2 and x6 magnification. For night-fighting a PERI Z 59 is used.
WIESEL with TOW 2 has the same weight but is slightly higher (1.87 m). Traverse is 45 degrees to both sides and 10 degrees elevation and depression as well. (TABLE)
The WIESEL 2 mortar version will probably be introduced by the German light infantry. The 120 mm Tampella mortar is muzzle-loaded. It has a soft recoil system. Prior to firing two stabilisers are lowered to the ground on either side of the rear.
A WIESEL 2 ambulance version is also under troop trials.

Role. 343 WIESEL 1 in service with the German Army (133 20 mm carriers and 210 TOW) another 54 WIESEL 2 C3 vehicles, 19 ambulance, 6 engineer recce, 23 ammunition transporter and 5 driving school tanks will be issued starting in 2001; WIESEL 1 also bought by the US Army (7).

Assessment. Small agile vehicle, very versatile and cost-effective.

WIESEL with 20 mm cannon

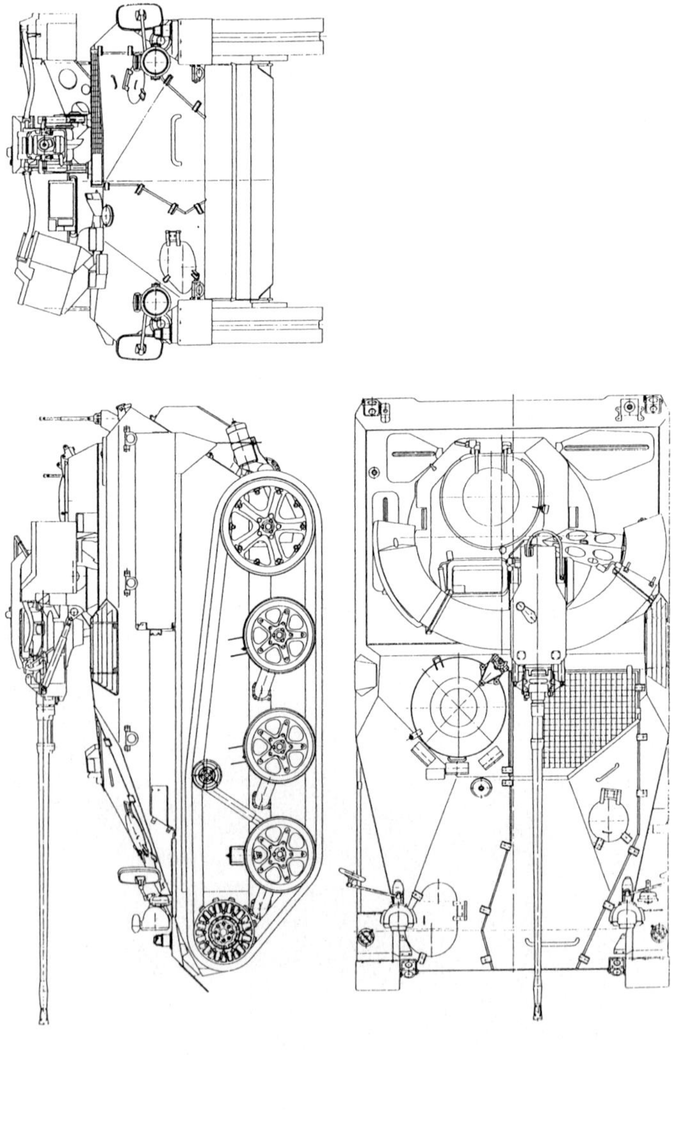

WIESEL airborne weapon carrier (scale 1:40)

WIESEL with TOW-2 ATGMS

WIESEL 2 with 120 mm mortar

Panzerhaubitze 2000 armoured artillery howitzer

Development and production.

1968 requirement for a M109 successor.
1972 trilateral SP70 project starts together with Italy and UK.
1986 SP70 project cancelled. Decision in favour of a national solution.
1993 combined troop and technical evaluation trials starts.
1996 contract of a first batch of 185 AAH.
1998 first deliveries.

Distinguishing features. Side. Seven roadwheel suspension using Leopard 2 components with front drive sprocket. Large box-shaped turret covering two third of chassis and protruding to the rear. Long gun tube extending half of its length over vehicle's front. Ripped muzzle brake, fume extractor at the end of first third of gun tube. Front. Sloped engine deck. Massive turret covering full width of chassis with sloped side walls. Rear. Turret covering whole width. Two large entrance doors.

Other details. 155 mm L52 gun fires all types of NATO standardized rounds up to a range of 30 km resp. 40 km using base-bleed rounds. 60 shell magazine in the hull, free-fly flick rammer, breech block with 32 round standard primer magazine, compartmentized charge container in the turret rear, on-board fire control equipment with ballistics computer and gun laying and navigation system to be used either fully autonomeously or controlled via fire control center. Rate of fire of 12 rounds per minute or 3 rounds in 10 seconds. ERA module roof protection against bomblet threat.

Role. On delivery for the German brigade artillery (185). Decision for introduction in Italy and the Netherlands pending.

Assessment. Highest performance of the time being according to reaction time and rate of fire.

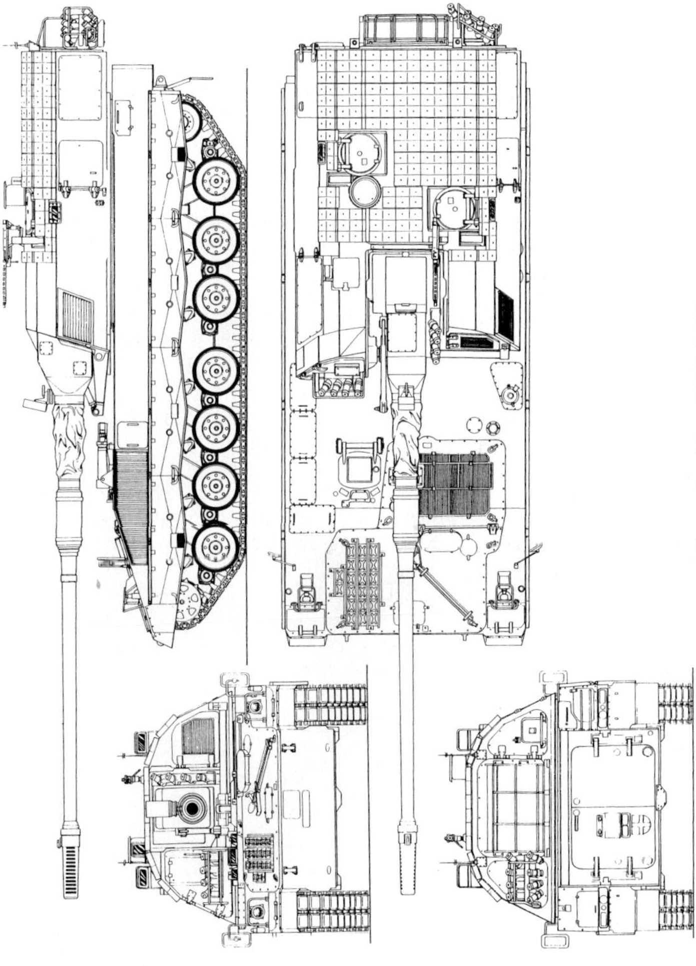

Panzerhaubitze 2000 AAH

Panzerhaubitze 2000 AAH

GEPARD armoured air defence gun

Variant.

PRTL AADG (Pantserrups tegen Luchtdoelen) former CA1 (NL)

Development and production.

1965 OR for a self-contained air defence system on LEOPARD 1 MBT hull. Development by Oerlikon-Contraves and Rheinmetall.
1968 prototype of Oerlikon-Contraves version (5PZF C5 CA1) with integrated target acquisition and tracking radar by Hollandse Signalapparaten, later ordered by NL.
1970 Rheinmetall version (MATADOR) is cancelled. The German Oerlikon-Contraves version (5PZF-B) is fitted with the Siemens MPDR12 tracking radar developed for the Rheinmetall version. Pre-production vehicles are ordered (12, NL 5).
1973 contract issued to Krauss-Maffei for 122 vehicles with an option on 298. 55 ordered by Belgium, 95 by the Netherlands.
1976 start of series production.
1980 final deliveries.
1984 combat improvement program initiated applicable also to the Dutch version.
1996 contract to Krauss-Maffei to upgrade 147 German and 60 Dutch vehicles; option for a further 191 resp. 10 tanks.
1999 first refurbished vehicles delivered to Romania.

Distinguishing features. Side. LEOPARD 1 hull. In azimuth and traverse movable tracking radar on turret front; rotating target acquisition radar above turret. Twin gun with longitudinally fluted barrels and flash eliminators on external mounting to either side of turret. Two stabilised rotating head sights on turret roof. Front. Narrow, high, heavily sloped turret with external guns at sides. Shell antenna of target acquisition radar (PRTL with beam-shaped antenna).

Other details. Computerised fire control system with self-monitoring facility. The fully coherent pulse-Doppler search radar (E/F-bands) has a range of 15 km and IFF capability. It can be operated while the tank is on the move. Enemy target information will be passed to the coherent pulse-Doppler tracking radar (K-band). The search radar is maintaining search for other targets. The fire-control computer calculates the lead angles taking into account continuously measured muzzle velocity and the cant of the tank. Wind speed and direction, ballistic air temperature/speed data are fed in manually. Opening fire is automatically at ranges between 3000 and 4000 metres with bursts of 20 to 40 rounds. Main armament consists of two Oerlikon 35 mm KDA cannons. Each cannon is provided with 310 rounds and 20 rounds of armour-piercing ammunition. The computer is provided with a built-in check system. In case of malfunction the crew

has a collective NBC system. An auxiliary generator is placed at front beside driver, exhaust on left hand side of hull.
Combat improvement program comprises the integration of a digital fire-control computer, a stabilized passive optronic sensor with thermal imaging, linkage to the battlefield air-defence control and monitoring system (HFlaFÅSys), an improved velocity measuring system, the use of new high-velocity frangible APDS ammunition increasing the range up to 5000 m and a self defence system. (TABLE)

Role. In service with Germany (338) and the Netherlands (70) and Romania (32 plus 11 to be cannibalised for spares).

Assessment. Fully self-contained system, at the time of its introduction the best of its kind. Combat improvement program will decrease susceptibility to ECM and improve effectiveness against fast very low flying targets and armoured combat helicopters.

GEPARD AADG

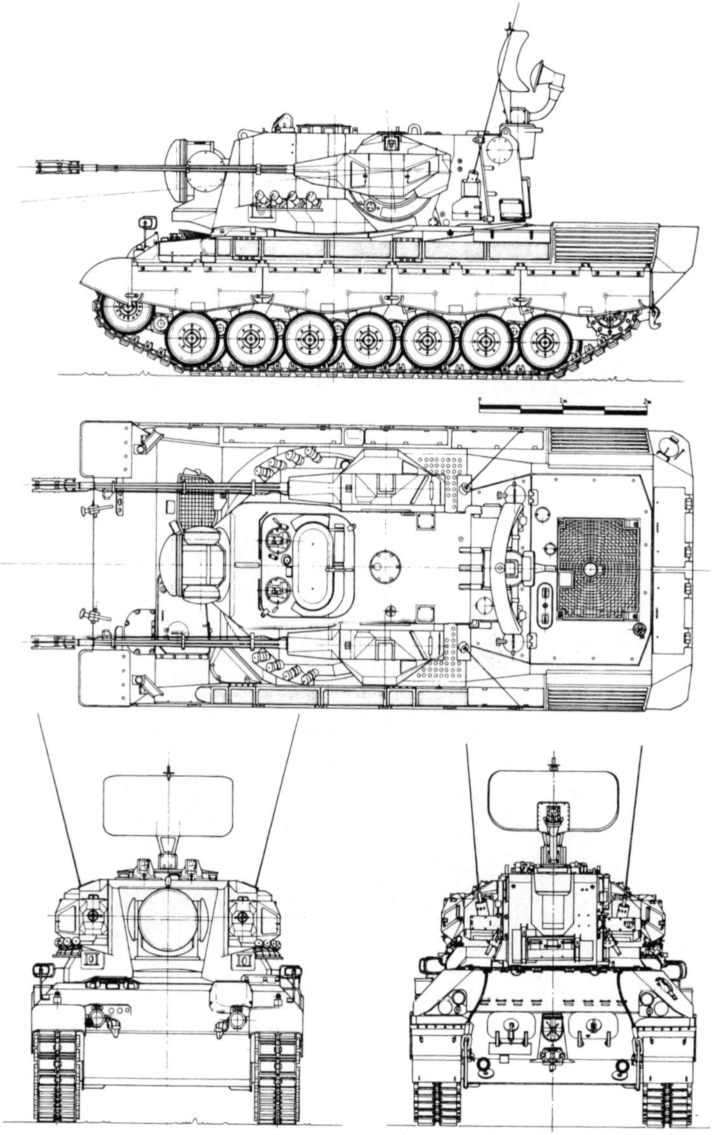

GEPARD AADG

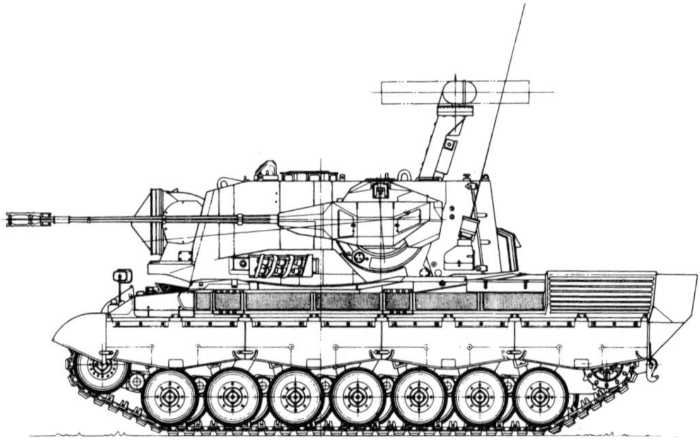

PRTL AADG

OZELOT short range air defence system

Models.

OZELOT launcher vehicle
OZELOT command post vehicle
OZELOT battery command post vehicle

Development and production.

1993 technical studies by Atlas E., DASA, Krauss-Maffei and Wegmann for an air-transportable missile system for the engagement of low-flying aircraft for the crisis reaction forces.
1995 the winning system ASRAD on WIESEL 1 chassis (Atlas short range air defence system) is awarded development and production contract.
1998 first deliveries.

Distinguishing features. WIESEL carrier hull slightly modified with high superstructure. Launcher unit resp. rectangular radar antenna on top of the rear hull roof.

Other details. The weapon system consists of the launcher and the combined sensor with the system electronics including the missile interface electronics, the platform electronics and the control and display console. The multipurpose launcher can fire different types of anti-aircraft missiles (Stinger, Mistral, Igla, RSB70/90, Starburst); four ready-to fire missiles.
The integrated sensor system consists of infra-red, TV and an eye-safe laser range finder. The system has automatic tracking and in-range and in-envelope processing. The system can be remote-controlled. It is equipped with a Pilkington Optronics passive Air Defence Alert Device.
The CP vehicle is fitted with a 3-D Helicopter & Aircraft Radio/Radar Detection system (HARD) from Ericsson in use with the Bofors RBS-90 system.

Role. In production for the German Army (50 launchers, 10 CP and 7 battery CP vehicles).

Assessment. Versatile light-weight SHORAD system which can be operated with different missile types.

OZELOT launcher vehicle

OZELOT CP vehicle

ROLAND armoured air defence missile launcher

Models.

ROLAND 1
ROLAND 3

Development and production.

1964 joint development of Bölkow and Nord-Aviation, Paris, of this missile system for the engagement of low-flying aircraft.
1971 trials with prototype on AMX 13 hull with trailer.
1973 trials on prototype of complete system on AMX 30 hull.
1975 option with the US Army for ROLAND 2 was taken up, collaboration with Hughes.
1977 start of series production for France.
1981 first deliveries to the German Army.
1988 production of ROLAND 3 starts.
1990 Roland Mach 5 missile offered by Euromissile with an effective range of 12 km

Distinguishing features. Side. MARDER AIFV hull slightly modified. Turret with brackets at either side for missile transit/launcher packs. One radar antenna at front side of turret, another (folding) at turret rear. Front. Narrow turret with elongated elliptical antenna right front, wider antenna on roof and at each side a large-diameter cylindrical transit/launch container on a bracket.

Other details. The weapon system consists of:
Acquisition system: radar and IFF system, Guidance system: optical sight, IR direction finder, tracking radar, fire-control computer and transmitter,
Launching system: turret and 2 launchers with 1 missile each (ROLAND 3 optional 2 x 2 launchers).
The weapon system is served by 2 men, the radar operator who is also the commander and the gunner. ROLAND 1 is a fair-weather system without tracking radar.
The ROLAND 3 missile has an extended range of 8000 m, a maximum speed of 600 m/s, a launch weight of 78 kg, a warhead weight of 9 kg.
(TABLE)

Role. In service with Brazil (4 ROLAND 1) and the German Army (143 ROLAND 3). Slovenia requested delivery of 6 systems.

Assessment. Successful attempt to combine launcher and guidance system on one armoured vehicle. Effective medium-altitude AADML with short reaction time.

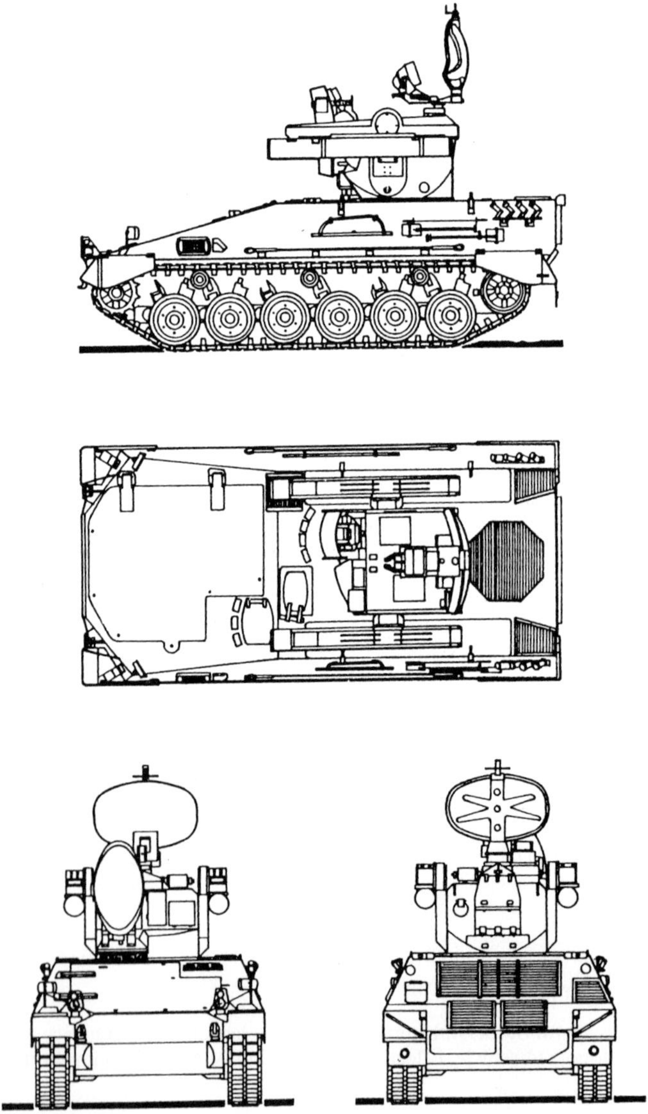

ROLAND AADML

ROLAND AADML

BIBER armoured vehicle launched bridge

Models

BIBER AVLB
LEGUAN AVLB

Non-serial version.

BrPz BIBER with SAS rapid bridging piers.

Development and production.

1967 request for an AVLB on the LEOPARD 1 hull.
1970 trials starts with 2 different prototypes, PT A built by Krupp MaK and PTB by Krauss-Maffei (bridges by Klöckner-Humboldt-Deutz).
1971 solution B is selected.
1973 production starts at MaK, bridges delivered by Eisenwerke Kaiserslautern.
1997 delivery of 9 LEGUAN MAN GHH bridge on LEOPARD 1 chassis to Norway.

Distinguishing features. Based on LEOPARD 1 MBT hull. Two-part bridge is carried horizontally and laid over front of tank. Dozer blade mounted at front under nose used as a support during laying of the bridge. LEGUAN with larger bridge of same appearance.

Other details. Bridge consists of two spans, lower half of bridge slides forward until its end is lined up with end of upper half. Total weight is 9.8 t, span is 22 m, MLC 60. The bridge is pushed forward with its nose free, the vehicle acting as counter-weight. The bridge can be retracted from both sides. Up to three bridges can be laid in an overlapping mode.
LEGUAN bridge has the same characteristics but a span of 26 m and a MLC of 70. Its weight is 10.9 t. The vehicle's weight is 49.5 t, the length 13.37 m, height 3.85, width 4.01 m. Tiltable dozer blade.
(TABLE)

Role. In service with Australia (5), Canada (9), Denmark (10 ex-Dutch), Germany (102), Greece (10 ex-German), Italy (63) and The Netherlands (15), Norway (9 LEGUAN).

Assessment. Novel laying system avoiding the scissors bridge high silhouette thus making it more difficult to detect.

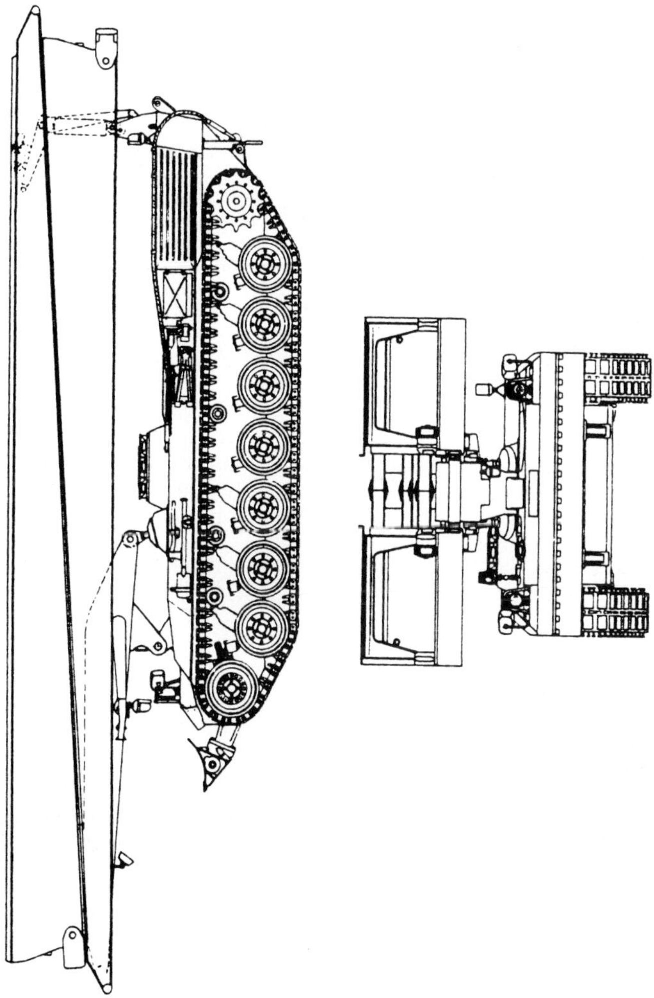

BIBER AVLB

BIBER AVLB

LEGUAN AVLB

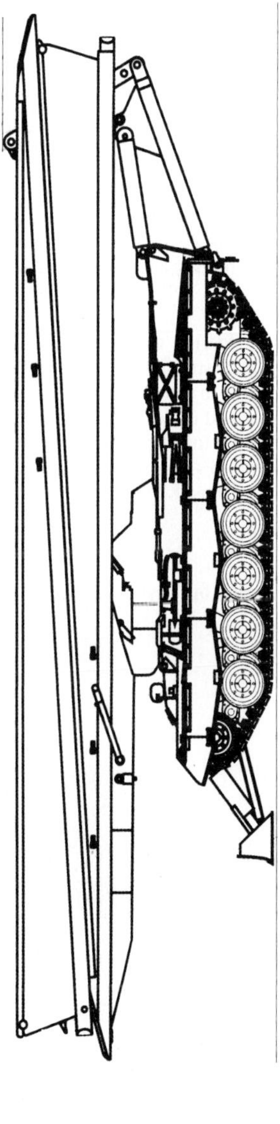

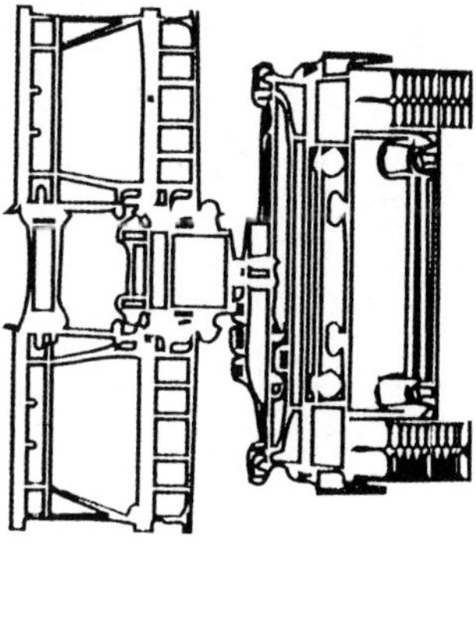

LEGUAN AVLB

KEILER armoured mine clearing vehicle

Development and production.

1971 trilateral requirement (with France and Italy).
1973 idea competition with 35 different solutions. The concept of Kaelble is selected.
1983 two prototypes built by MaK.
1997 first serial version delivered. 24 on order.

Distinguishing features. Reconstructed hull of M48A2C tanks with a rotatable and tiltable flail arm. In the road march condition it is positioned longitudinally on top of the vehicle.

Other details. When preparing for operation the clearing mechanism swings forward hydraulically. At the two part clearing mechanism a total of 24 steel flail hammers weighing 15 kg each are mounted on chains in a staggered fashion. In operation the shaft is height controlled by a sensor and offset in a 20° position against the lateral axis to the left. It rotates as 400 rpm in the opposite direction of vehicle travel. The ploughed depth is adjustable to -25 cm. The maximum clearing speed is 4 km/h, the width of the cleared path is 4.7 m.

Role. 24 on delivery for the German Army.

Assessment. The world-wide only reliable mine-clearing system to cope with any type of mines.

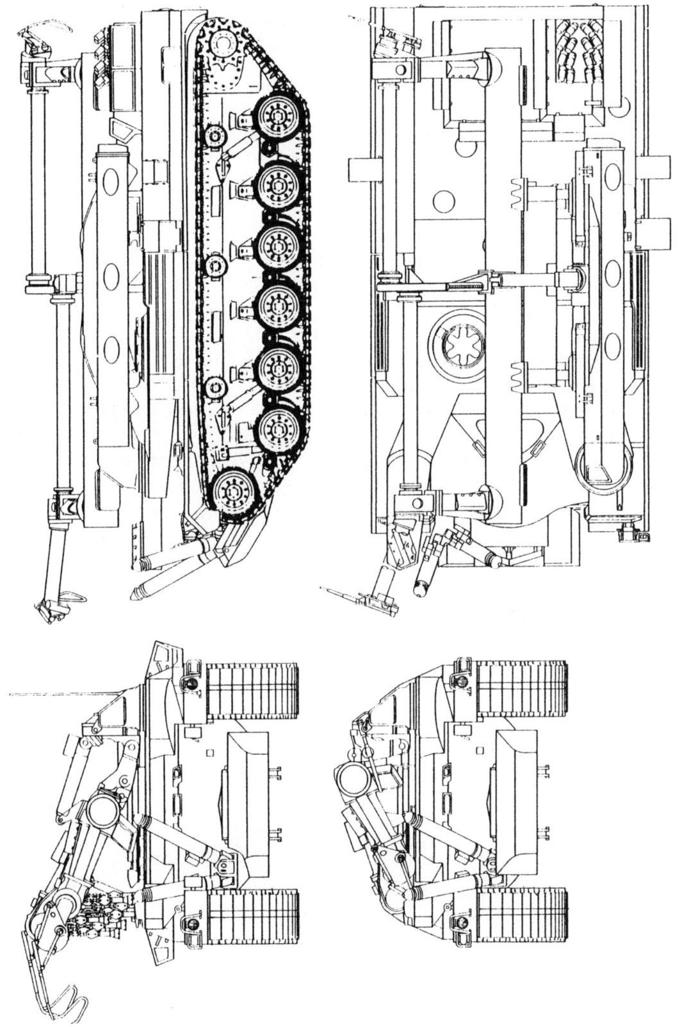

KEILER AMCV

KEILER AMCV

KEILER AMCV in operation

Pionierpanzer armoured engineer vehicle

Models.

Pionierpanzer 1
Pionierpanzer 2 DACHS
BADGER AEV (Canada)
Mine clearing tank (Netherlands)
Demining Tank (Norway)

Development and production.

1966 OR for a LEOPARD based AEV. Prototype from Krupp MaK (Porsche development).
1968 series production at Krupp MaK starts.
1972 development work starts on PiPz 2 for the roles of approach/ exit improvement and recovery on water obstacles. Prototype 1 (MaK) with 2 excavator units; PT 2 (EWK) with single excavator unit on turret. Project cancelled owing to financial stringency.
1981 re-start of PiPz 2 development in a re-built version. Testing of 3 prototypes.
1985 production order is placed.
1988 retrofitting of 36 PiPz 1 and 104 Bergepanzer 2 to PiPz 2 DACHS starts.
1990 final deliveries.
1997 Dutch conversion of 8 AEV and 9 ARV to mine clearing tanks.

Distinguishing features. All main assemblies other than the supplementary equipment needed for the engineer role are identical to those of the BPz 2 ARV.

Other details. The heavy duty dozer blade can be widened to 3.75 m by side extensions and/or fitted with four small grubbing teeth with a working depth of 50 mm or eight large teeth with a working depth of 400 mm. Dozing/grubbing speed is 8 km/h giving a capacity of some 200 m3/h. An auger stowed on the vehicle can be fitted to the crane jib; it is equipped with a hydraulic radial piston motor and can dig foxholes with a diameter of 700 mm and up to a depth of 2 m. The vehicle is also provided with special stowage for explosives, and with electrical welding equipment.
PiPz 2 DACHS with a traversing telescopic excavator in place of crane operated either by the commander or by the driver (also via remote-control and when tank is operated submerged). The vehicle is optimized for the amphibious use and has now a stronger hydraulic system of 300 bar. The excavator is 1.5 m wide and has a capacity of 1.1 m3. It can be operated up to a height of 8.3 m and a depth up to 5.05 m.
(TABLE)
Dutch conversion to mine clearing vehicle with Pearson mine plough, electro-magnetic influence mine clearing system, path marking system.

Bofors and Swedish Demining developed a 9 t roller system with tungsten-carbide teeth which is mounted on the front of a LEOPARD 1 chassis and has a width of 4 m. Clearance speed is about 5 km/h.

Role. In service with Belgium (6), Italy (39, of which 28 were produced under licence) and the Netherlands (22, 7 converted to mine clearing, 6 to breaching tanks). Germany uses 140 PiPz 2 DACHS, Canada 9 called BADGER. Norway has 7 Demining tanks.

Assessment. A practical and versatile vehicle with special equipment well suited to forward engineer roles.

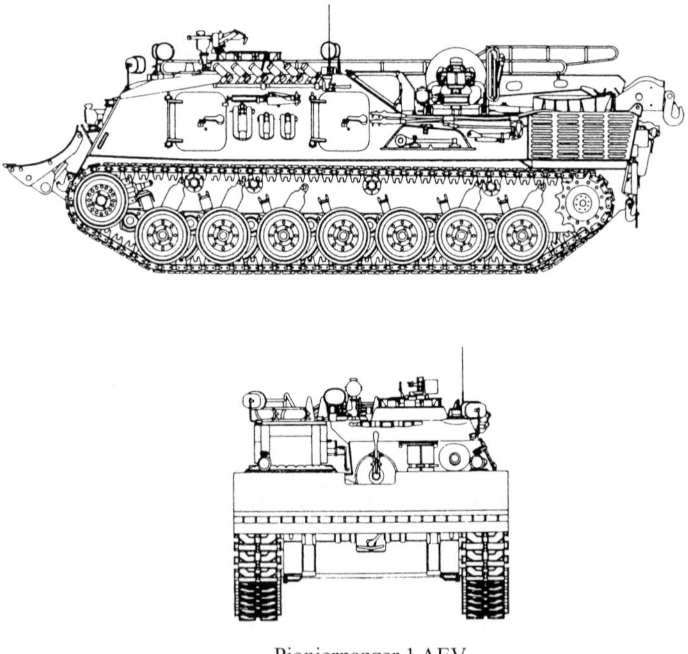

Pionierpanzer 1 AEV

Pionierpanzer 1 AEV

DACHS AEV

BADGER AEV

BERGEPANZER 2 armoured recovery vehicle

Models.

Bergepanzer 2
Bergepanzer 2 A2

Development and production.

1962 OR for an ARV with as many assemblies as possible common to the LEOPARD 1 MBT.
1964 two prototypes delivered from Jung-Jungenthal (Porsche development).
1966 start of production at MaK.
1977 improved model (BPz 2 A2) to meet needs of GEPARD AADG.
1997 Dutch conversion of 9 ARV to mine clearing tanks.

Distinguishing features. Front. Sloping fully sponsoned hull side. Full width dozer blade. On right hand side turntable with crane (jib stowed rearwards when not in use). Commander's hatch with ADMG ring. Side. LEOPARD 1 running gear. High crew compartment at front with dozer blade under forward part of hull which has flat sides with distinctive step down from crew to engine compartment at vertical rear. Crane jib in stowed position runs whole length of right hand side.

Other details. Driver's hatch at front centre of crew compartment roof. Rotating periscope left near periscope ring and ADMG ring-mounting. Periscope for hull MG left front. Platform for replacement power-pack at rear. Two hatches in left side wall of crew compartment. Smoke dischargers (6) left side. Collective NBC protection. Snorkels to 4 m depth with full width tube on cdr's hatch. Main winch with cable tensioning and wind-on devices powered by two hydraulic motors. Welding and cutting equipment, power spanners, chain saw, ground tackle, fuel system filling and emptying equipment.
BPz 2 A2 has newly developed subsystems. Double rotary pump for two hydraulic systems (dozer blade and spade). Rear spade, supplementary spade arm and plate right side swing upward in travel position. New valve block. Dozer blade lock. Strengthened traversing crane jib with extended range (20 t load). Wt 40.58 t, L 7.68 m. (TABLE)

Role. In service with Australia (8), Belgium (36), Brazil (1), Canada (8 A2 named TAURUS), Denmark (14 ex-NL), Germany (340 plus 100 BPz 2 A2), Greece (5), Italy (127 of which 10 are for sale), Netherlands (28, 9 mine-clearing tanks), Norway (17) and Turkey (12).

Assessment. A versatile, manoeuvrable and highly mobile tank with very well designed recovery equipment. The capability for quick in-situ power-pack change on breakdowns reduces total down-time.

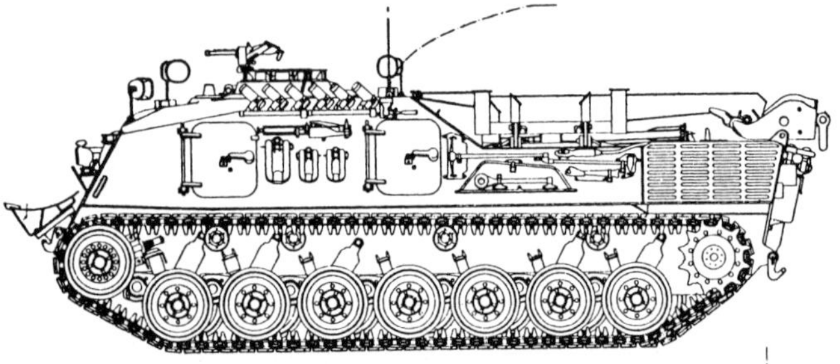

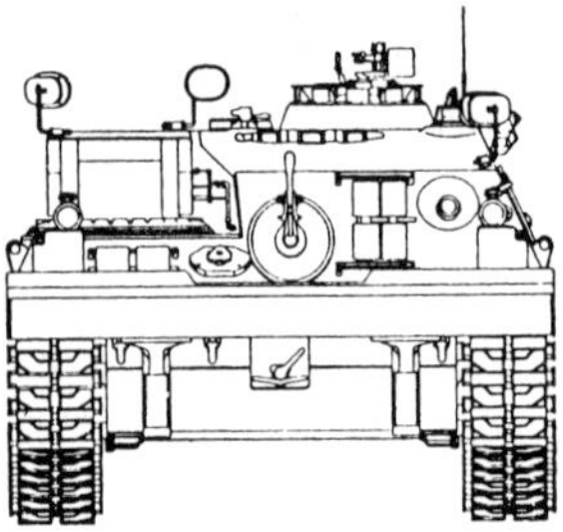

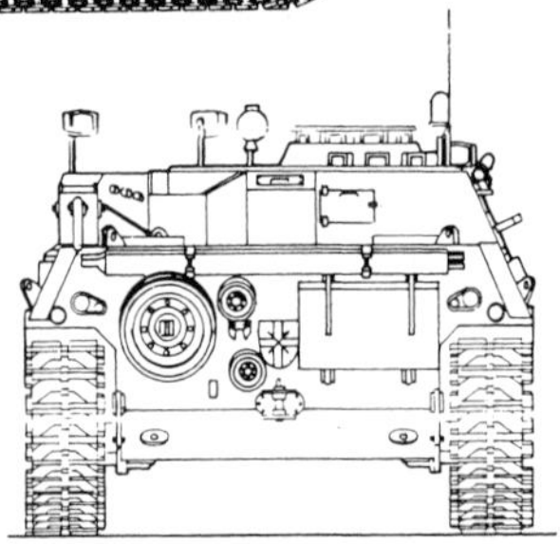

Top: Bergepanzer 2 ARV
Down: Bergepanzer 2 A2 ARV

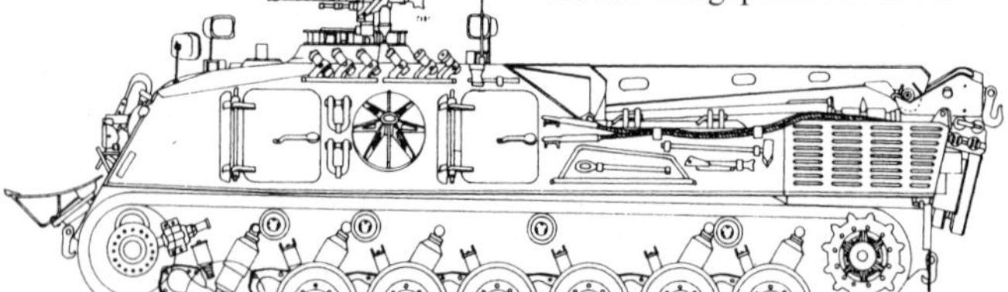

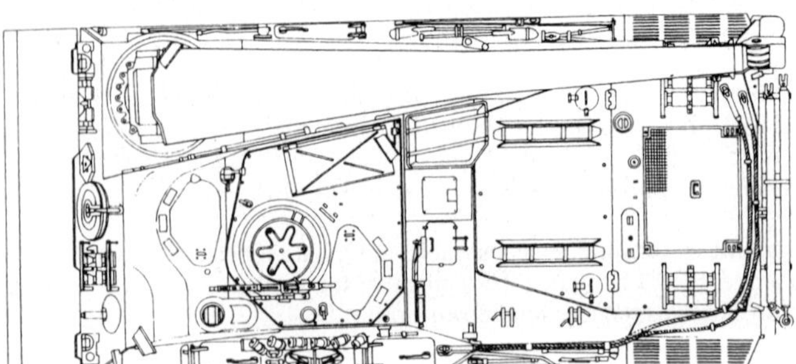

Bergepanzer ARV (Netherlands)

Bergepanzer 2 A2 ARV

BERGEPANZER 3 BÜFFEL armoured recovery vehicle

Development and production.

1977 concept studies on MLC 60 ARV on LEOPARD 2 MBT chassis starts.
1982 development of BPz 3 commenced, construction of an experimental vehicle.
1987 end of definition phase, two prototypes built.
1989 troop trials.
1991 start of delivery.

Distinguishing features. Side. LEOPARD 2 MBT chassis. High crew compartment at front with distinctive step down to engine compartment, dozer blade under forward part. Three large doors on the left side of the crew compartment. Crane jib in stowed position runs whole length of right hand side. Front. Left part of the tank with steep glacis with driver's position and flat cdr's cupola with ADMG ring behind. On right side turntable with crane. Full width dozer blade raised when not in use.

Other details. Main components of the recovery equipment include the traversing crane unit with a capacity of 30 t maximum (electronic load momentum limiter) and a dual capstan winch with an effective cable length of 160 m (total length 180 m, diameter of 33 mm) of a tractive capacity of 35 t with a guide pulley for double traction. An auxiliary winch is also used to extract resp. recover the main winch cable. There is also cutting and welding equipment and a rear power pack transportation device, various couplings and tow bars including a quick-recovery bar. The tank is also provided with a fire suppression system both in the crew and in the engine compartment, has a NBC protection system and has submerged driving capability.
(TABLE)

Role. In service in Germany (75), the Netherlands (25) and Switzerland. 5 German AFVs are on lease in Spain which has a requirement of 23 to be produced from 2000 onwards. Sweden will field between 28 and 32 ARVs starting in 2002.

Assessment. Most advanced ARV currently available with outstanding performance data.

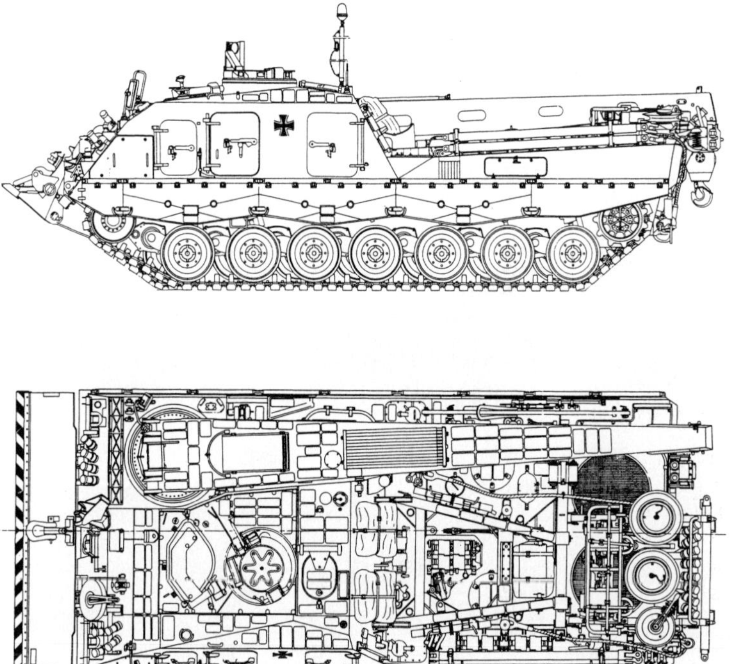

Bergepanzer 3 BÜFFEL ARV

BÜFFEL ARV

CVR(W) FOX FV721 armoured scout car
Combat Vehicle Reconnaissance (Wheeled)

Development and production.

1965 by Daimler as successor of FERRET scout car.
1972 series production starts.
1986 production ceases.

Distinguishing features. Side. 4x4 wheeled vehicle with two large road wheels each side with stowage boxes between. Low octagonal turret with slender gun. Front. Sloped glacis plate with driver's hatch opening to the right side in upper part. Turret on small ring with lower and upper sides sloping inwards. Two large periscopes on turret roof.

Other details. Driver has one wide angle vision block. Commander/loader at left side with peri-binoculars and seven periscopes, gunner right side with binocular sight (x1 and x10 magnifications) and two periscopes. To the right side of the gun is a Rank Precision Industries SPAV L2A1 passive night side with x5.8 or x1.6 magnification. The 30 mm RARDEN gun fires single shots or bursts of six at a cyclic rate of some 90 rds/min. From 1985 APDS-T projectiles with 300 g penetrator against 40 mm steel at 1500 m are in use. Commercial 4.2 litre Jaguar engine as in SCORPION with few modifications. Original FOX had erectable flotation screen later removed in British service.

Role. In service with Malawi (22) and Nigeria (50). Phased out of British service, turrets used for SABRE modification.

Assessment. A very handy, fast, quiet vehicle. Main gun of high performance against light armoured targets.

FOX AVR

VR(T) SCORPION armoured vehicle reconnaissance
Combat Vehicle Reconnaissance (Tracked)

Models.

SCORPION FV101 AVR
SCORPION 90 ARFSV
SABRE AVR
STRIKER FV102 TDM
SPARTAN FV103 APC
SPARTAN MCT (MILAN compact turret) APC
SAMARITAN FV104 AAV
SULTAN FV105 ACCC
SAMSON FV106 ARV
SCIMITAR FV107 AVR

Development and production.
1965 development by Alvis as SALADIN replacement.
1969 first prototype completed.
1971 first SCIMITAR prototype constructed.
1972 start of series production of SCORPION.
1974 first deliveries of SCIMITAR.
1975 first STRIKER delivered.
1977 SULTAN enters service use.
1978 service enter of SPARTAN, SAMSON and SAMARITAN.
1981 export order for SCORPION 90 by Malaysia.
1985 SPARTAN MILAN accepted for service use in the BAOR.
1993 British SCORPION conversion to SABRE standard.
1997 AFV family still in production (for Indonesia).

Distinguishing features. Side. Five roadwheels, drive sprocket front, no top rollers. Low, sloped glacis, vertical lower front and rounded nose. Horizontal hull roof, vertical hull rear with oblong stowage box. Octogonal turret with lower portion undercut. Wedge-shaped mantlet with short gun. Front. Driver's hatch in upper left of sloped glacis, louvres on right side. Sponsoned hull with sloped sides. SCIMITAR identical, smaller but longer gun tube.
Variants with enlarged crew compartment and different external appearance.

Other details. Flotation screen removed from British vehicles. Jaguar J60 6-cylinder petrol engine of 140 kW is replaced in late production export vehicles by Perkins T6-3544 turbo-charged 6-cylinder diesel engine with 147 kW.
SCORPION has two roof-mounted sights (x1 and x10 magnifications) and a Rank Precision Ind. passive night sight (x1.6 and x5.8 magnifications). Irish SCORPIONS have an addtional 12.7 mm M2 HB MG on the turret roof. Export SCORPION differ in many details from British type (i.e. lack of NBC protection system). (TABLE)
British SCORPIONS are either phased out or received the 30 mm gun turret of

the formerly used FOX (4x4) AVR; new designation is SABRE. The L37A2 7.62 mm coaxial MG has been replaced by the chain gun of the same calibre in use with the WARRIOR.
SCORPION 90 is fitted with a 90 mm Cockerill Mk III gun (34 rounds). The gun has an elevation of +30 down to -8 degrees. It has power traverse/elevation. Nigerian SCORPION 90 have Belgian OIP-5 fire-control system. Barr & Stroud laser rangefinder/gunner's sights were partly refitted.
STRIKER is equipped with a rear elevatable launcher box with five SWING-FIRE ATGM, five reload missiles are stowed inside the tank. The gunner has a monocular sight (x1 and x10 degrees magnifications) traversable through 55° left and right. The missiles range is up to 4000 m. Vehicle's weight is 8.35 t, height is 2.28 m.
SPARTAN can carry four additional soldiers. It is also used carrying a ZB 298 ground surveillance radar on top of the roof. The weight is 8.17 t, height 2.26 m
SPARTAN MILAN has the MCT with two ready-to-launch MILAN, plus eleven missiles carried inside.
SAMARITAN can carry four stretcher or five sitting patients, or two stretcher and three sitting. Crew of two, weight is 8.664 t.
SULTAN has additional radio equipment and a pintle-mounted 7.62 mm MG on top of a No 27 cupola. Weight is 8.66 t, height 2.56 m
SAMSON has an internal winch of 4.1 t pull with a 229 m rope. An A-frame can be erected at the hull rear. Various tool equipment. Weight is 8.74 t, height 2.25, crew of three.
SCIMITAR has the same turret as SCORPION but is armed with a 30 mm RARDEN cannon (165 rounds carried plus 3000 7.62 mm). Weight is 7.8 t.
For the vehicles deployed in Bosnia the SPIRE sighting system was procured comprising of a Hughes thermal imaging module, one display for the commander and the gunner each and an electronics unit.
Export versions partly lack NBC protection system. Several countries retrofitted SCORPIONs with laser rangefinders.

Role. In service or ordered by Belgium (43 STRIKER, 230 SPARTAN, 47 SULTAN, 41 SAMARITAN), Bolivia (12 SCORPION), Botswana (36 ex-Belgium SCORPION), Brunei (16, 2 SULTAN, 1 SAMSON), Chile (30 ex-British SCORPION), Great Britain (127 SABRE, 315 SCIMITAR; 58 STRIKER, 553 SPARTAN + 75 SPARTAN with MCT, 265 SULTAN, 50 SAMARITAN, 86 SAMSON), Honduras (12, 3 SCIMITAR, 1 SULTAN), Indonesia (121 SCORPION 90), Iran (50), Ireland (14 SCORPION), Jordan (19 SCORPION donated by Iraq as war booty), Malaysia (26 SCORPION 90), New Zealand (26 SCORPION), Nigeria (17, 33 SCORPION 90, 5 SCIMITAR), Oman (30 plus SULTAN, SPARTAN, 3 SAMSON), Philippines (28 SCORPION), Spain (17 SCORPION), Tanzania (36 SCORPION), Thailand (144 SCORPION), Togo (12 SCORPION 90), UAE (80 SCORPION), Venezuela (84 SCORPION 90, SULTAN, SAMSON).

Assessment. Versatile AFV family, light weight due to aluminium armour. Highly mobile with long range when using diesel engine.

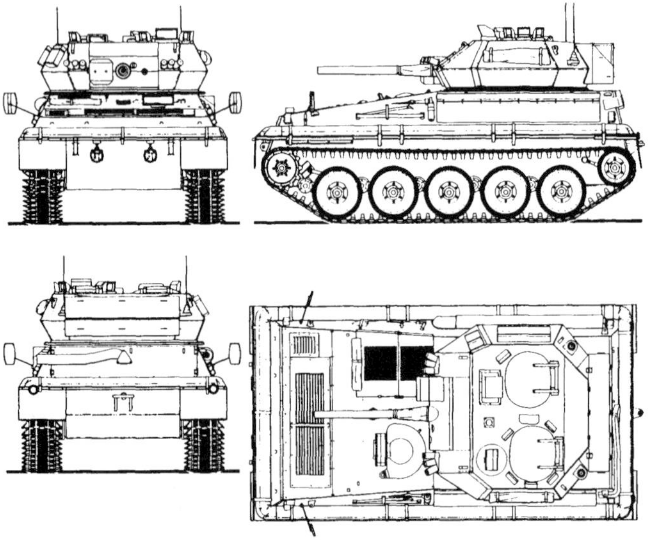

SCORPION AVR

SCORPION AVR

SCORPION 90 AVR

SABRE AVR

STRIKER TDM

SPARTAN MILAN APC

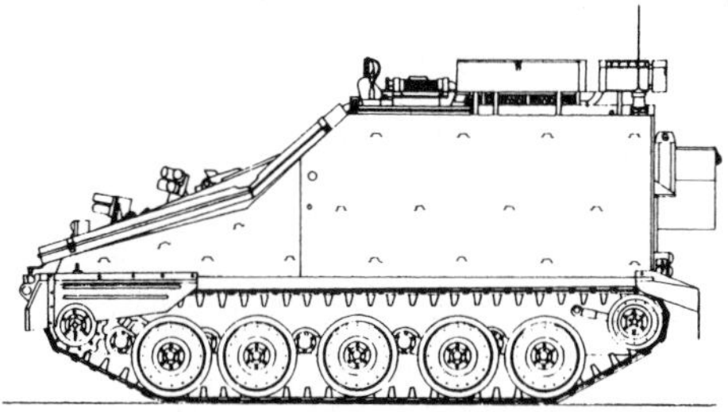

SPARTAN APC

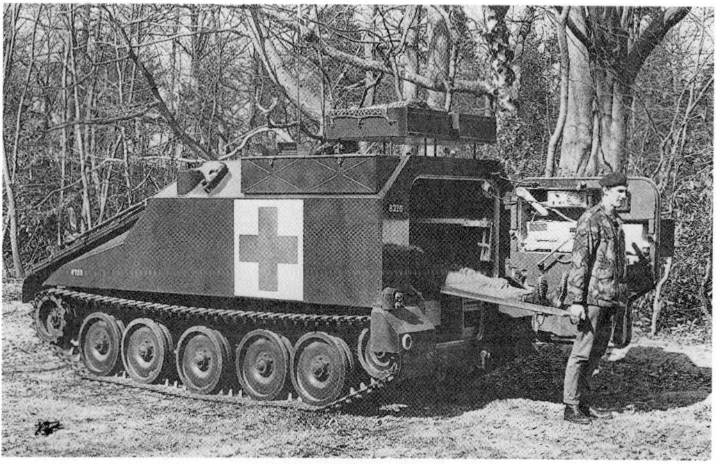

SAMARITAN AAV

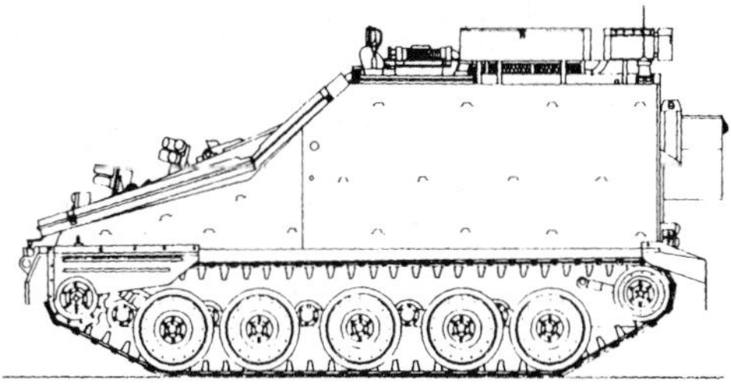

SAMARITAN AAV

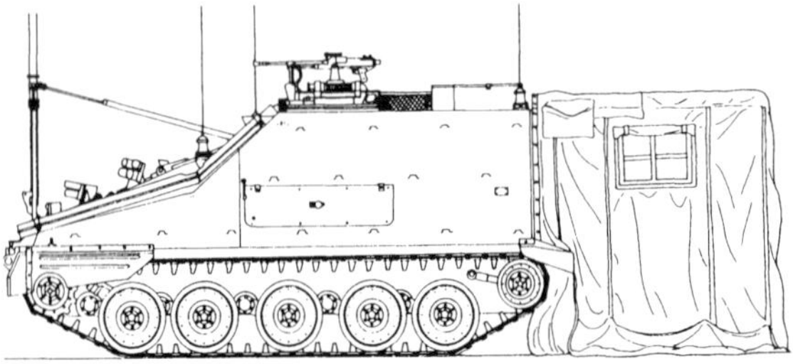

SULTAN ACCC

SULTAN ACCC

SAMSON ARV

SCIMITAR AVR

CENTURION main battle tank

Wt 51.8 t L 9.86 W 3.39 H 3.0 551 kW 45 km/h 500 km Crew 4

Models.

CENTURION Mk 8/1
CENTURION ARV Mk 2
OLIFANT MBT (South Africa)
OLIFANT ARV
TARIQ MBT (Jordan)
SHOT-KAL D MBT (Israel)
NAGMASHOT/PUMA APC (Israel)
NAKPADON APC (Israel)
MAR 290 AARL (Israel)

Development and production.

1944 A41 design with 76.2 mm gun.
1945 start of series production (Mk 1).
1948 Mk 3 with 83.4 mm gun produced.
1959 Mk 9 with 105 mm gun produced as an upgunned Mk 7.
1962 production ends after 4.423 MBT having been delivered, many older versions in British and foreign use were upgraded.

Distinguishing features. Side. Six roadwheels, four return rollers, sprocket rear. Upper part of suspension covered by metal skirts. Sloped front, horizontal hull roof and raised engine deck, almost vertical rear. High rounded turret with bustle and wide low-profile cupola. Long gun with fume extractor near to centre. Front. Sloped glacis with split driver's hatch on the right. Cylindrical turret with built-on boxshaped stowage boxes on both sides, wide mantlet, cupola right. Vertical hull sides.

Other details. Main armament of all versions still in service is the L7 105 mm gun. A 12.7 mm ADMG (600 rounds) and a 7.62 mm coaxial MG (4750 rounds) are secondary armament.
The OLIFANT is the South African modernisation of former Mk 3 or Mk 5 with new diesel engine (Continental AVDS-1790-2A copy), the L7 105 mm gun (72 rounds stowage) with Armscor fire control system. With its increased fuel capacity of 1280 litres the OLIFANT has a range of 500 km. Its maximum road speed is 45 km/h. An upgraded version OLIFANT Mk 1B has the 706 kW V-12 turbocharged engine coupled to a new AMTRA III automatic transmission. Suspension was also upgraded. The Mk 1B has special armour on the glacis, the front, front sides and front top of the turret. The OLIFANT ARV has a new lower ratio final drive, a new dozer blade and some equipment changes.
The Jordanian Mk 7/2 and Mk 5/2 were fitted with a Belgian SABCA fire control system with laser rangefinder, the AVDS-1790 diesel engine, a Cadillac

Gage turret drive and stabilisation system, hydropneumatic suspension, and a Graviner explosion suppression system. Its designation is now TARIQ.
An ambitious refurbishment was made by Israel including the integration of the L7 105 mm gun and the AVDS diesel engine. An Elbit fire control system was retrofitted. From 1982 the Israeli CENTURION are equipped with external explosive reactive armour (BLAZER). Israeli CENTURIONs can also be equipped with the front mounted RAMTA mine clearing device.
In 1985 CENTURIONs with their turrets removed are in use as APC for the infantry (NAGMASHOT) or the engineer corps (PUMA). The NAGMASHOT is armed with two or three pintle mounted MGs while the PUMA is equipped with the Rafael OWS (overhead weapon station). The NAKPADON is also provided with ERA around the upper part of the troop compartment.
The MAR 290 is a CENTURION chassis with a four-tube 290 mm rocket launcher with a firing range of 25 km. The vehicle's weight is 50.8 t, its height is 3.6 m.

Role. In service with Israel (several hundred MBTs and APCs), Somalia (28 Mk 8/1 from Kuwait) and South Africa (154).

Assessment. Totally obsolete heavy tank even when extensively upgraded. Israeli APC versions represent efficient low-cost solutions.

OLIFANT MBT

SHOT-KAL MBT

NAGMASHOT APC

NAKPADON APC

MAR 290 AARL

CENTURION ARV

CHIEFTAIN FV4201 main battle tank

Models.

Mk 3/3P (Persia)
Mk 5/2K (Kuwait)
Mk 5/5P (Persia)
Mk 15 QAYID (Oman)
FV4030/2 Khalid
CHIEFTAIN AVRE

Development and production.

1956 requirement to replace CONQUEROR and CENTURION, three prototypes (FV4202) built by Leyland Motors.
1958 modification of concept led to FV4201, one PT built.
1961 further six PTs built, adoption for British Army.
1963 production start.
1974 Iranian order for 1200 FV4030/1, only 187 delivered.
1981 deliveries of KHALID (former SHIR 1) to Jordan starts.
1984 delivery of Mk 15 for Oman starts.
1986 conversion of 12 CHIEFTAIN AVRE at BAOR.
1991 48 CHIEFTAIN AVRE ordered.

Distinguishing features. Side. Six roadwheels and three return rollers, drive sprocket rear. Upper part of suspension covered by skirts. Well sloped glacis with reclining driver, raised engine compartment. Low-profile turret with heavily sloped front, sloped sides and bustle. Long gun with fume extractor and thermal sleeve in mantletless mounting. Commander's cupola with ADMG and IR spotlight. Front. Large stowage bin on right hand side of turret; permanently built-in searchlight on left (removed after installation of TOGS). Cdr's cupola on the right side.

Other details. Stabilised 120 mm gun using seperate ammunition, fitted with a muzzle reference system. NBC protection system with filter pack in the turret rear. Infra-red detector mounted on the turret roof. Part of the tanks are fitted with dozer blades.
KHALID with Perkins Condor 12V 1200 diesel engine (882 kW) coupled to a TN37 transmission. Bogie type suspension. Tank is fitted with a Pilkington PE Condor cdr's day/night sight.
CHIEFTAIN AVRE with provision to carry Class 60 trackway and pipe facines and/or attachment of Pearson Engineering Mine Plough (EMP) or Pearson Engineering Combat Dozer Blade (UDK1); gun retained.

Role. In service with Great Britain (60 AVRE plus 79 MBT in storage), Iran (60 - 100, status unknown), Jordan (274 KHALID), Kuwait (45 in storage) and Oman (27 in storage).

Assessment. Good firepower, mainly with KE. Markedly below-average mobility.

KHALID MBT

CHIEFTAIN AVRE

CHALLENGER main battle tank

Models.

CHALLENGER 1
CHALLENGER 2

Non-serial version.

CHALLENGER 2E with MTU 883 engine.

Development and production.

1971 experimental vehicle (FV4211) with new Chobham special armour.
1979 projected production start of SHIR 2 for Iran; contract cancelled, vehicles FV4030/3) completed as CHALLENGER prototypes.
1983 issue of first CHALLENGER MBT for the British Army.
1986 work starts on CHALLENGER 2 as private venture.
1989 CHALLENGER 1 production stops. Last effort contract for Vickers to develop an indigenous British MBT (CHALLENGER 1 Mk 2) with new L/30 120 mm CHARM gun to be ready in 1990.
1991 decision in favour of CHALLENGER 2.
1993 start of series production.
1997 first tanks issued to field units.

Distinguishing features. Side. Six roadwheels, four return rollers, sprocket rear. Upper part of suspension covered with skirts. Sloped glacis plate with front idlers projecting ahead of nose. Low hull with slightly raised engine deck, pointed rear. Large, angular turret with sloped front and rear bustle. Cdr's cupola with ADMG. Front. High lower front with driver's hatch recessed in centre. Large overhanging turret with vertical sides. Cdr's cupola on right side. CHALLENGER 1 with opening in right side of turret for TOGS. CHALLENGER 2 has the thermal sight mounted in a box above the gun.

Other details. Main armament is the standard L11A5 120 mm gun firing seperate ammunition. Marconi IFCS fire-control system with laser rangefinder. After the decission to phase out CHALLENGER 1 the CHALLENGER Improvement Program (CHIP) has been reduced to the exchange of the ill-fated TN37 transmission by the TN54 type with six forward and two reverse gears. New Perkins auxiliary power unit already integrated in late production vehicles, new generator.
MBT serving in the 1991 Gulf campaign were fitted with Vickers passive side armour and reactive front armour (designation CHALLENGER 1 Mk 2). 20 vehicles received Gallay air-conditioning systems.
The last 156 vehicles produced have a revised internal layout and armoured charge bins for better protection of the propellant charge; their designation is Mk. 3.
(TABLE)

CHALLENGER 2 has a new-designed turret with the RO 120 mm L30 rifled gun firing seperate ammunition. The fire-control system is derived from the M1A1 ABRAMS type. The roof-mounted SFIM stabilised day-sight periscope is identical that of the LECLERC MBT. The gunner is provided with a roof-mounted stabilised day and night sight. The hull is almost identical to CHALLENGER 1. A new designed double pin track is used.
The CHALLENGER 2 ordered by Oman have an improved cooling system for the engine, a single pin track and has the 7.62 mm loader's MG replaced by a 12.7 mm pintle-mounted one.

Role. In service with the British Army CHALLENGER Driver Training Tanks (CTT). Jordania (288 CHALLENGER 1). CHALLENGER 2 in production for the British Army (386 on order) and for Oman (18, 20 more ordered).

Assessment. The development and procurement process for both versions suffered from political and economic pressure. This irregularity added to many unreliabilities. CHALLENGER 2 has good armour protection, average fire-control system, low mobility.

CHALLENGER 1 MBT

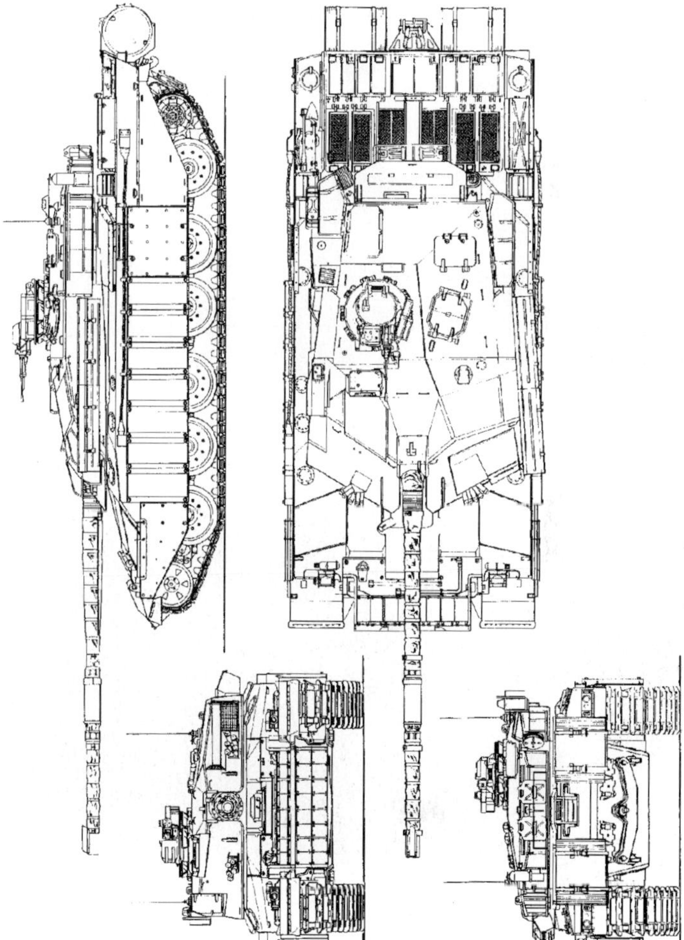

CHALLENGER 1 Mk 2 MBT

CHALLENGER 1 Mk 2 MBT

CHALLENGER 2 MBT

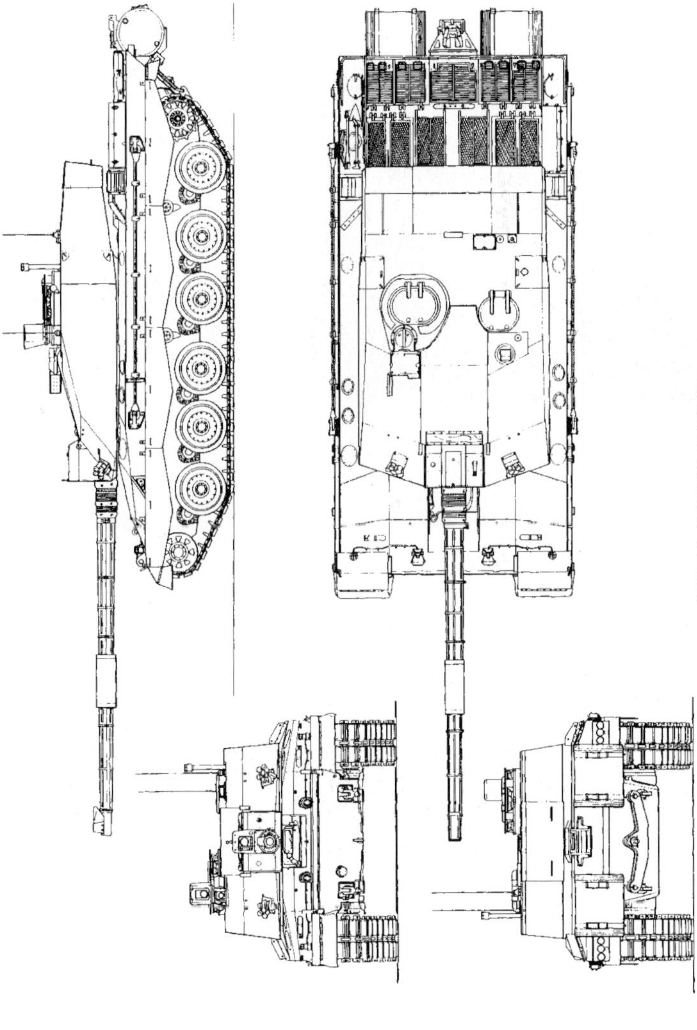

CHALLENGER 2 MBT

CHALLENGER 2 MBT (Oman)

Vickers main battle tank

Wt 40 t L 9.79 W 3.17 H 2.64 529 kW 50 km/h 530 km crew 4

Models.

Mk 1 (India: VIJAYANTA)
Mk 3
CATAPULT AAH (India)
Vickers AVLB
Vickers ARRV
VIJAYANTA ARV

Non-serial versions.

Vickers Mk 3(M) MBT with ERA add-on armour
155 mm GBT 155 turret on Mk 3 chassis
MARKSMAN 2x35 mm anti-aircraft system on Mk 3 chassis
KARTIK AVLB (India)
VIJAYANTA with GBT 155 turret

Development and production.

1963 two PT private venture by Vickers. First export order by India.
1965 licence production of Mk 1 starts in India, designation VIJAYANTA.
1970 order by Kuwait.
1977 order of Mk 3 plus ARV by Kenya.
1981 Indian conversion of locally built VIJAYANTA chassis'.
1983 export order by Nigeria incl. ARV and AVLB.

Distinguishing features. Side. Six roadwheels, large idler front, three return rollers, sprocket rear. Upper part of track and suspension normally covered by sheet metal skirts ribbed horizontally. Sloped glacis plate, horizontal hull top with slightly raised engine deck, vertical rear. Main gun with fume extractor and thermal sleeve. Front. Driver's hatch on the right side. Turret with external mantlet with stowage boxes each side of turret rear. Cdr's cupola right side of turret roof with prominent day/night sight in forward part.

Other details. Technical data shown above refer to Mk 3 MBT. Both versions have electric turret power control and are fitted with one 12.7 mm ranging MG. Fuel capacity is 1000 litres.
Mk 1 MBT has a weight of 38.6 t and is equipped with the L60 Mk 4B Leyland engine of 478 kW. Ammunition stowage is 44 rounds for the 105 mm L7A1 main gun, 3000 for the two 7.62 mm MG and 600 for the RMG. Gun elevation is +20 up to -7 degrees. Gun stabilisation.
Mk 3 has a cast turret front and is equipped with a General Motors 12V-71T diesel. Ammunition stowage for the main gun is 50 rounds, 2600 rounds for the two 7.62 mm MG. India has upgraded about 1100 vehicles with the integration of

the Marconi SFCS 600 fire-control system, an new locally built 550 kW diesel engine and Kanchan appliqué armour.
Mk 3 is also offered with a laser range-finder instead of the RMG. Maximum depression of the gun is increased to -10 degrees.
CATAPULT is built on an elongated hull with seven roadwheels and four return rollers. The 130 mm M-46 gun is integrated in an open mounting pointing to the rear. The crew compartment is covered by a canvas top. About 30 rounds can be carried.
AVLB carries a bridge of 13.41 m length.
ARRV is produced with or without a crane. The main winch can pull 25 t. The crane can lift up to 4 t to 3.62 m. The vehicle's weight is 36.8 t.
Indian ARV has an A-frame boom attached on the front hull roof which is folded backward in transportation mode.

Role. In service with India (1250 Mk 1 in reserve, 100 CATAPULT), Kenya (76 Mk 3, 7 ARV, 6 AVLB), Nigeria (108 Mk 3, 15 ARV, 18 AVLB), Tanzania (3 ARV with GM engine).

Assessment. Light armoured, average mobility and firepower. Mk 1 quite unreliable propulsion system.

VICKERS Mk 1 MBT

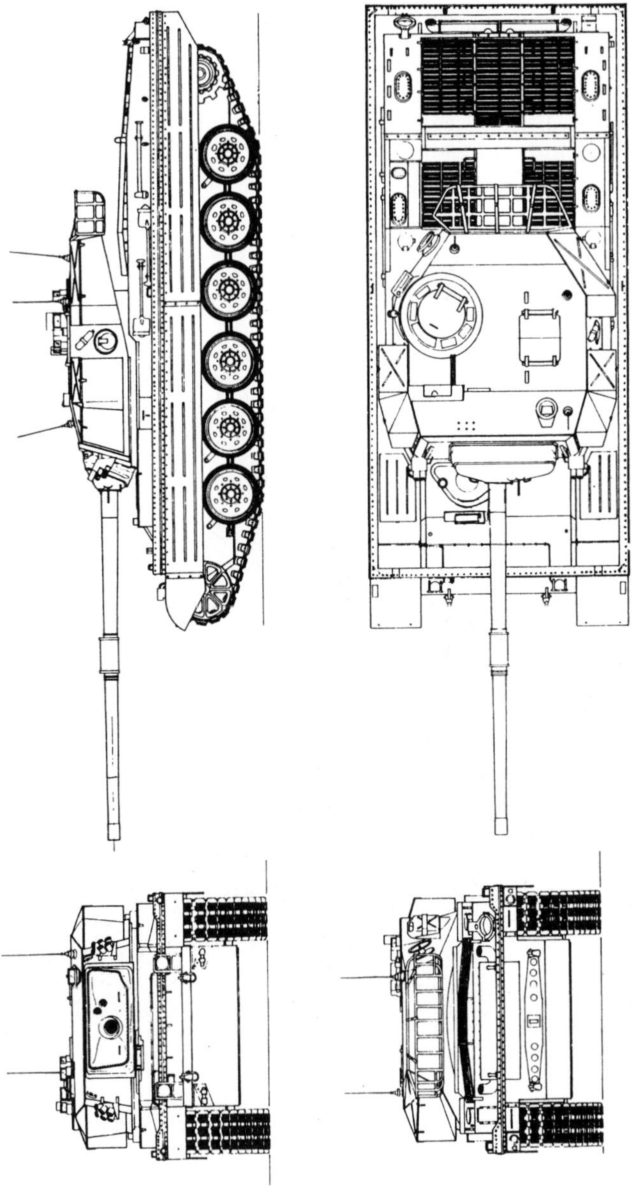

Vickers Mk 1 MBT

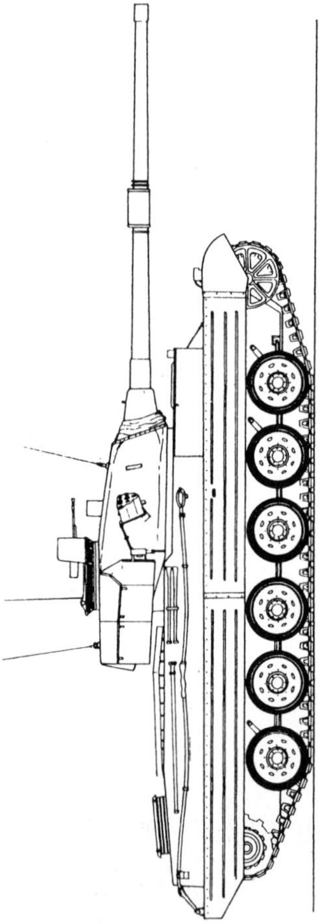

Vickers Mk 3 MBT

VICKERS Mk 3 MBT

CATAPULT AAH

VICKERS AVLB

VICKERS ARRV

WARRIOR FV510 Mechanized Combat Vehicle

Models.

WARRIOR AIFV (Section vehicle)
WARRIOR with add-on armour
WARRIOR with MILAN ATGM
DESERT WARRIOR
WARRIOR AICV
WARRIOR Mechanised Artillery Observation Vehicle (MAOV)
WARRIOR Battery Command Vehicle (BCV)
WARRIOR Mechanised Combat Repair Vehicle (MCRV) FV 512
WARRIOR Mechanised Recovery and Repair Vehicle (MRRV) FV 513

Non-serial versions.

WARRIOR Recce version
WARRIOR DFV with 25 mm M242 chain gun in Cockerill CM 2 turret
Low Profile WARRIOR
WARRIOR DFV with 25 mm Oerlikon-Bührle KBB gun in CM 25 turret
WARRIOR with Euromissile HCT (HOT compact turret)
WARRIOR with Cockerill 90 mm Mk III gun in CM 90 turret
WARRIOR with LMT 105 turret
WARRIOR with ADATS AD/ATGM launcher

Development and production.

1969 first OR for replacement of FV432 APC.
1977 full project definition is undertaken by GKN Sankey, prime contractor for 12 prototypes (MCV 80).
1985 production contract awarded for 1048 vehicles.
1987 first series vehicle is handed over to the army.
1988 first prototype of WARRIOR MAOV completed.
1990 delivery of WARRIOR MRVR starts.
1994 first deliveries of DESERT WARRIORs to Kuwait.

Distinguishing features. Side. Six roadwheels, front sprocket, three return rollers covered by skirt. High hull with sloped front plate. Horizontal hull roof with centered turret. Vertical rear. Front. Grill front plate, driver in upper left side. Turret has vertical sides and is positioned slightly to the left on raised portion of hull, sloped upper side to right. Long barrel cannon with flash eliminator mounted in external mantlet. Distinctive roof-mounted periscopes on the front of cdr and gunner. Both are provided with a hatch opening to rear. Single door in the hull rear with stowage boxes on each side.

Other details. Perkins CV8 TCA V.8 diesel engine coupled to Detroit Diesel Allison X300-4B transmission (four forward and two reverse gears). The

Vickers turret has power traverse. Pilkington PE Raven day/night sights are used (x1 and x8 magnifications, in the light intensification mode it has magnifications of x2 and x6). NBC protection system. Non amphibious. (TABLE)
Prior to Operation "Desert Storm" most of the WARRIORs were upgraded with frontal and side spaced add-on armour increasing the weight by approx. 4 t.
The WARRIOR can be equipped with a front mounted counter-mine dozer and a rear mounted Pathfinder marking launcher.
The anti-tank version is equipped with a pintle mounted MILAN launcher on the right turret roof side. It is planned to replace it by a TRIGAT (MR) launcher later.
The command vehicles are equipped with different radio equipment according to level of command.
DESERT WARRIOR AIFV with Delco LAV-25 two-man turret with a stabilised M242 25 mm Chain Gun, coaxial 7.62 mm MG and one single TOW ATGM launchers on either turret side. This version has a thermal sight, air conditioning system and additional hull and turret add-on armour.
The MAOV is externally similar to the section vehicle but has a dummy cannon. The observer is equipped with a Pilkington combined day/night thermal sight with integrated laser rangefinder. The vehicle is also fitted with an inertial land navigation system and the Battlefield Artillery Target Engagement System (BATES). The tank has a crew of six. The crew compartment contains a Thorn EMI MSTAR battlefield surveillance radar.
The repair/recovery versions have a hydraulically operated crane with a capacity of 6.5 t and a hydraulic stabiliser leg. The MRRV addionally has a rear winch with a capacity of 20 t and a rear spade. Both tanks are operated by a crew of five. Armament consists of a 7.62 mm MG in a small turret. Vehicle's weight is 30 t. These versions may tow a T4 trailer.

Role. In production for Kuwait (254 DESERT WARRIORS) and in service with the British Army (789 including 387 section and 84 command vehicles, 105 with MILAN, 105 MCRV, 19 BCV, 52 OPV and 39 MRRV).

Assessment. IFV without provisions for fighting from the vehicle. Non-stabilized main gun, no thermal imaging. Excellent mobility.

WARRIOR AIFV

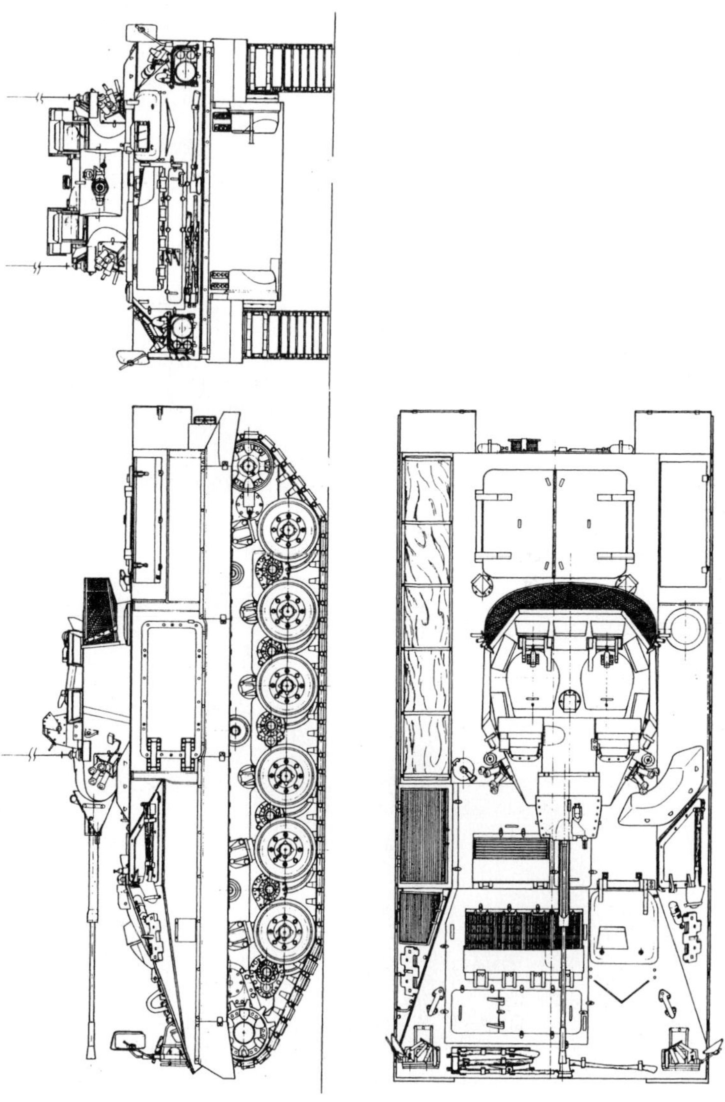

WARRIOR AIFV

WARRIOR AIFV with add-on armour

DESERT WARRIOR AIFV

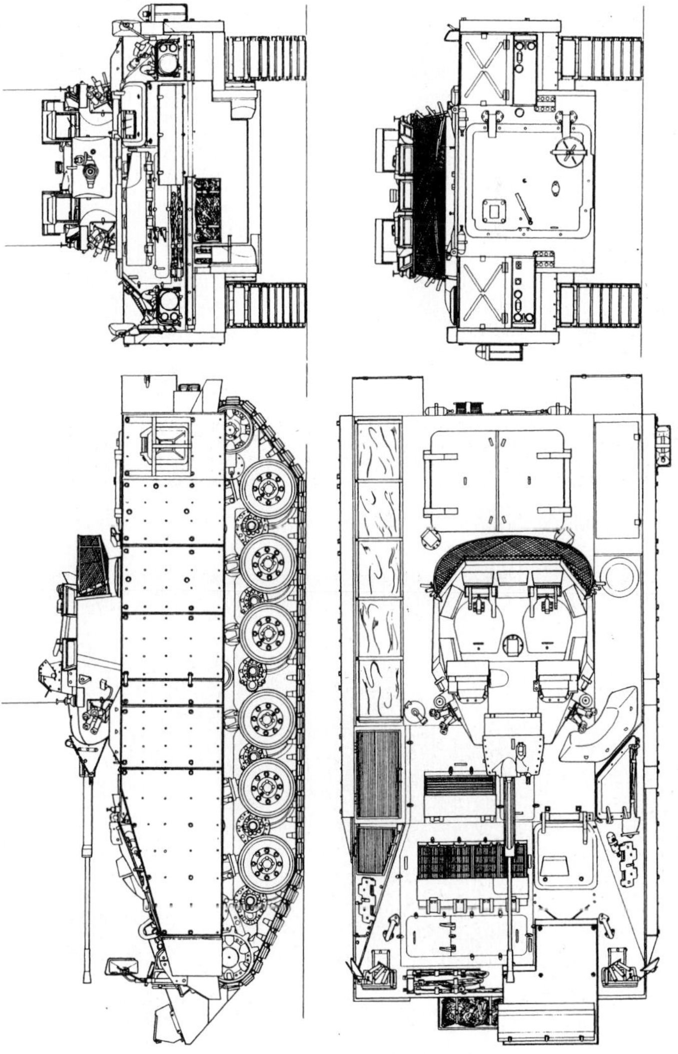

WARRIOR AIFV with add-on armour

WARRIOR AICV

WARRIOR MAOV

WARRIOR MCRV

WARRIOR MRRV

FV432 armoured general purpose carrier

Wt 15.3 t L 5.3 m W 2.8 m H 2.3 m 176 kW 52 km/h 424 km

Models.

FV432 Mk 1 APC
FV432 Mk 1/1 APC
FV432 Mk 2 APC
FV432 Mk 2/1 APC
FV432 ACCC
FV432 EW vehicle
FV432 AMC with 81 mm mortar
FV432 AACPV
FV432 with Cymbeline mortar locating-radar
FV432 with ZB 298 ground surveillance radar
FV432 AEV
FV432 AAV
FV432 recovery vehicle
FV434 maintenance carrier
FV438 Wavell ACCC
FV439 Signals vehicle

Development and production.

1961 first prototype following a requirement for a light AFV family derived from the FV420 series.
1963 first production vehicles completed by GKN Sankey (initially named TROJAN).
1971 production ends (about 3000 built).
1996 delivery of the EW version.

Distinguishing features. Side. Five roadwheels, two return rollers, drive sprocket front, boxed-in top run of track. Sloping front, horizontal hull roof, vertical rear. Front. Large square front plate, vertical hull sides. Driver front right, commander behind him. Large door in rear opening left.

Other details. Flotation screen removed from all vehicles. Engine (Rolls-Royce K60 No 4 Mk4F 2-stroke) in the front left.
FV432 Mk 1 have exhaust on the left hull side and NBC pack protruding on right side.
FV432 Mk 2 with exhaust on top of hull, NBC pack almost flush against hull. Mk 2/1 fitted with Peak Engineering turret.
ACCC with different radio arrangements and stowage boxes on hull top.
AMC with 81 mm L16 mortar (range 5.7 km) on a turntable in the hull rear, traversable at 360 degrees. 160 mortar bombs carried. Wt 16.4, crew of six.
AACPV with FACE (Field Artillery Computer Equipment).

AARV with roof-mounted Thorn EMI Cymbeline mortar-locating radar (range 20 km). Former FV436 GREEN ARCHER was phased out.
AIRV is fitted with Marconi Avionics ZB 298 radar (10 km range).
The electronic warfare carrier is equipped with a Racal RDF 3725 ESM DF system on an extractable mast.
AEV is used to tow Bar minelayer (RANGER anti-personnel minelaying system (on hull top) or Giant Viper mine-clearing trailer.
AAV is unarmed and carries four stretchers or two stretchers and five seated patients, crew of two.
FV432 recovery version has a winch installed in the rear crew compartment with a maximum pull of 18 t.
FV434 has a HIAB crane on the right side with a lifting capacity of 1.25 t. Wt 17.5 t, crew of four.
FV438 was former TDGM with Swingfire now fitted with Wavell Automatic Data Processing (ADP) system. The tank has a large external box on the roof. Wt 16, H 2.83 m.
FV439 in different configurations recognizable by extensive antenna arrays on the hull roof.

Role. In service with the British Army (1253 incl. 192 AMC, 210 ARV), AACPV also used by India.

Assessment. By now obsolete vehicle of average mobility. Impossible to fight mounted.

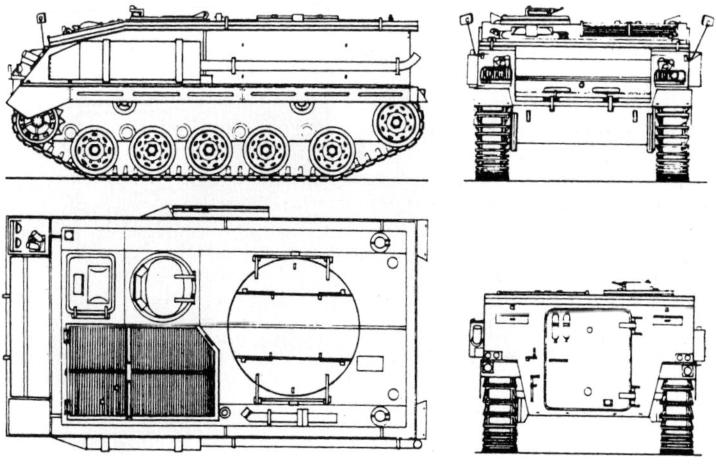

FV 432 Mk 1 APC

FV 432 Mk 1 APC

FV 432 Mk 2 APC

FV 432 Mk 2/1 APC

FV 432 ACCC

FV 432 with Cymbeline radar

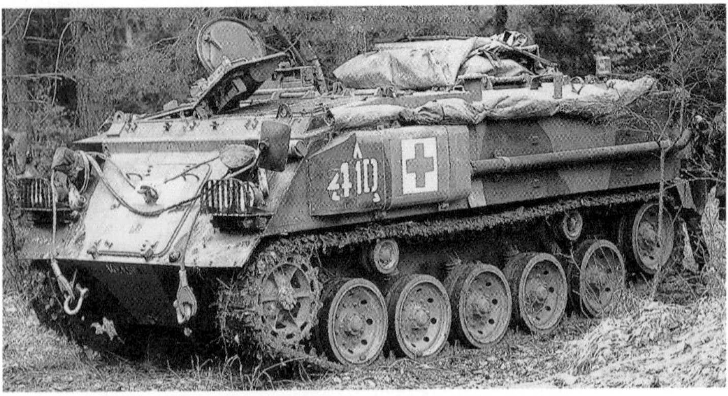

FV 432 AAV

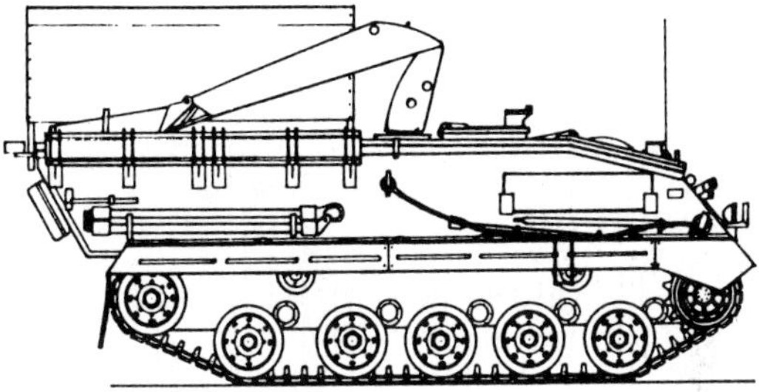

FV 432 ARV

FV 432 ARV

STORMER armoured general purpose carrier

Models.

STORMER IFV
STORMER APC with Helio FVT 900 turret
STORMER APC with Thyssen Henschel TH-1 turret
STORMER TRV (Troop Reconnaissance Vehicle)
STORMER ACCC
STORMER AAV
STORMER with STARSTREAK high velocity missiles (HVM)
STORMER VLSMS with MINOTAUR
STORMER AVLB
STORMER ARV
STORMER ACC

Development and production.

1978 first appearance of the so called FV4333 APC.
1980 Alvis purchased the production rights (former design by a British research and development group).
1982 STORMER enters production, three sold to the USA to participate in the LAV competition (armed with M242 turret). Export order from Malaysia.
1986 STORMER selected by the British Army (STARSTREAK carrier)
1990 contract awarded to Alvis for the production of the French MINOTAUR scatterable antitank mine system carrying STORMER sent to the Gulf.
1991 production of STARSTREAK starts.
1997 last deliveries to Indonesia.

Distinguishing features. Side. Six road-wheels, drive sprocket front, no return roller. Flat front, boxlike crew compartment. Horizontal hull top and vertical hull rear. Front. Blunt nose with sloped glacis, engine right and driver left. Sides vertical, chamfer between sides and roof.

Other details. STORMER can be equipped with a flotation screen. In water it is propelled by the tracks (speed 8 km/h), an appliqué propeller kit is optional. Fuel capacity is 405 litres. The engine of the first vehicles (Malaysian order) was a Perkins T6/3544 water-cooled 6-cylinder turbo-charged diesel of 187 kW coupled to a ALVIS/SCG T300 transmission.
A whole variety of weapon applications and roles for this family is proposed resulting in different one/two-man cupolas for the APC versions. Up to eight infantry soldiers can be carried in the rear compartment which is optionally provided with side and rear looking firing ports. The vehicles have NBC protection systems and with the exception of the recovery and the flatbed versions are fully amphibious using a front flotation screen. Driven by the tracks the speed in water is 6.5 km/h.

Recently produced vehicles are powered by the Perkins Phaser diesel engine with 184 kW retaining the T300 transmission.
(TABLE)
The command and the ambulance versions have a raised superstructure. The recovery version has a rear spade and is equipped with a roof mounted crane.
The MINOTAUR system is installed on a flatbed version of the STORMER and consists of the MSM scatterable mine system developed by GIAT. The system resembles the German SKORPION and has six launch batteries with 20 tubes each. Each launch tube fires five anti-tank mines. Vehicle's weight is 14 t. It is operated by a crew of two.
The cargo carrier and the AVLB are also flat-bed carriers. The bridge-layer carries a MLC 30 scissors bridge of 15 m length.
STORMER with STARSTREAK carries eight ready-to-fire missiles in two protected servo-controlled containers on the vehicle's roof. Gunner's surveillance, firing and target tracking turret is adjacent fitted with a Avimo thermal imaging sight and the Pilkongton passive infra-red Air Defence Alerting Device (ADAD). Inside the sight there is a guidance beam transmitter collimated to the target sight line. The missile is boosted by a second stage within 300 m to its maximum speed of Mach 4 plus. It contains three winged darts with a small HE war head. They are homed in on the target using laser beam riding technology and a grid matrix. Range is beyond 5000 m. The TRV acts as a missile resupply and a battery command post vehicle.

Role. In service in Indonesia (20, different versions), Malaysia (12 with FVT 900, 13 with TH-1 turrets) and Oman (4 ACCC). The British Army operates 135 STARSTREAK carriers, 10 TRV and 6 VLSMS.

Assessment. Slightly enlarged SPARTAN derivative with only limited AIFV qualities. Versatile vehicle family concept.

STORMER IFV

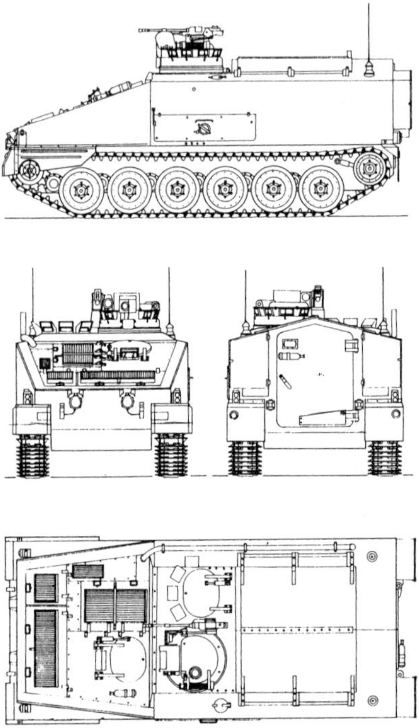

STORMER APC

STORMER APC with FVT 900 turret

STORMER ACCC

STORMER AAV

STORMER STARSTREAK AADML

STORMER VSLMS

SAXON AT105 armoured personnel carrier

Models.

AT105-P APC without armament
AT105-E APC with two 7.62 mm MG
AT105-Q ACCC
AT105-A AAV
AT105 ARV
AT105 IS (internal security) vehicle
SAXON Patrol

Development and production.

1972 development of AT104 as a private venture by GKN Sankey Ltd.
1975 improved successor prototype (AT105) completed.
1976 production starts for export.
1983 initial order by the British MoD (vehicle now called SAXON).
1989 end of series production.
1992 production resumes with the patrol version.

Distinguishing features. Side. 4x4 wheeled vehicle with very high, sloped front, box-shaped hull and vertical rear. Stowage bins on left side of hull and wire stowage box on roof. Grill over first axle. Single door in right side of hull (or in left side at right-hand drive model). Flat roof with commander's fixed cupola on forward part with rear-opening hatch and vision block in each side. Front. Grilled front plate high above beam axle, driver's position offset to the left (resp. right). Vertical sides. Double doors at rear.

Other details. Box-like driver's compartment with three windows. Many automotive components compatible with standard 4 t truck. Optional equipment includes air-conditioning system, front-mounted winch, auxiliary power unit, grenade launchers, search-lights and replacement of Bedford engine by Perkins T6.3544 diesel with 143 kW.
AT105-P has four sockets for MG pintle.
AT105-E is fitted with a turret with twin 7.62 mm MG with an elevation of +50 and a depression of -15 degrees.
AT105-A with internal fittings for four stretchers or two stretchers plus five sitting wounded. Crew 3.
AT105 ARV has a 5 t winch on the left side of 40 m cable length. Crew 4.
Internal security version can be equipped with dozer blade and water cannon.
Patrol version has a barricade removal device at the front, an anti-wire device, roof-mounted searchlights and an armoured radiator cover. (TABLE)

Role. In service with Bahrain (10), Brunei Malay Regiment (24 AT104), Great Britain (624 incl. 20 Command Post Vehicles for the STARSTREAK system; 100 SAXON upgraded with Cummins 6BT 5.9 litre turbocharged diesel engine of 118 kW), Malaysia (44), Netherlands (AT104), Nigeria (75), Oman (20), UAE (10).

Assessment. Sub-standard APC, only useful in internal security role.

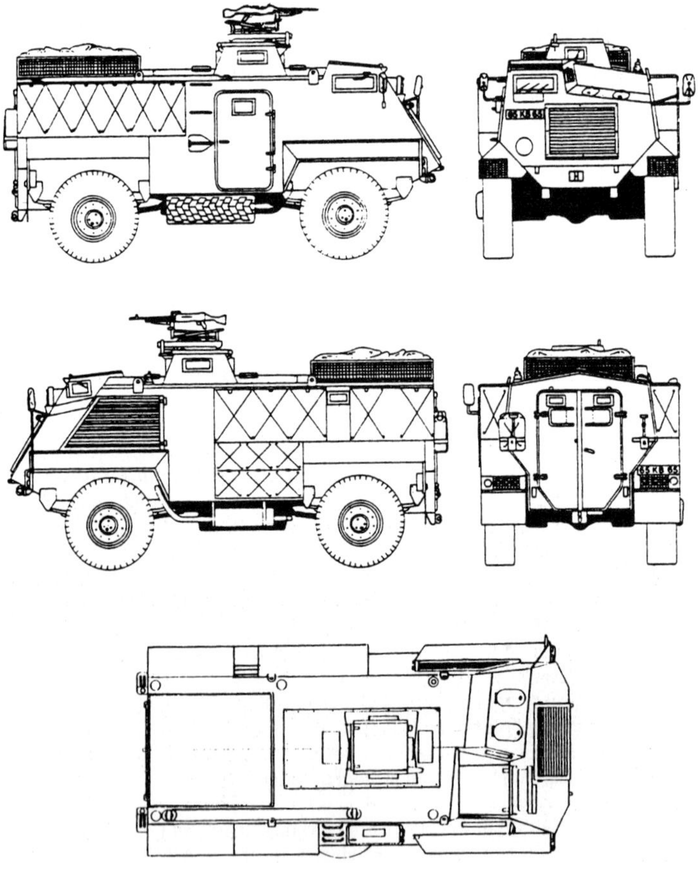

SAXON APC

AT 105-P APC

AT 105-E APC

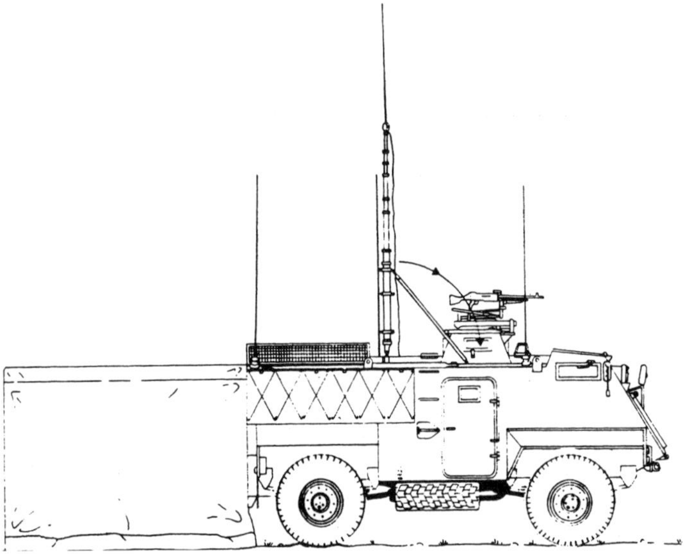

SAXON ACCC

AT 105-Q ACCC

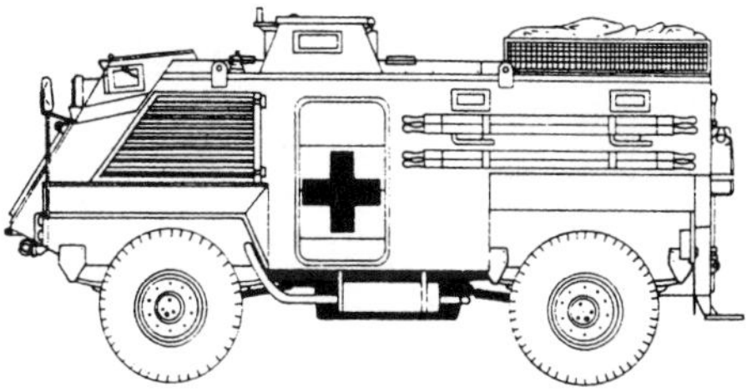

AT 105-A AAV

AT 105-A AAV

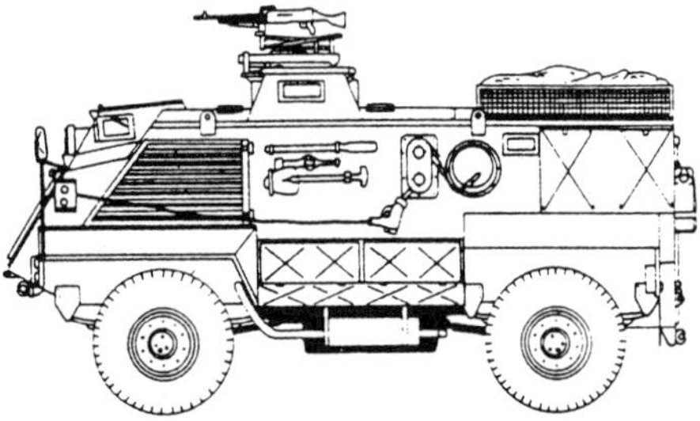

AT 105 ARV

AT 105 ARV

AT 105 IS

SAXON PATROL

SIMBA armoured personnel carrier

Wt 11.2 t L 5.35 m W 2.5 m H 2.19 m 154 kW 100 km/h 660 km

Models.

SIMBA APC
SIMBA IS (internal security) vehicle
SIMBA ACCC

Development and production.

1982 private venture concept by GKN Sankey.
1992 selection by the Philippine Armed Forces.
1994 first deliveries.
1998 order placed by Malaysia.

Distinguishing features. Side. Two large wheels on each side with door between them. Coffin-shaped hull with nose sloping back under hull and sloped glacis; horizontal roof. All hull corners angled. Vertical hull rear.Front. Sloped glacis with driver's bay on the left side. Hull sides sloped inwards top and bottom.

Other details. APC versions have a crew of 8 to 10 infantry soldiers plus commander and driver. The basic APC is fitted with a GKN turret armed with 7.62 mm MG (12.7 mm optional). Vision blocks and firing ports are provided on both sides and in the upper part of the two rear doors. The vehicle is powered by a Perkins 210Ti turbocharged 6 litre diesel engine coupled to a Clark 13.7 LHR 28422 semi-automative transmission. Fuel capacity is 296 litres. The SIMBA lacks a NBC protection system and is not amphibious.
The IS version has a front-mounted barricade remover, side riot screens, a CS and smoke grenade launcher and a collective CS filtration system.

Role. Licence produced by the Philippines (150), in production for Malaysia.

Assessment. Agile anti-riot vehicle with low armour protection.

SIMBA APC

SIMBA IS

AS90 BRAVEHEART armoured artillery howitzer

Versions

AS 90
Chrobry (Poland)

Development and production.

1982 private venture of GBT 155 turret by VSEL for the installation in several MBT chassis. Development of a special AAH chassis starts in cooperation with Verolme Estaleiros Reunidos do Brazil after abortion of the SP-70 project.
1986 first prototype (AS90) revealed (suspension without return rollers, 441 kW engine and a weight of 36 t).
1987 firing trials with AS90 with 52 calibre barrel.
1988 issue of AS90B in competition with the RO M109UK and the Vickers Defence System VSP-91.
1989 adoption by the British Army as ABBOT replacement.
1991 start of series production.
1995 final deliveries.
1999 Contract with Marconi Electronic Systems to develop L52 upgrade for 96 vehicles until 2004. Polish decision in favour of AS90 turret.

Distinguishing features. Side. Six roadwheels with equal spacing, three return rollers, sprocket front. Sloped front with raised engine deck, horizontal hull line. Vertical rear sponson boxes. Recessed vertical rear door. Large box-like turret covering more than rear half of the vehicle with sloped front. Long barrel in cylindrical mounting with muzzle brake. Smoke recuperator in rear half of the ordnance. Small access door on the left side. Front. Vertical hull bow with curved final drive housing. Sloped glacis with driver's hatch on the left side and raised engine grill on the right. Wide turret with sloped sides covering the whole width of the vehicle.

Other details. Main armament is the 155 mm 39-calibre RARDE gun. A bustle magazine holds 31 shells, an additional 17 are are stowed within the hull. The AAH is fitted with an autonomous navigation and gun laying system (AGLS). Main component is an inertial dynamic reference unit (DRU) derived from the US MAPS. The gunner's interface for gunlaying, navigation and AGLS diagnostic functions is the layer's display unit (LDU). The rate of fire is 3 rds/10s (burst), 6 rds/min (intense) and 2 rds/min (sustained). With a shell transfer arm (STA) the projectile is shifted from the magazine to the breech. The second round is on the flick rammer, a third one can be transported by the STA prior to firing. The roadwheels have Airlog hydro-pneumatic transmission units.
(TABLE)
Upgrade will include RO L52 barrel extending range up to 40 km and the modular charge system (MCS) from Denel (SA).

AS90 AAH

Chrobry (the Brave) is the Polish designation of a hybrid system using AS90 turrets and the indegenious Kalina chassis derived from the SUM minelayer built by OBRUM. The first six turrets will be delivered in 2000 by Marconi MLN Systems. The Polish company HSW will be prime contractor for licence production. The initial batch will amount to 83 systems.

Role. In service use with the British Army (179).

Assessment. Up-to-date AAH with high-performance armament. Under-powered chassis.

MARKSMAN twin 35 mm AADG KDA turret

Model.

MARKSMAN on T-54/55 chassis

Non-serial versions.

MARKSMAN on Vickers Mk 3
MARKSMAN on CENTURION
MARKSMAN on CHALLENGER
MARKSMAN on T-59
MARKSMAN on M48

Development and production.

1984 prototype by Marconi as private venture. Turret by Vickers, guns from Oerlikon-Bührle.
1985 production standard turret for firing trials.
1989 order from Finland.
1992 follow-on order.

Distinguishing features. Turret with twin in drums externally mounted guns. Small dish antenna on rear roof. Two roof-mounted sights.

Other details. Self-contained turret (Wt 11 t) with own power supply. Elevation +85 - -10 degrees. Power drive, traverse up to 90°/sec. Marconi 400 combined surveillance and tracking radar (X/J band) with a range of 12 km in the surveillance mode resp. 10 km.
Radar and optical sights (with laser range finder) are stabilised.
Digital fire-control system. The 35 mm cannon has a cyclic rate of 550/sec, each gun has 230 rounds plus 20 AP ready. Effective range is 4000 m maximum. The turret is operated by a crew of two sitting side by side. Built-in test facility, IFF system optional. Reaction time is under six seconds, target speed up to 400 m/sec.

Role. In service in Finland (12).

Assessment. Comparable to GEPARD AADG, but no acquisition whilst tracking.

MARKSMAN (T-55) AADG

TRACKED RAPIER armoured air defence missile launcher

Models.

Mk 1B TRLV (Tracked Rapier Launch Vehicle)
Tracked Rapier Support Vehicle (TRSV)
Forward Area Support Team (FAST)

Development and production.

1974 contract placed by Iran on BAC to put existing towed RAPIER British surface-to-air missile system on a fully tracked AFV. US M548 cargo carrier was selected, first prototype built.
1981 50 ordered for British Army (Iranian order cancelled).
1984 first deployment.
1985 prototype on Mk 1B, all Mk 1A to be refitted.

Distinguishing features. Side. M548 type suspension with five road wheels, idler rear and drive sprocket front. Fully enclosed armoured cab at front, commander's hatch in roof. Blast shield behind cab. On the flat load platform is the launcher with eight tubes in two elevating housings. In firing position tracking radar is raised above launcher. Front. Wide high cab with vision slits, antenna mountings at the corners and raised right side of roof with periscope head. TRSV and FAST vehicles with canvas hood above load platform.

Other details. In the light alloy armour bin below turntable is the guidance computer. Command transmitter antenna arranged on two traversing arms, automatically controlled by target acquisition radar (pulse-Doppler F-band, range 11.43 km). A thermal imager mounted on the Tracker rotary head (TOTE - Tracker Optical Thermally Enhanced, called Darkfire) can be used in an automatic passive search program in addition to its day and night tracking role. Targets are tracked either optically or using TOTE. The J-band command antenna is mounted on an elevating mechanism. The missile (Mk 1) has an effective range of 247 to 6858 m up to levels of 3000 m.
Each TRLV is accompanied by an TRSV with 20 reload missiles.
The FAST has a front-mounted crane, test equipment and spare parts. The TRSV and the FAST are non-armoured M548 cargo-carriers.
(TABLE)

Role. In service with the British Army (70 TRLV/TRSV plus 12 FAST).

Assessment. Lightly armoured self-contained AD system operating in an entirely passive mode when surveillance radar is switched off by TOTE.

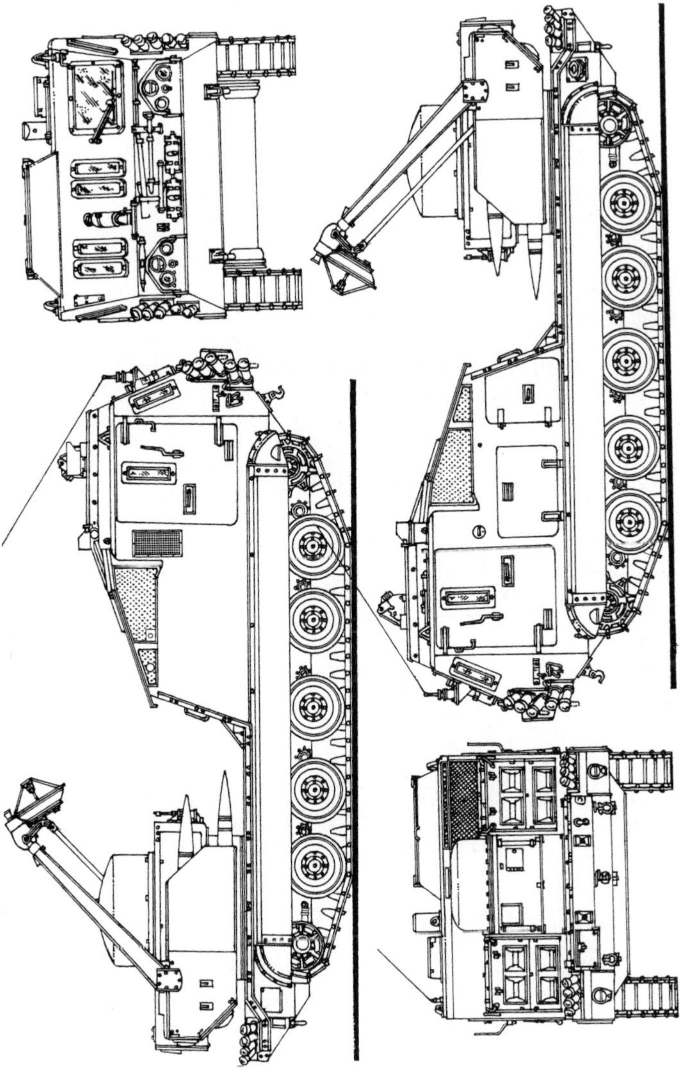

TRACKED RAPIER AADML

TRACKED RAPIER AADML

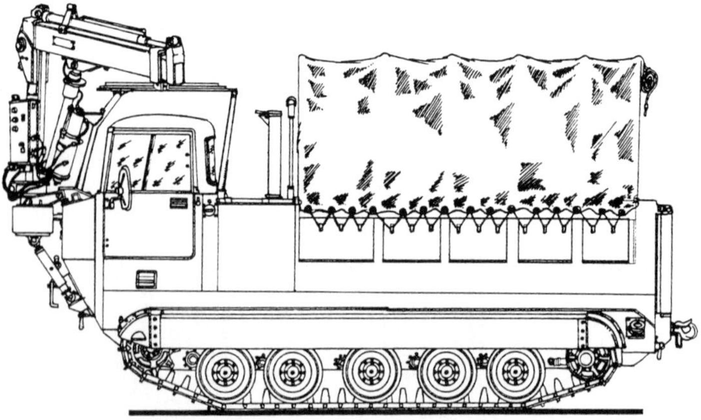

FAST vehicle

CHIEFTAIN armoured vehicle launched bridge

Models.

CHIEFTAIN AVLB (FV4205)
CHIEFTAIN Mk 6 AVLB

Development and production.

1971 development of CENTURION AVLB successor on CHIEFTAIN hull.
1974 delivery of first production vehicle.
1984 conversion order awarded to Vickers of 11 CHIEFTAIN Mk 1/4 into AVLB.
1986 vehicles handed over to Royal Engineers.

Distinguishing features. CHIEFTAIN MBT hull with a high bridge superstructure. Different types of bridge carried.

Other details. The AVLB carry either No. 8 or No. 9 bridge (MLC 60).
The No. 8 Tank Bridge is carried folded and launched in the scissors manner over the vehicle's front. It has a length of 24.38 m, a weight of 12.2 t and can span a gap of about 22 m.
The No. 9 bridge is carried horizontally and is swung vertically through 180 degrees over the front of the vehicle. It is 13.41 m long, has a weight of 9.14 t and can span about 12 m. Both bridges are 4.16 m wide and have an overall depth of 0.91 m.
CHIEFTAIN Mk 6 AVLB has a more powerful hydraulic system and carries the Track Width Mine Plough System.
Currently under development is the BR 90 system include three new MLC 70 bridges (No. 10 scissors bridge of 26 m , No. 11 of 16 m and No. 12 of 13.5 length). These bridges are due to replace the old No. 8/9 bridges in the next years.
(TABLE)

Role. In service with Great Britain (37) and Iran (14 delivered, status unknown).

Assessment. Despite the use of a scissors type bridge this is a rugged and operationally effective bridgelayer.

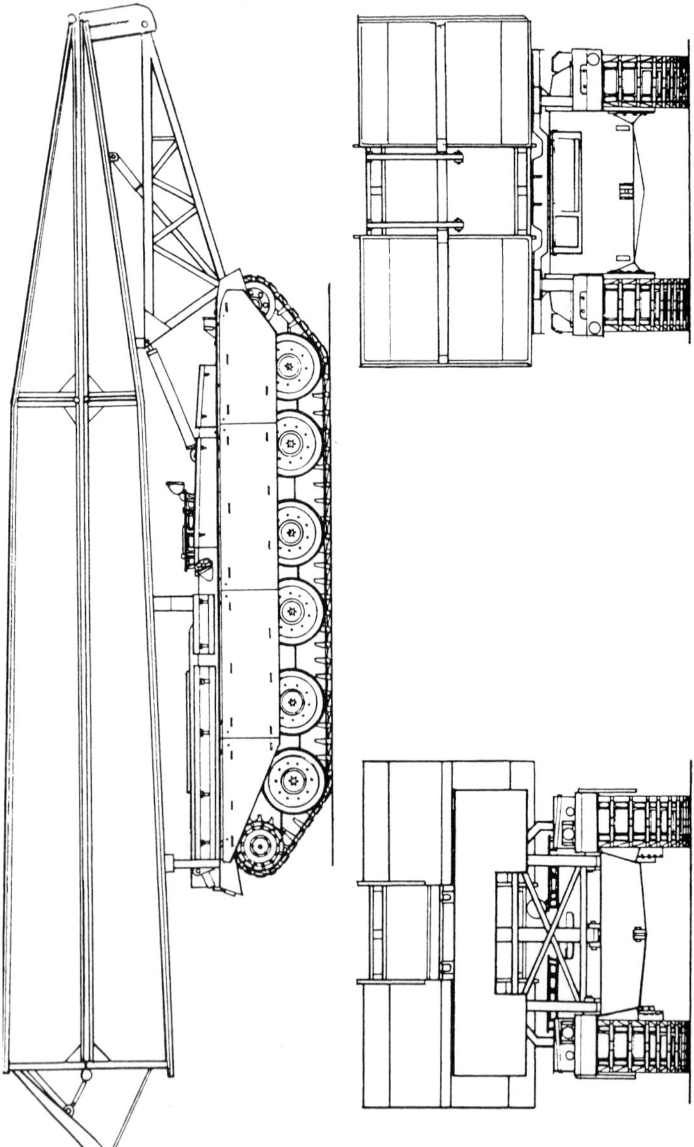

CHIEFTAIN AVLB

CHIEFTAIN AVLB

CET FV180 armoured engineer vehicle
Combat Engineer Tractor

Development and production.

1968 first two prototypes completed by RO Nottingham.
1973 complete redesign, seven further prototypes built.
1978 start of serial production running until 1981.
1990 delivery of 6 new vehicles to India, 9 taken from British stock.
1993 Singapore order of 18 CETs.

Distinguishing features. Side. Five roadwheels, no top rollers or idler, rear sprocket. Flat front and glacis plate sloping up to roof, sloped hull rear. Prominent rear bucket (march configuration). Front. Hull with vertical sides. Driver and bucket operator on left with swivel seats.

Other details. Rocket-launched grapnel. Splashboard with buoyancy bag. Two hydrojets for amphibious operation. Two-speed winch. Shovel capacity 1.72 m^2, maximum height of bucket 1.8 m, lifting capacity 67 kN, earth-moving capacity 336 m^2/h. (TABLE)

Role. In British (141), Indian (21) and Singapore (36) service.

Assessment. Agile vehicle, yet lacking in capacity as earth-moving plant.

CET AEV

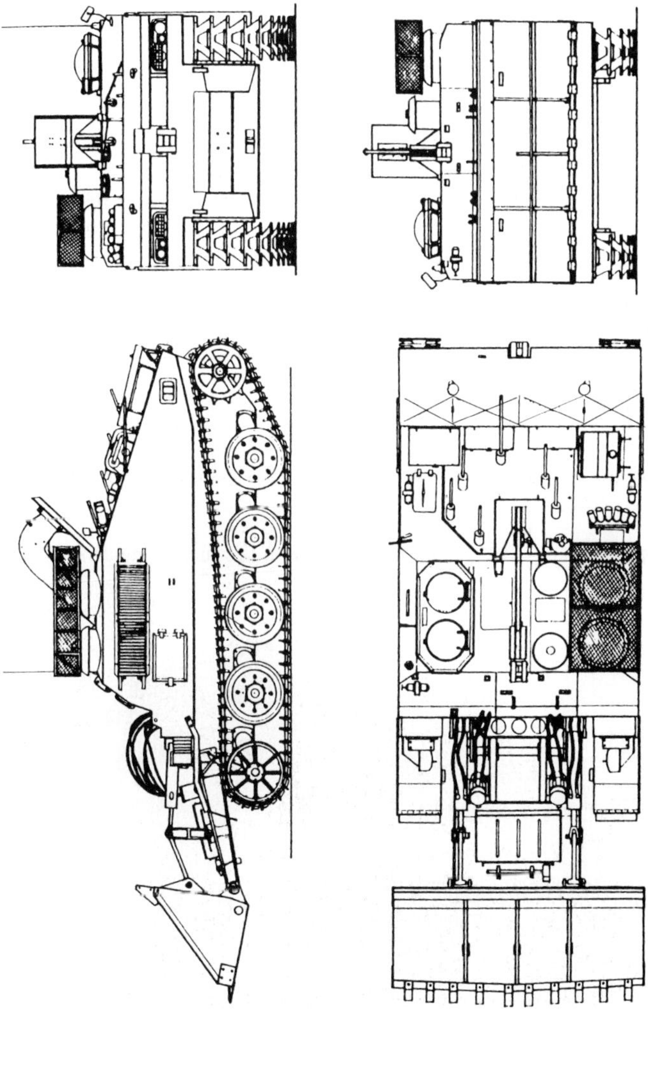

FV180 AEV

CHIEFTAIN (FV4204) armoured recovery vehicle

Models.

CHIEFTAIN ARV
CHIEFTAIN ARRV

Development and production.

1971 first appearance, developed by FVRDE Chobham.
1976 date of service enter.
1983 modification of ARV to ARRV standard.

Distinguishing features. Side. CHIEFTAIN Mk 5 MBT chassis with modified hull. Box-shaped superstructure with sloped front. Large stowage boxes at each side. Flat roof with commander's cupola centre. Front. Front dozer blade across full width of tank. Well sloped glacis plate with driver's hatch in upper part. Cdr central in newly developed cupola with inclined periscope. Tanks often carry replacement power pack on the hull rear.
ARRV with additional crane on the left side of superstructure.

Other details. ARRV is equipped with ATLAS AK 6000M crane which lifts 5.8 t at a reach of 3.62 m. NBC protection system. (TABLE)

Role. In service with Great Britain (186), Iran (41 delivered, status unknown) and Jordan (30).

Assessment. Heavy ARV, lack of crane makes it fall short of modern requirements.

CHIEFTAIN ARV

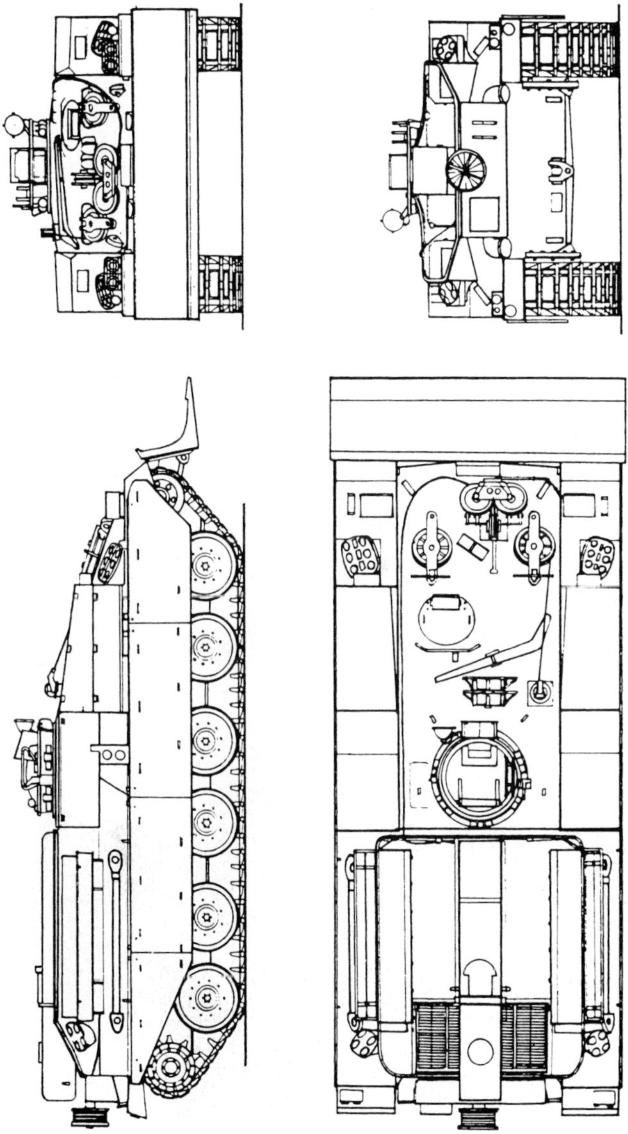

CHIEFTAIN ARV

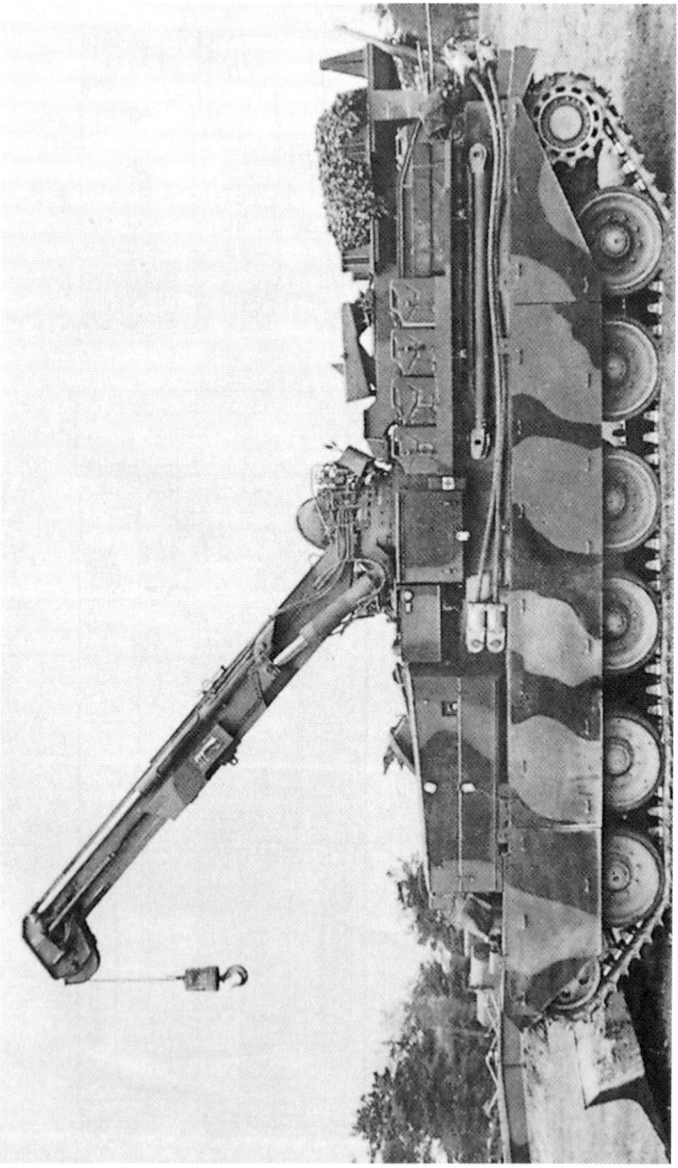

CHIEFTAIN ARRV

CHALLENGER Armoured Repair and Recovery Vehicle
(CR ARRV)

Development and production.

1984 competitive bids put in for CR ARRV by Royal Ordnance Leeds and Vickers Defence Systems.
1985 fixed price contract awarded to Vickers for the development and production of 30 vehicles.
1987 roll-out for the first of six pre-production vehicles (four of them brought to series standard later).
1989 start of main production.
1993 production complete.

Distinguishing features. Chassis and automotive components from CHALLENGER MBT, new TN-54 automatic transmission. High crew compartment at front with dozer blade under forward part of hull which has flat sides with step down from crew compartment. Short telescopic crane jib on left centre part of hull stowed in lowered rear position when not in use. Front. Steep glacis with winch exit and sloped upper section with driver's position on the left side, cdr's cupola behind.

Other details. The hydraulically driven double capstan main winch has a maximum line pull of 52 t (510 kN) and a maximum usable rope length of 150 m. The auxiliary winch (300 m usable rope length) has a capacity of 15 kN. The Atlas AK 6000 M8 hydraulic crane has a maximum lifting capacity of 6.5 t at 4.9 m reach. In addition to carrying a complete CHALLENGER powerpack on its engine deck the CR ARRV also carries welding equipment, an air compressor and a wide range of tools. The cdr has a modified No 32 Mk 1 cupola fitted with a 7.62 mm MG that can be fired from inside.
(TABLE ->)

Role. In service with the British Army (80).

Assessment. Efficient ARV with high standard equipment, below average mobility.

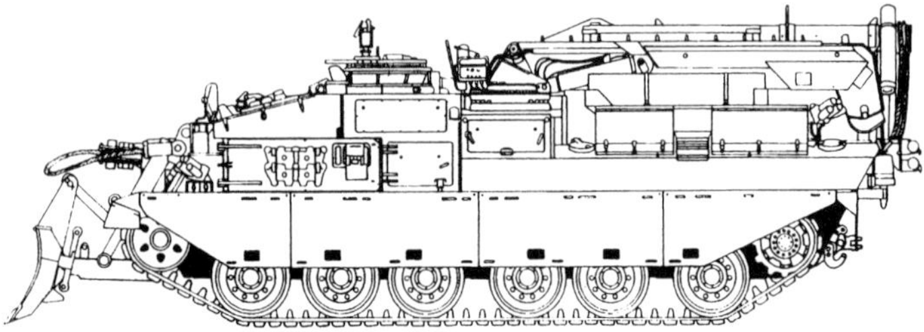

CHALLENGER ARRV

CHALLENGER ARRV

FUG D-422 armoured vehicle reconnaissance
Felderito Uszo Gepkocsi

Wt 7 t L 5.79 W 2.5 H 1.91 74 kW 87 km/h 600 km crew 2+4

Models.

FUG D-422 (Czech OT-65)
OT-65A
OT-65DP AVR
FUG D-442MRP AADFDV
FUG D-442VS/OT-65Ch ANBCRV
OT-65ChV logistic command version
OT-65DPP AAOP
OT-65R-2 ACCC
OT-65RL AARV
FUG D-422M AAV

Development and production.

1964 AVR then called FUG-63 enters service. Production at Györ.
1966 export to Czechoslovakia and Poland.

Distinguishing features. 4-wheeled vehicle with two auxiliary axles which can be raised. Rear engine compartment. two hydrojets at rear below engine. High lower front with folding wave deflector on lower glacis. Low superstructure closed in, roof almost parallel to crew compartment. Single exhaust box on hull right rear. Foot steps cut into the hull sides.

Other details. Amphibious, speed in water 9 km/h. NBC collective ventilation system. Adjustable tyre pressure. Pivot mounting for 7.62 mm MG (1250 ammunition stowage).
OT-65A with the turret of the OT-62B APC with 7.62 mm M59T MG. Externally mounted 82 mm T-21 TARASNICE recoilless gun.
NBC version with racks of lane marking poles on each side of the hull rear.

Role. In service with the Czech Republic (185), Hungary (161) and Slovakia (33).

Assessment. Poor firepower but good mobility. Swimming performance improved as compared with Soviet BRDM by installation of second hydrojet.

FUG D-422 AVR

FUG D-422 VS ANBCRV

PSzH D-944 armoured personnel carrier
Pancelozott Szallito Harcjarmü

Wt 7.6 t L 5.69 W 2.5 H 2.3 74 kW 80 km/h crew 3+6

Models.

PSzH D-944 (Czech OT-66)
PSzH D-944Z.PK ACCC
PSzH D-944Z.TÖF turretless ACCC
PSzH D-944M AAV

Development and production.

1970 first appearance (originally called FUG-70).

Distinguishing features. Similar to OT-65A with increased sloping of the hull front. Forward-opening door in each side of hull. Splashboard no longer in front of lower front but on upper glacis. No auxiliary wheels. Turret with 14.5 mm MG with a stepped top with forward-facing periscopes in the top part, no roof hatch. Ventilator fan on left side of the turret.

Other details. Coaxial 7.62 mm MG left of main gun. Ammunition stowage is 2000 resp. 500 rounds. Turret has manual controls, elevation -5o to +30o. Power-winch of 40 kN.

Role. In service with Bulgaria (50), Czech Republic (182), Hungary (420) and Slovakia (40).

Assessment. Light armoured vehicle with similar combat value as Soviet BRDM-2.

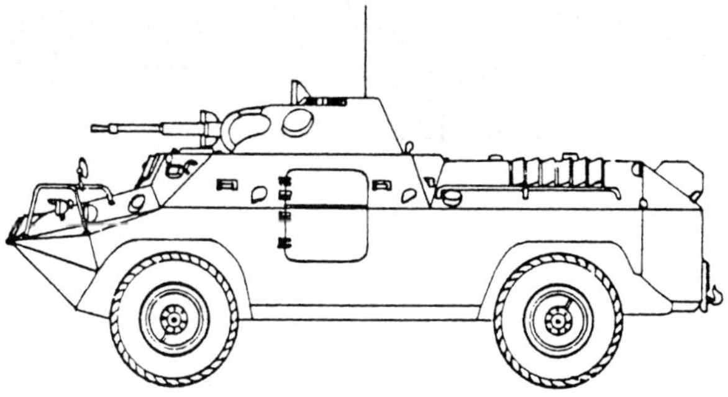

PSzH D-944 APC

PSzH D-944 APC

PSzH D-944Z.TÖF ACCC

ZULFIQAR main battle tank

Development and production.

Developed probably during the late 80s using components from the US PATTON family and the Soviet T-72 MBTs.
1994 first public appearance in TV.

Distinguishing features. Side. Suspension with six roadwheels, idler at the front and five return rollers. Raised engine deck like the M48/M60 MBT series. Long turret covering two thirds of front section. Long gun tube with central fume extractor. Front. Box shaped hull, square turret not covering whole width with slightly sloped sides.

Other details. There are no official specifications available except the gun calibre 125 mm and the approximate engine (735 kW) and weight (40 tons) data. Since Iran successfully obtained M60 fire-control systems slipping through embargo regulations and has also been equipped with Russian T-72S tanks an integration of a second generation FCS must be considered.

Role. Estimated to be in low-scale production.

Assessment. Present information do not allow serious estimation.

ZULFIQAR MBT

Timoney armoured personnel carrier

Wt 9.9 t L 4.95 W 2.5 H 2.75 132 kW 95 km/h 1000 km

Models.

Timoney Mk 4 with twin 7.62 mm MG in Technology Investments turret
Timoney Mk 6 with Creusot-Loire TLi 127 turret
BDX APC with 7.62 mm MG in ring-mount (Timoney Mk 5)
BDX without armament (partly equipped with a dozer blade)
BDX AMC with 81 mm mortar

Several non-serial versions.

Development and production.

1973 private venture by Irish firm Timoney.
1975 Belgian trials of No. 3 prototype. Manufacturing licence granted to Beherman Demoen.
1977 licence production starts in Belgium.
1979 export to Argentina, first BDX handed over to Belgium Gendarmerie.
Beherman-Demoen starts development of a new, diesel powered BDX-2 as a private venture.
1981 deliveries to the Irish Army. Production for Belgium completed.

Distinguishing features. Side. 4x4 vehicle. High sloped glacis with raised driver's hatch and side windows. High, completely vertical rear. Small angular turret central. Front. Vertical sides, upper part sloped. Large front screen. Timoney and BDX differ externally in the arrangement of the front windows and the firing-ports on both sides.

Other details. Slight differences between the Timoney and the BDX according dimensions and weight (Timoney Mk 6 9.98 t, BDX 10.7 t). Timoney Mk 6 is powered by the 4-53T Detroit diesel engine, the BDX with Chrysler V-8. Crew is 3+9 resp. 2+10. Fuel capacity is 300 resp 248 litres, range 1000 resp. 500 km.

Role. In service with Argentina (5), Belgium (Gendarmerie 41 APC, 26 with dozer blade, 13 mortar carriers, Air Force has 43 APC with 7.62 mm MG), Ireland (5 Timoney Mk 4 and 5 Mk 6), Luxembourg, Mexico (18 BDX) and Uruguay (22 BDX).

Assessment. A light armoured vehicle for police and internal security roles.

TIMONEY Mk 4 APC

TIMONEY Mk 6 APC

BDX APC

RBY/RAM scout cars

Wt 5.4 t L 5.02 W 2.03 H 1.72 97 kW 96 km/h 600 km crew 6

Models.

RBY Mk 1
RBY Mk 1 with 106 mm M40 recoilless rifle
RAM V-1
RAM V-1 command and communications version
RAM V-1 with 106 mm M40 recoilless rifle
RAM V-1 with TOW ATGM system

Non-serial version.

RAM V-2 closed top version

Development and production.

1975 first appearance of RBY Mk 1 as a private venture project by Israel Aircraft Industries' subsidiary RAMTA.
1979 RAM family announced.

Distinguishing features. Side. Low, open-topped 4x4 vehicle. Two large wheels at extreme end of the vehicle, both axles on leaf springs. Rear engine. Spare wheel over front axle. Different weapon arrangements. Extensive external stowage. Front. Low hull with slightly sloped sides. Driver's and co-driver's vision screens with flaps.

Other details. The RAM series is offered in two different wheelbases (3.4 m or V-1/V-2L with 3.8 m). The RAM V-1L has a weight of 5.75 m and a length of 5.52 m. Ground clearance is 0.31 m (RBY has 0.27 m). RBY Mk 1 is powered by a Chrysler 225-2 6-cylinder water-cooled petrol engine of 88 kW (RAM with Deutz air-cooled 6-cylinder diesel). Fuel capacity is 140 l (RBY), 120 (RAM V-1) resp. 160 l (RAM V-1L).
The AT versions with 106 mm RR carry 16 resp. 18 rounds/missiles. The rifle/launcher can be traversed through 360 degrees.

Role. Known users beside Israel (12) are Botswana (RAM), Guatemala (10 RBY upgraded with 107 kW diesel engine and overhead cover), Honduras (10 RBY), Lesotho (19 RBY), Morocco (RAM V-1 incl. 106 mm) and Zaire (RAM).

Assessment. Mixture of reconnaissance vehicle and carrier with poor protection. General cross-country performance is slight, but it is suitable for desert conditions. Optimized for anti-mine protection.

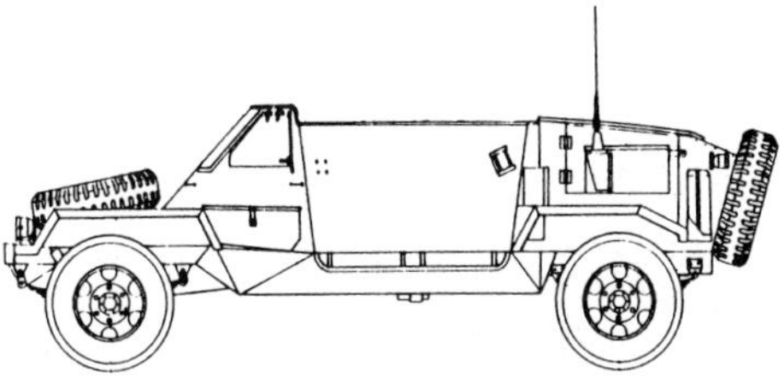

RBY Mk 1 ARSC

RBY Mk 1 ARSC

RAM V-1 ARSC

RAM V-1 ARSC with TOW ATGMS

MERKAVA main battle tank

Models.

MERKAVA MARK 2
MERKAVA MARK 3

Non-serial version.

MERKAVA 4 technology program with 140 mm gun
155 mm AAG on modified MERKAVA chassis (SLAMMER)

Development and production.

1970 development of genuine Israeli MBT starts with priority on protection.
1974 first prototype issued partly financed by the USA.
1978 MERKAVA enters production.
1983 MERKAVA MARK 2 delivered to the army.
1989 MERKAVA MARK 3 revealed.
1994 all tanks receive turret roof add-on armour and laser sensors (MERKAVA 2 B resp. 3 B).

Distinguishing features. Side. Six roadwheels, second and third one closer together. Four return rollers, front sprocket. Upper part of suspension covered by skirts with wavy rim. Long low, nearly horizontal glacis. Horizontal hull roof with slightly bulged rear deck. Nearly vertical rear with entry hatch. Turret rear of mid-point of vehicle with distinctive pointed front. Tapered mantlet cover, long turret bustle with stowage rack which extends to hull rear. Gun with fume extractor half-way along barrel with thermal sleeves. Up to three external MG. Front. Almost vertical hull front, sloped glacis with distinct bulge in right side above engine. Very narrow upper turret on dome-shaped lower part. Mantletless mounting. Recess right for sight. Small persicope head in front of (right) cdr's hatch.

Other details. Mk 1 and 2 are armed with the 105 mm M68 gun. Ammunition stowage is 62 or more. One 7.62 mm MG is mounted coaxially on the left side of the gun. Two 7.62 mm ADMG are mounted externally and occasionally for training purposes or during urban fighting a 12.7 mm MG is mounted on top of the gun mantlet. 10.000 rounds of 7.62 mm ammunition are carried. A 60 mm SOLTAM mortar is carried externally with a stowage of 30 mortar bombs. Mk 1 was protected by laminated armour, Mk 2 has additional external armour plates on the turret and special armour on the hull front. The gunner's periscope has a magnification of x1 or x8 and incorporates a laser rangefinder. Mk 1 was equipped with the Matador Mk 1 digital fire control system. Mk 1 and 2 have the CENTURION type helical spring suspension. They are powered by the Teledyne Continental AVDS-1790-6A V-12 diesel engine of 662 kW. MERKAVA MARK 2 is equipped with the more advanced Matador Mk 2 fire-control system. All Mark 1 tanks were upgraded to this standard.

The MERKAVA MARK 3 is armed with a 120 mm smoothbore gun with 50 rounds of stowage. Turret controls are now electric. The gunner's fully stabilised sight has a magnification of x12. Turret and hull have special armour. This version is powered by the Teledyne AVDS-1790-9AR diesel of 882 kW coupled to a new Ashot transmission. The suspension consists of roadwheels mounted independently on trailing arms and pairs of concentric coil springs. The tank is provided with a threat warning system developed by Amcoram. (TABLE)
The MERKAVA Mk. 3 BAZ has a revised armour package, an improved suspension and an auto-tracking system based on the video output from a TV camera (either day or IR light). The next version MARK 4 is under development.

Role. In service with the Israeli Army (900).

Assessment. Good protection from configuration and type of armour. Mk 2 average firepower. Below-average mobility. Large rear compartment is suitable for transport of ammunition and personnel.

MERKAVA Mk 2 MBT

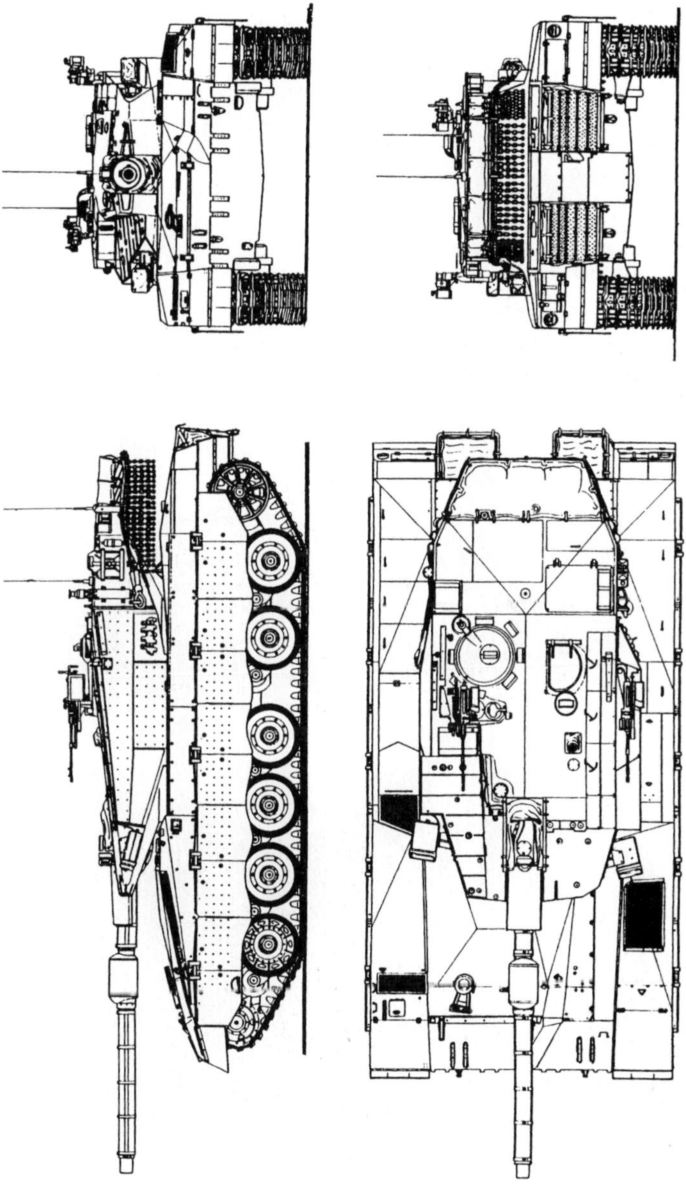

MERKAVA Mk 3 MBT

MERKAVA Mk 3B MBT

ACHZARIT armoured personnel carrier

Wt 44 t L 7.0 W 3.64 H 2.3 478 kW 60 km/h (provisional data)

Development and production.

1987 first prototype built.
1988 start of series production.
1997 improved Mk. 2 version under development.

Distinguishing features. Side. T-55 hull and suspension covered by thick add-on armour layer with pointed hull bow. Right rear hull section slightly higher. Upper part of track covered by thick metal skirts. Front. Sponsoned hull with apliqué armour sections extending on both sides. Sloped upper and lower hull bow portions.

Other details. Completely rebuilt T-55 chassis covered with massive special armour segments providing a high degree of protection against mid-calibre HEAT charges and any type of AIFV cannon fire. In the front hull the driver (left), the commander (centre) and the gunner sit side by side, the latter one provided with a 7.62 Rafael OWS (Overhead Weapon Station) with day and light intensification sights. The original turret space accomodates up to seven infantry who are provided with vision blocks and two roof hatches. The original engine is replaced by a General Motors 8V-71 TTA diesel engine which is mounted transversely, coupled to an Allison XTG-411 4 automatic transmission. The engine is moved to left side leaving a narrow corridor leading to a hydraulically operated clamshell rear exit with a ramp which opens downward. Three pintle mounts for 7.62 mm MG are positioned on the hull roof. The vehicle has a fire supression system and individual NBC protection systems.
Recently produced vehicles are powered by a Detroit Diesel 8V-92 TA engine (549 kW) coupled to an Allison XTG-411-5 transmission.
Command tanks have additional radio equipment.
The next Mk. 2 version will be provided with improved special armour, a MERKAVA-type long-travel suspension and a Rafael OWS25 (overhead weapon station) with a remotely controlled M242 25 mm Bushmaster cannon.

Role. In service with Israel (approx. 400).

Assessment. Excellent low-cost solution of an infantry companion vehicle capable for combined operations with MBTs with a high degree of armour protection. At the time being the best protected infantry fighting carrier.

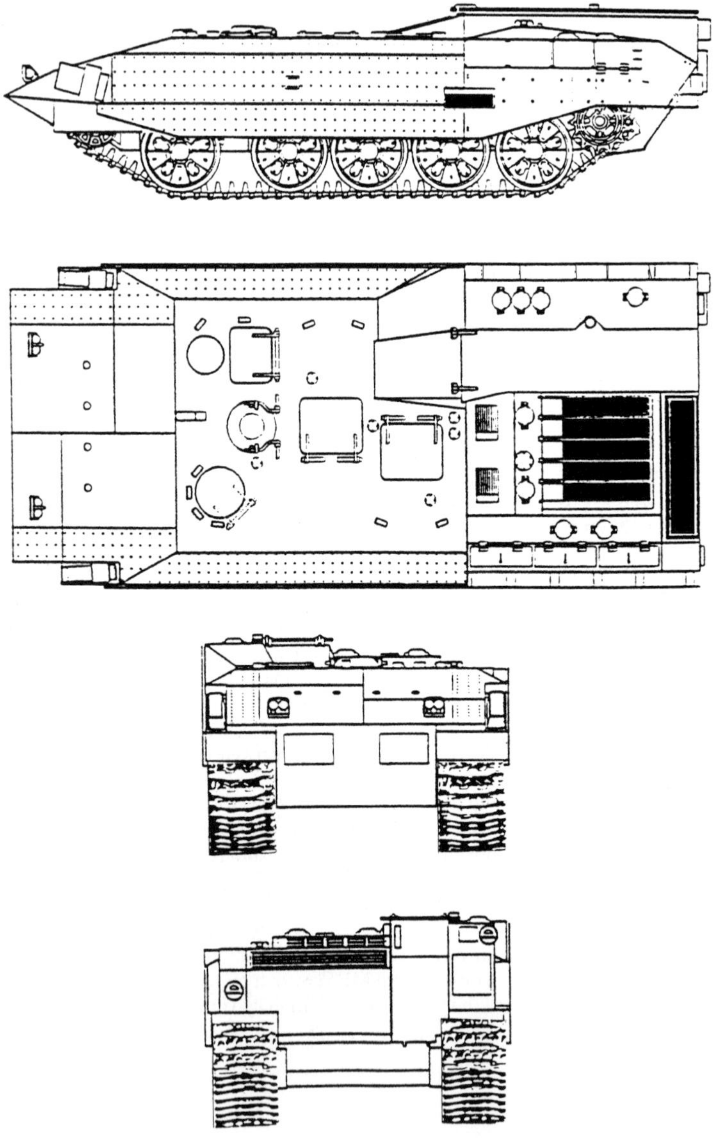

ACHZARIT APC

ACHZARIT APC

6616 armoured vehicle reconnaissance

Wt 8 t L 5.37 W 2.5 H 2.04 118 kW 100 km/h 700 km Crew 3

Development and production.

1970 private venture by Fiat.
1972 prototypes by OTO Melara built.
1977 order for Carabinieri placed, first export orders.
1985 further development by IVECO into 6634 AVL, 6633 AVM and 6636 AVH.

Distinguishing features. Side. 4-wheeled vehicle. Hull front slopes back under hull, shallow glacis with raised driver's hatch. Horizontal louvres above rear wheel stations, small hatch in lower part of hull side between wheels. Large angled turret in centre with gun in sloped front and MG above it. Front. Sponsoned, heavily sloped hull. Well sloped glacis with driver's half-circular position on the right. Wide turret with sides sloping slightly inwards. Commander's cupola on left protrudes in turret side. Armament in drum mantlet.

Other details. Rh 202 20 mm cannon with 250 ready-use rounds and a coaxial MG above with 300 rounds. A further 150 resp. 700 rounds are carried inside. Two-men turret with cdr on the left. Gunner equipped with Aeritalia P204 periscope (magnification x1 and x8 replacable by a P194 image intensifier (magnification x8). The periscope head has e depression of -10 and an elevation of +60 degrees. Turret mowement is servo-controlled. NBC system is optional.

Role. In service with Italy (40), Libya (100), Peru (20), Somalia (30, status unknown), and other countries.

Assessment. A fast vehicle with average armament, lightly armoured.

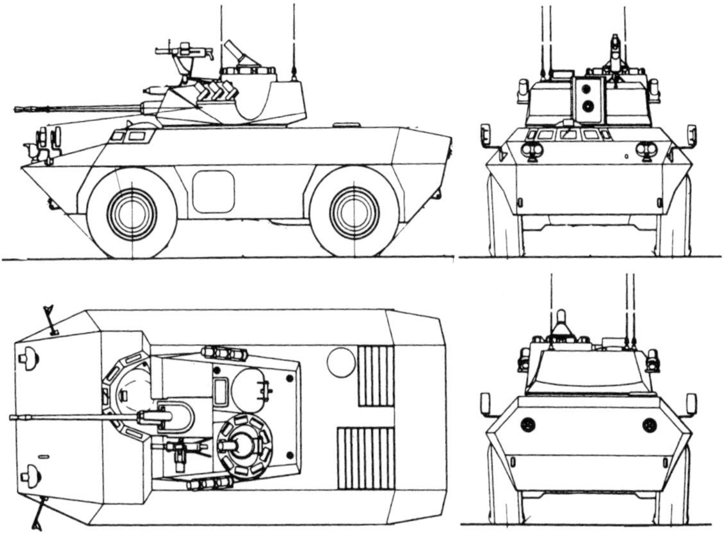

6616 A AVR

CENTAURO B-1 armoured vehicle reconnaissance

Non-serial versions.

CENTAURO VBC APC with OTO Breda TC-25 turret
CENTAURO 155 mm AAH

Development and production.

1983 private venture design by OTO Melara of a new family of wheeled armoured fighting vehicles including the 6x6 AVH6636.
1984 requirement of a wheeled TDG armed with a 105 mm gun.
1987 first prototype (8x8) issued, adopted as B-1.
1990 start of series production as M-47 replacement.
1996 last deliveries. Development contract for the VBC version.

Distinguishing features. Side. 8x8 vehicle with equal spacings between wheels, slightly bigger between first and second pair. Edged bow with inclined bottom and sloped glacis. Horizontal hull roof. Sharp edge above wheel arches. Slightly sloped hull rear with door. Flat long turret moved to the rear of the tank with flat cupola and roof mounted periscope. Pointed mantlet with main gun extending over vehicle's front, thermal sleeve and muzzle brake. Front. Boat-like front bow with sloped front plate. Driver's position on the left side, engine compartment to his right. Hull sides sloping inwards above wheel level. Flat turret with slightly sloped sides. Main sight on the right front, rotating periscope on the left turret roof.

Other details. Main armament is an OTO Melara designed 105 mm gun with long recoil and 52 calibre length. Ammunition stowage is 40 rounds. One coaxial and a 7.62 mm ADMG are available. Gun elevation is -6 to +15 degrees. The tank is fitted with the Officine Galileo TURMS fire control system with digital computer, sensors, muzzle reference system, laser rangefinder and stabilised cdr's periscope. Gunner's main sight is also stabilised, has a magnification of x5 and is provided with a thermal image unit. The tank is powered by an IVECO FIAT MTCA V-6 turbocharged diesel engine coupled to a German ZF SHP-1500 transmission. The run-flat tyres are provided with a central tyre pressure regulation system. NBC system installed and fire surpression systems for the crew and the engine compartments. Standard equipment includes a front mounted winch.
For 150 vehicles a kit of tools and accessories are delivered to accomodate an infantry squad of four in the hull back. This requires the removal of of the two rear 12-rounds ammunition racks decreasing the stowage to only 16 rounds. This compartment is fitted with a spall liner.
(TABLE)
As a stop-gap solution CENTAUROs deployed in Somalia were provided with Royal Ordnance ROMOR A ERA add-on armour to provide more HEAT protection.

CENTAUROs operating in Bosnia received the German IBD MEXAS add-on armour.

Role. In service with the Italian Army (400), 22 on order by Spain.

Assessment. Mobile, fast tank destroyer with adequate firepower not up-to-date against latest MBT armour protection. Insufficient armour when fighting enemy MBT.

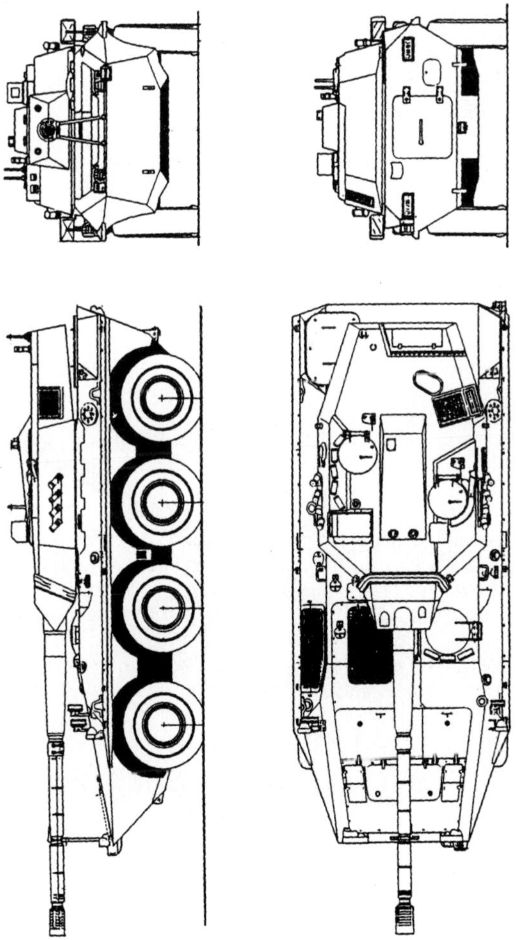

CENTAURO AVR

CENTAURO AVR

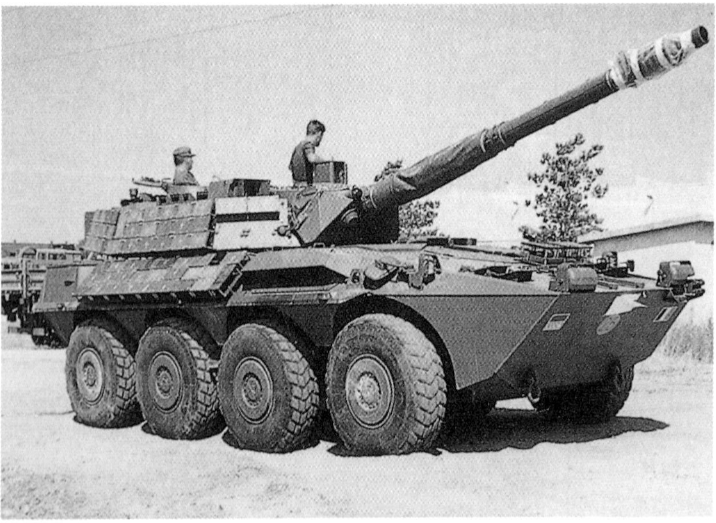

CENTAURO AVR with add-on armour

OF-40 main battle tank

Wt 45.5 t L 9.22 W 3.51 H 2.76 610 kW 60 km/h 600 km Crew 4

Models.

OF-40 Mark 2
OF-40 ARV

Development and production.

1977 copy of LEOPARD 1 MBT with several modifications.
1980 issue of a prototype.
1981 export order for initial version (OF-40 Mark 1).
1984 production of OF-40 Mark 2 and ARV.

Distinguishing features. Side. Hull similar suspension identical to LEOPARD 1 (but five instead of four return rollers). Turret similar to LEOPARD 1A3. Rotating-head periscope in front of cdr's hatch. Front. Appearance like LEOPARD 1A3, wedge-shaped gun mantlet. Rear. Exhaust grills above tracks. Vertical sides.

Other details. Transmission as LEOPARD 1. Track of the rubber-bushed connector type. 105 mm 52-calibre rifled gun with falling wedge breech-block by OTO Melara (62 rounds stowage) firing NATO standard ammunition. Mark 2 with Galilco OG14L2A fire-control system. The gunner's sight has magnifications of x7 and x14 degrees incorporating a laser rangefinder. The cdr's stabilised day/night is self-stabilised during panoramic search and slaved to the weapon stabilisation. A LLLTV camera is mounted externally on the right side of the mantlet. One 7.62 mm ADMG and a 7.62 mm coaxial MG (5700 rounds). OF-40 ARV almost identical to German Bergepanzer 2.

Role. In service with Libya (100), United Arab Emirates (36, 18 of which former Mark 1 brought to Mark 2 standard, 3 ARV).

Assessment. Standard similar to LEOPARD 1A3 but with increased first round hit capability due to laser rangefinder.

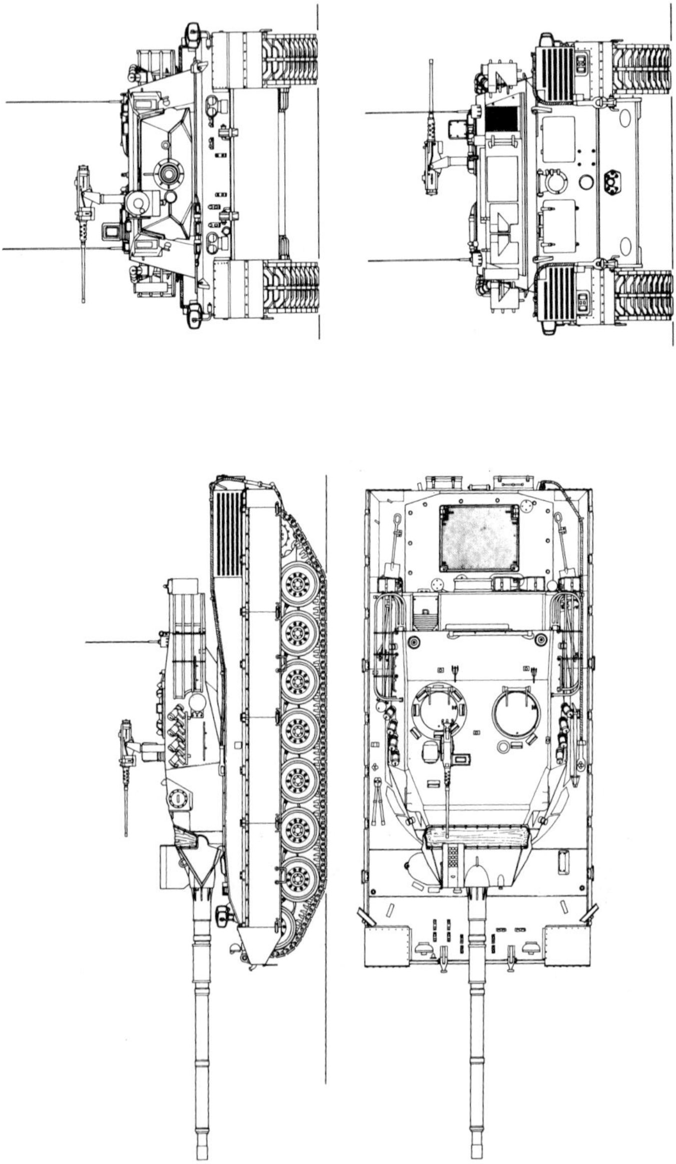

OF 40 Mk 2 MBT

OF-40 Mk 2 MBT

OF-40 ARV

C-1 ARIETE main battle tank

Wt 48 t L 10.54 W 3.7 H 2.46 895 kW 65 km/h 550 km Crew 4

Development and production.

1984 development starts by OTO Melara and IVECO.
1987 first prototype (Tricolore) issued.
1988 six prototypes available for testing.
1995 series production starts. Studies on an advanced model with automatic loader and 1100 kW engine (ARIETE 2).

Distinguishing features. Side. Seven roadwheels, four return rollers, sprocket rear. Upper part of suspension covered by skirt. Long hull with wedge-shaped bow. Heavily sloped front. Horizontal forward half of hull top, raised engine deck with side grills. Long turret with sloped front and prominent rear bustle and vertical end. Distinct roof-mounted periscope. Front. Flat sponsoned hull with driver's position on the front right. Large box-like turret with small mantlet. Main sight recessed into right front roof section with periscope head behind it.

Other details. Main armament is a 120 mm smooth-bore L44 gun built by OTO Melara with 42 rounds stowage. Elevation arc is from -9 to +20 degrees. One 7.62 mm MG is mounted coaxially another one is externally mounted as ADMG (2500 rounds). The fire-control system. The gunner has a combined primary stabilised day/thermal sight (magnification x5) with integrated Neodym-YAG laser rangefinder. The commander uses a Galileo/SFIM panoramic binocular periscope with primary stabilisation. The day channel has two magnifications (x2.5 and x10), the thermal image of the gunner's sight is projected into the periscope, too. A COSMO digital ballistic computer is installed. Armour protection includes special armour in the frontal parts. The tank will possibly be fitted with explosive reactive add-on armour. The tank is powered by an IVECO MTCA V12 supercharged direct-injected diesel engine coupled to a ZF LSG3000 automatic transmission.

Role. In production for the Italian Army (200).

Assessment. Scaled-down development of a modern MBT intentionally below latest MBT technology.

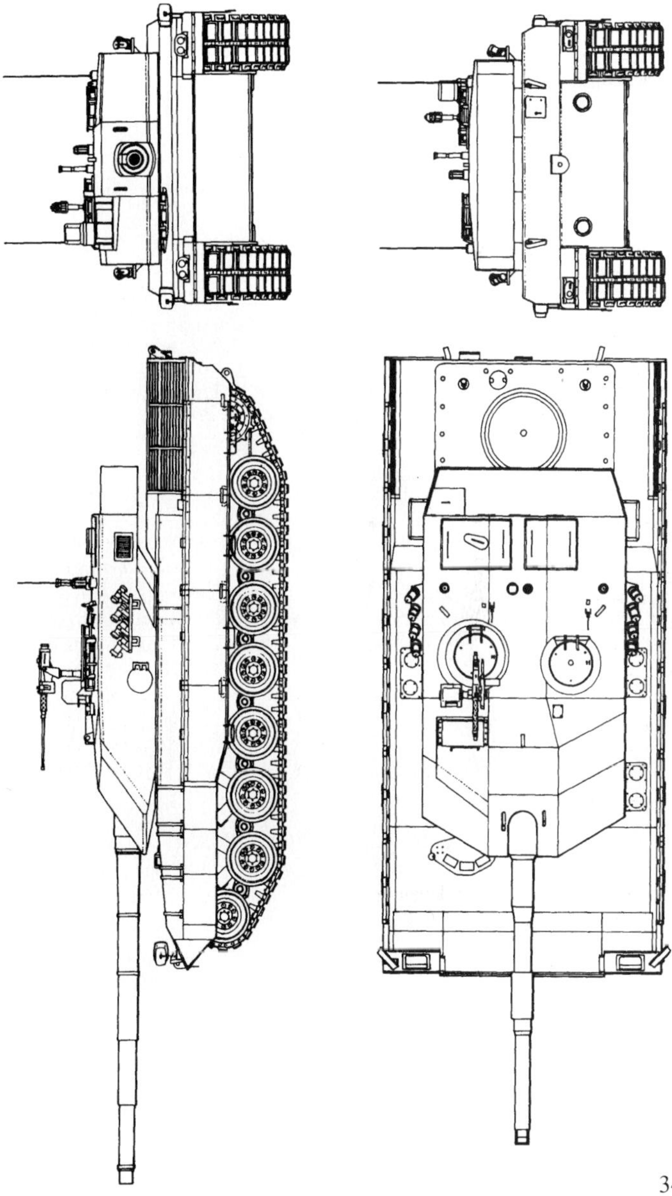

ARIETE MBT

ARIETE MBT

DARDO AIFV

Wt 23 t L 6.705 W 3.0 H 2.64 382 kW 70 km/h 600 km Crew 2+7

Development and production.

1982 development contract awarded to OTO Melara and IVECO Fiat.
1983 first prototype of private venture OTO C13 vehicle family issued.
1985 first of three VCC-80 prototypes issued.
1994 OTO Melara proposes a down-scaled version.
1998 first production order for the Italian Army expected.

Distinguishing features. Side. Six roadwheels with sprocket front, three return rollers, upper part of suspension covered by side skirts. Well sloped front, horizontal roof, hull rear slopes inwards with integral ramp. Troop compartment rear with two outward-opening roof hatches, circular roof hatch between. Two spherical firing ports with vision blocks above in each side.
Turret central with sloped front and rear, prominent periscope heads. Front. Sloped glacis plate with driver's hatch in upper left side, large horizontal air louvres on forward right side.
Hull sides above tracks slope inwards. Turret centre with sides sloping inwards. Cdr's cupola on left with periscopic sight to his immediate front. Main sight to his immediate front. Main sight protruding right turret roof.

Other details. Main armament of the DARDO is the Oerlikon-Bührle 25 mm B02 cannon in the OTO TD 25 HITFIST two-man turret. The gun has a rate of fire of 550 rpm. Muzzle velocities are 1355 m/s for the APDS round resp. 1160 for HEI. Traverse and elevation (-10 to +60 degrees) are electric. The gunner has a roof-mounted P204 periscopic sight (x8 magnification). The panoramic periscopic sight for the commander proposed for the VCC-80 is deleted and the stabilisation system to reduce production costs. The fire-control system includes laser rangefinder and thermal image night vision.
A 7.62 mm MG 42/59 is mounted coaxially. 200 ready rounds for the cannon and 700 7.62 mm are available. On both turret sides a TOW ATGM launcher is externally mounted. The vehicle is powered by a FIAT V-6 turbo- charged diesel engine coupled to a ZF 20 HST 500 automatic transmission. The hull is all aluminium armour with an additional layer of steel. The vehicle can ford 1.5 m. The tank is provided with a collective NBC system.

Role. In production for the Italian Army, C 13 offered for export.

Assessment. Sub-standard AIFV with a flat compact appearance. Good mobility and comparably low weight. Average protection and firepower.

DARDO AIFV

6614 armoured personnel carrier

Wt 8.5 t L 5.86 W 2.5 H 1.78 118 kW 100 km/h 700 km

Models.

6614 APC
6614 AMC with 81 mm mortar
6614 ACCC
6614 AAV

Development and production.

1970 development starts as a private venture by Fiat.
1972 prototypes by OTO Melara built.
Licence production in South Korea (Asia Motors) under the designation KM900.

Distinguishing features. Side. 4-wheeled vehicle with box-shaped superstructure sloped all round. Pointed hull front with engine grill above front wheel. Hull roof in line with driver's roof. Doors between wheel stations. Rear with reverse slope and large ramp-type door. Four weapon ports at either side of hull. Front. Engine right front beside 4-periscope driver's hatch. Hull sides above wheel arches slope inwards. Central commander's cupola with 7.62 mm MG.

Other details. The vehicle uses many components from the 6616 AVR.
An infantry section of 10 can be carried. The carrier is fully amphibious being propelled in the water by its wheels.
Italian vehicles deployed in Bosnia received appliqué armour.

Role. In service with Italy (Army 57, Air Force 12 and Police 40), South Korea (275 two versions KM-901 and KM-902 with a wrap-around MG mount), Libya (200), Peru (15, also AMC), Somalia (270, status unknown) and Venezuela (24).

Assessment. Good configuration, very lightly armoured. The firing ports offer only limited facilities for fighting mounted, poor firepower. Suited for low-intensity conflicts.

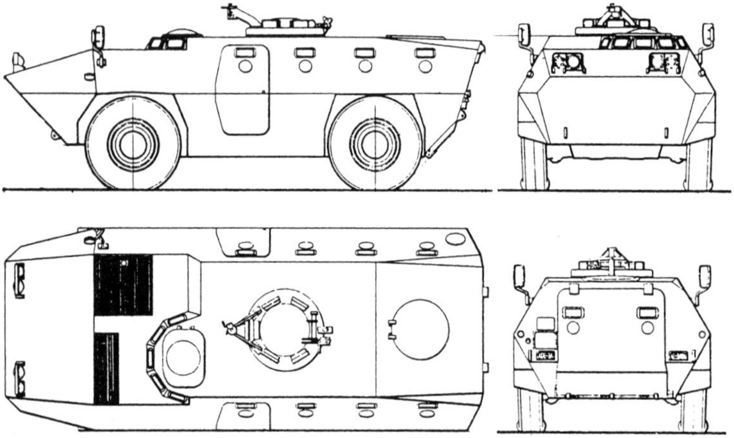

6614 APC

6614 ACCC

KM 900 APC

PUMA (6634 G) armoured general purpose carrier

Wt 5.5 t L 4.65 W 2.08 H 1.67 134 kW 105 km/h 800 km Crew 1+6

Models (provisional).

PUMA APC
PUMA ACCC
PUMA with MILAN ATGM
PUMA with TOW ATGM
PUMA with 81 mm mortar
PUMA with MISTRAL SAM
PUMA AAV

Development and production.

1982 development starts of a FIAT private venture wheeled armoured fighting vehicle family, 6634 (4x4 Light), 6633 (4x4 Medium) and 6636 (6x6 Heavy).
1988 two prototypes of a slightly bigger 6634 were built to meet the requirements of the Italian Army for a light AGPC family.
1999 contract signed for 300 4x4 and 150 6x6 vehicles.
2001 projected start of series production.

Distinguishing features. Side. Two large wheels with entrance door in front of second one. Heavily sloped glacis with vertical bow front. Horizontal hull roof, vertical rear with door. Front. Sloped front engine access hatches. Driver at forward left. Hull side with small vertical section heavily sloped to top and bottom ends, rounded edges.

Other details. The 4x4 version is 5.1 m long and can carry up to six personnel plus the driver, while the 6x6 version is 5.5 m and carries eight soldiers plus driver. The PUMA is powered by an IVECO four-cylinder turbocharged TCA diesel engine (132 kW), coupled to an automatic five- speed transmission, that provides a maximum speed of 100 km/h and a range of 800 km for the 5.7 t 4x4 variant resp. 700 km for the 7.5 t 6x6 version. The APC versions mount a 12.7 mm MG in an overhead version with electric servos for traverse and elevation, day/night sight with light intensifier.

Role. Preparation for series production for the Italian Army.

Assessment. Carrier family born out of the stringency to procure less complex and cheap fighting vehicles due to budget restraints. Questionable employment value inside Europe, only suitable for reconnaissance missions.

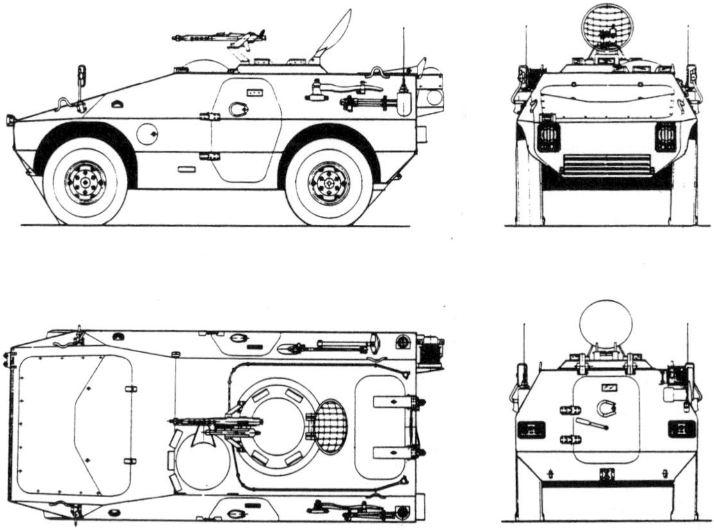

PUMA APC

PUMA APC

PUMA APC (6x6)

PALMARIA armoured artillery howitzer

Development and production.

1977 developed by OTO Melara based on OF-40 MBT hull.
1981 prototype completed.
1982 start of series production for export only.

Distinguishing features. Side. Suspension identical to OF-40, hull similar but forward part in line with rear engine deck. Two horizontal louvre sections on left forward hull sides. Large flat-sided turret on hull centre with howitzer with double muzzle brake and fume extractor. Front. Sponsoned hull with vertical sides. Sloped glacis, driver front right. Large wide turret with slightly sloping sides extending almost the full width of hull. Buffers arranged diagonally in wide conical drum mantlet.

Other details. 155 mm howitzer (41 calibres long) is designed by OTO Melara. The power-operated turret can be traversed 360 degrees. In the turret rear there is the automatic loader (23 rounds). The charge is loaded manually. Seven additional rounds are stowed inside the hull.
(TABLE)

Role. In service with Libya (210) and Nigeria (25); 20 turrets were delivered to Argentina.

Assessment. 155 AAH of above average standard due to advanced loading system.

PALMARIA AAH

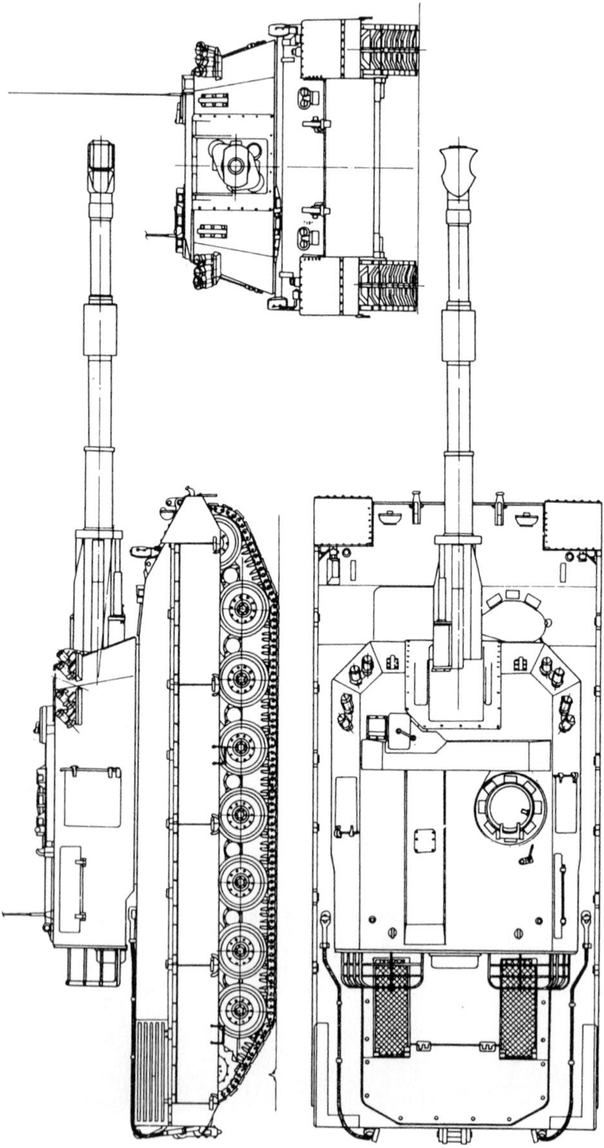

PALMARIA AAH

Type 87 armoured vehicle reconnaissance

Wt 14 t L 5.99 W 2.48 H 2.8 224 kW 100 km/h 500 km Crew 5

Models.

Type 87 AVR
Type 82 ACCC
Type 87 ANBCRV

Development and production.

1974 development of 4x4 and 6x6 ARV starts.
1979 prototype turret with Rh 202 20 mm cannon.
1982 series production of Type 82 ACCC starts at Mitsubishi.
1984 prototype with 25 mm KBA cannon.
1985 prototype with 35 mm KDE cannon.

Distinguishing features. Side. 6x6 vehicle with three large equally spaced road wheels each side with forward-opening door in left side between second and third wheels. Hull with flat nose, heightened engine compartment at rear with vertical end. Turret central. Side. Well sloped glacis plate with driver's position on right (three periscopes) and radio operator left (one periscope). Turret with eight sides all of which vertical with gun in drum mantlet. Cdr right with six periscopes, gunner left with one large sight forward and one periscope left side. Type 82 ACCC similar to Type 87 AVR but with markedly raised rear crew compartment. Pintle mounted ADMG. Large front window with shutter.

Other details. Type 87 AVR has 80 % commonality of parts with Type 82 ACCC. Serial version armed with Oerlikon-BÅhrle 25 mm KBA cannon and one 7.62 mm coaxial MG to the left. Left side of after hull used as an aisle. The vehicle has no NBC system and is not amphibious.
Type 82 ACCC has a slightly lower weight of 13.6 t, a length of 5.72 and a height of 2.38 m.
Type 87 ANBCRV with chemical detector arm mounted on the right rear of the vehicle.

Role. In production for the Japanese Army, 70 Type 87 resp. 250 Type 82 on order. 19 ANBCRV in service.

Assessment. Relatively large and high vehicle.

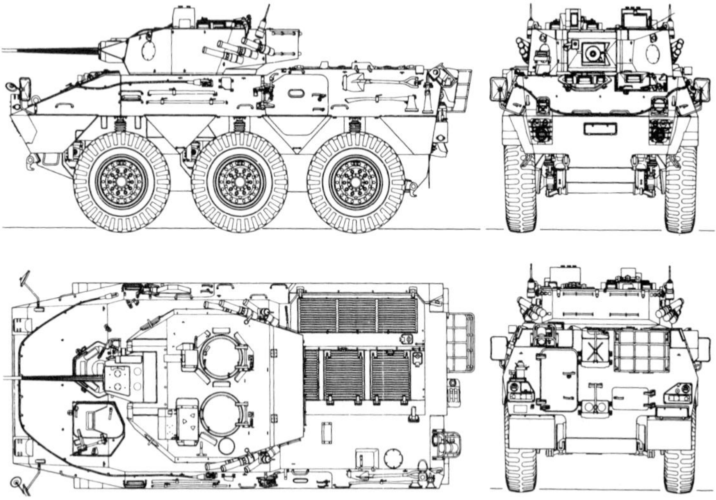

Type 87 AVR

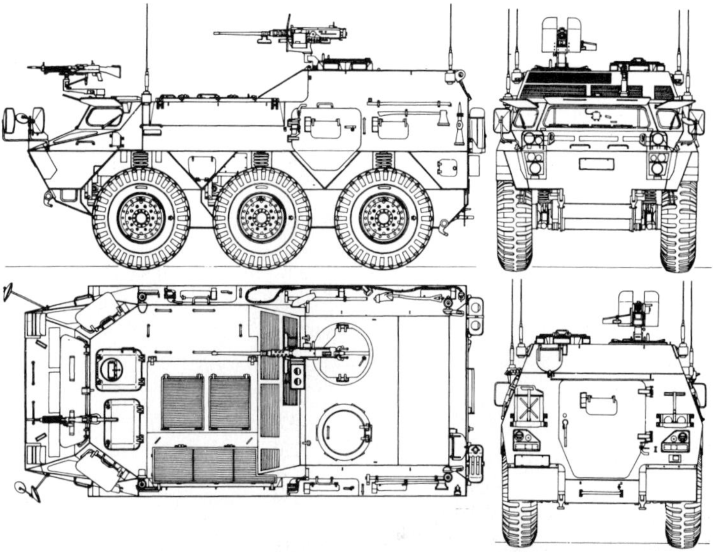

Type 82 ACCC

Type 87 ANBCRV

Type 74 main battle tank

Development and production.

1962 project definition studies for a Type 61 successor started.
1969 first two prototypes (STB-1) completed with automatic loader.
1971 STB-3 issued without auto-loader and turret with longer bustle.
1973 appearance of last prototype (STB-6), production contract placed.
1975 start of series production, running in small numbers up to 1988.

Distinguishing features. Side. Five large roadwheels, rear sprocket, no top rollers. Low-profile turret with compound-sloped mantlet. Gun with fume extractor half-way along barrel. Front. Sponsoned hull with heavily sloped sides and well sloped glacis plate. Driver's hatch upper left side. Heavily sloped turret sides. Commander's cupola right with high rotating head periscope.

Other details. Weapon stabilisation in both axis. Simple ballistic computer with laser rangefinder housed in the cdr's sight (x1 and x8 magnification). Gunner sitting in front of cdr with J2 infra-red periscope of x1 and x8 magnification. 105 mm L7A3 main armament (55 rounds), coaxial 7.62 mm MG (4500 rounds) and 12.7 mm ADMG outside armour with external ammunition feed (660 rounds). Raised driver's hatch with three periscopes protruding over glacis. Hydropneumatic suspension which can be raised or lowered to give a ground clearance of 0.2 to 0.65 m. Mitsubishi 10ZF Type 22 WT 10-cylinder air-cooled diesel engine and manual Mitsubishi MT75A transmission (6 forward, 1 reverse gears). External exhaust on both rear sides. Deep wading capabilities. NBC collective protection system.
Later upgraded with gun thermal sleeves, add-on turret roof and front armour, passive thermal infra-red night vision system.
(TABLE)

Role. In service with Japanese Self Defence Forces (873).

Assessment. Though characterised by excellent turret and hull configuration not more than average armour protection. Good cross-country performance thanks to adjustable supension. Very cramped fighting compartment.

Type 74 MBT

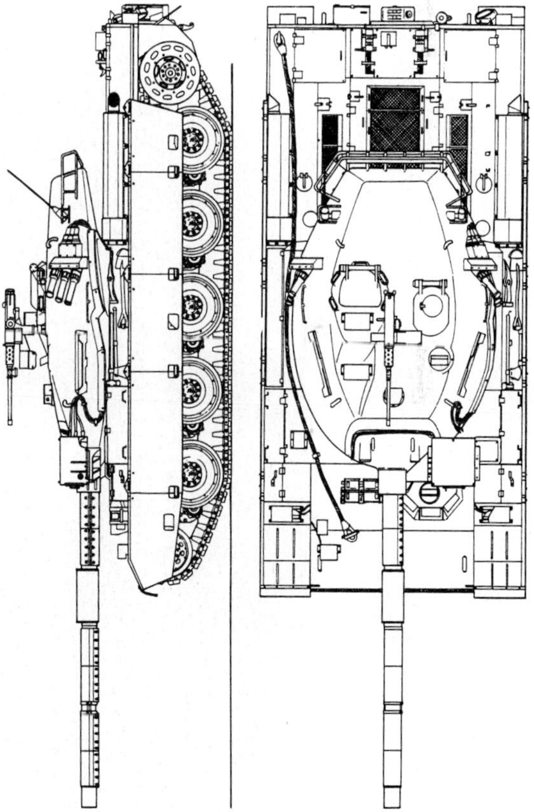

Type 74 MBT (modified)

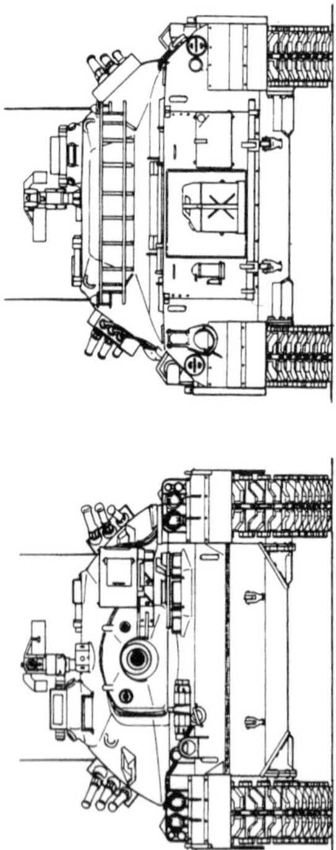

Type 74 MBT (modified)

Type 90 main battle tank

Models

Type 90 MBT
Type 90 ARV

Development and production.

1974 request for replacement of Type 61 MBT.
1977 development of components starts.
1982 test bed (STC) without turret constructed.
1986 four prototypes (TK-X) built by Mitsubishi Heavy Industries.
1991 series production starts.

Distinguishing features. Side. Six roadwheels with centre pair closer together. Upper part of suspension covered by skirts. Three return rollers, sprocket rear. Wedge-shaped front bow with heavily sloped glacis. Horizontal forward half of hull roof, raised engine deck, vertical rear with recessed rear hull bottom. Long box-like turret with vertical front and rear part. Prominent roof mounted sight in front of flat cdr's cupola. Relatively short-barreled gun with several steps and fume extractor offset half-way point. Front. Flat sponsoned hull with nearly horizontal glacis. Driver on left forward end of hull. Slightly overhanging turret with vertical sides. Rectangular mantlet. Large roof mounted sight on the left side and rotating roof mounted sight in front of cdr's hatch.

Other details. Main armament is the Rheinmetall 120 mm smooth-bore gun as used in the LEOPARD 2. The gun is fitted with an auto-loader in the rear turret bustle with blow-up roof plates. A 7.62 mm coaxial MG and one 12.7 mm ADMG are in use. The computerized fire control system includes a Neodym-YAG laser rangefinder, a digital ballistic computer and a mast mounted cross wind sensor. The gunner's sight is provided with a thermal sight. The gun controls are either electric or manual. Behind the mantlet a laser warning sensor is placed on the turret roof. The hybrid suspension consists of four hydro-pneumatic units and two torsion bars for the two centre wheels, it enables longitudinal hull adjustments (±15 cm). NBC system installed, no deep-wading capability.
(TABLE)
ARV on MBT hull of similar layout as the German BÜFFEL ARV with 25 t crane on the right vehicle side. Pintle mounted cal. 50 MG. The vehicle's length is 9.2 m, height 2.7 m and its weight is 50 tons.

Role. Under production for the Japanese Army, 220 MBTs and 5 ARV issued.

Assessment. Firepower comparable with M1A1 ABRAMS, excellent mobility and good protection. The low production numbers result in relatively high costs.

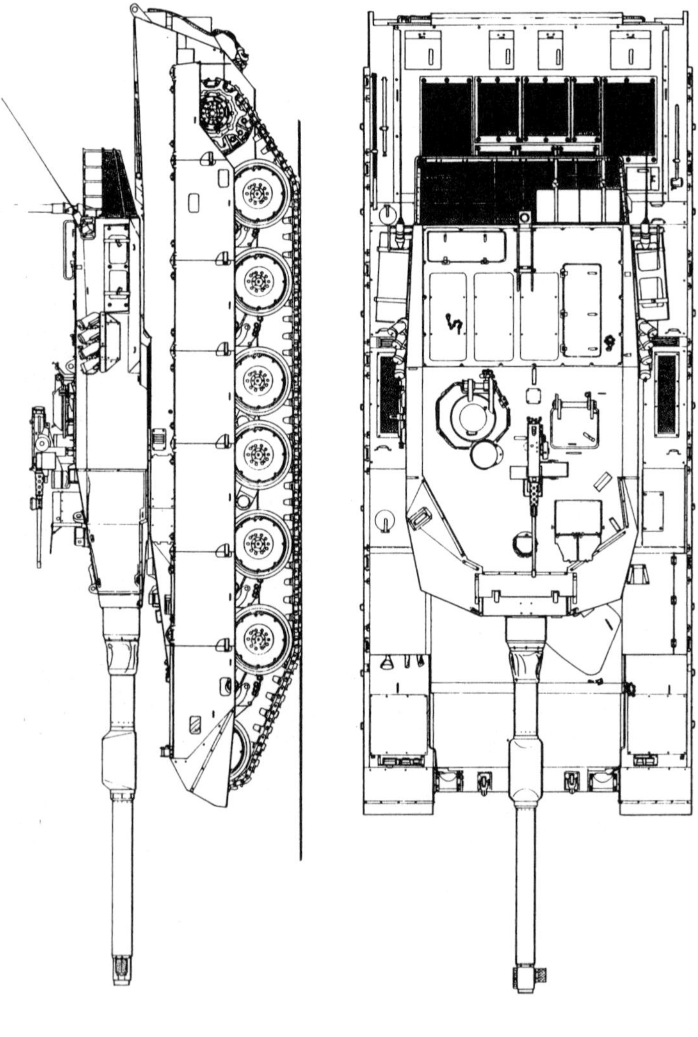

Type 90 MBT

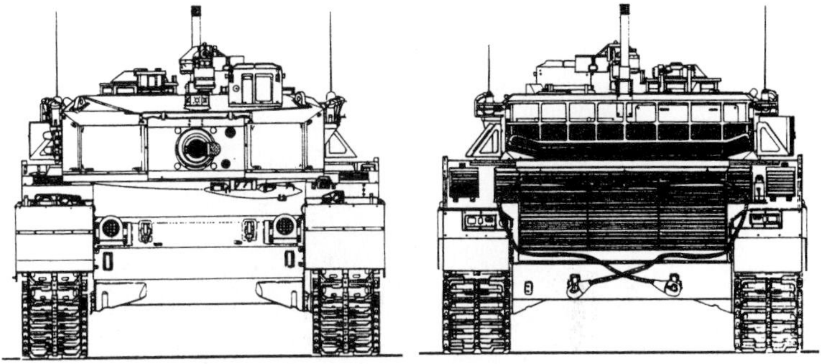

Type 90 MBT

Type 90 MBT

Type 90 ARV

Type 89 armoured infantry fighting vehicle

Development and production.

1981 request for the development of a new AIFV, four prototypes built by Mitsubishi Heavy Ind.
1989 start of production.

Distinguishing features. Side. Six roadwheels, three return rollers, sprocket front. Upper part of suspension covered by skirt. Sloped glacis with nearly vertical front bow. Horizontal hull top with raised troop compartment deck, steep rear end. Small exhaust grill in front left side, three spherical firing ports in each rear hull side. Large turret with pointed mantlet, cannon with flash suppressor extending vehicle's front. Front. Sloped front section with engine on the left and driver's place on the right side. Sponsoned hull with sloped sides. Box-like turret with vertical sides on full width of hull roof. Two prominent periscopes on top of the turret roof. Two rectangular launcher pods on either side. Small box housing thermal sight on the right side of the cannon.

Other details. Main armament is the Oerlikon-Bührle KDE 35 mm cannon with a cyclic rate of fire of 200 rpm. The gun fires APFSDS rounds up to 1000 m and HEI rounds up to 3000 m. One 7.62 mm MG is mounted coaxially on the left side of the cannon. On both turret sides a Jyu-MAT Type 79 medium range wire-guided anti-tank missile is mounted to be reloaded manually. Through the weapon ports 5.56 mm submachine guns can be fired. The infantry squad dismounts via two large rear doors opening to both sides. The vehicle does not have amphibious capabilities. NBC system is fitted.
(TABLE)

Role. In production for the Japanese Army as a Type 61 APC replacement, 148 on delivery.

Assessment. Versatile AIFV with powerful main armament and good capability for fighting under-armour; average armour protection.

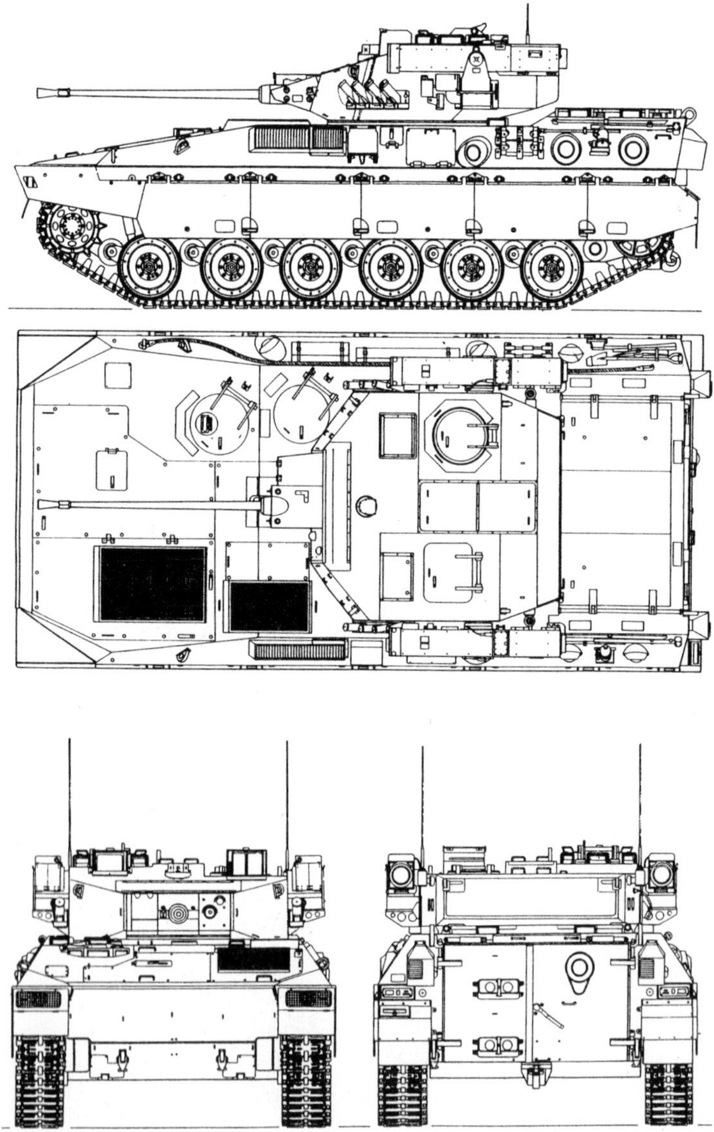

Type 89 AIFV

Type 89 AFV

Type 60 armoured personnel carrier

Wt 11.8 t L 4.85 W 2.4 H 2.31 162 kW 45 km/h 230 km Crew 10

Models.

Type 60 APC
Type 60 AMC with 81 mm mortar
Type 60 AMC with 107 mm mortar
Type 60 ANBCRV
Type 75 armoured dozer

Development and production.

1956 development of a fully tracked APC.
1958 first prototypes appear (SU-I by Komatsu and SU-II by Mitsubishi).
1959 eleven second series prototypes issued including AMC versions
1960 start of series production until 1972.

Distinguishing features. Side. Five large roadwheels, three top rollers, front sprocket. Steep glacis, box-like superstructure with vertical rear. Front. Glacis with ball mounted MG in left side, engine compartment on the left side to rear of bow MG operator and commander. Vertical hull sides. Rear troop compartment with roof hatches opening left and right. Twin doors in rear.

Other details. Main armament consists of a 12.7 mm MG in open mounting with a small metal shield. 7.62 mm bow MG. Rear compartment for six infantry men. No provision for fighting mounted under armour. No NBC system, not amphibious, no night vision equipment. Some vehicles equipped with two KAM-3D anti-tank guided missiles on hull roof at rear. Also two carriers with front mounted dozer in service.
AMC versions firing to the rear, 107 mm vehicle with modified rear compartment. Weight of the two vehicles is 12.1 (81 mm AMC) resp. 12.9 t. 24 resp. 8 mortar rounds are carried. Maximum range is 3000 resp. 4000 metres.
NBC detection vehicle has device at rear for taking soil samples.
Type 75 armoured dozer has a weight of 19.2 t, has a length of 6.84, a width (working order) of 3.45 and a height of 2.79 m. Dozer blade is hinged vertically in centre for use as a V-blade. The hydraulically operated blade is to the rear when travelling.

Role. In service with JSDF in the APC role (341), both AMC versions (18 each) and ANBCRV.

Assessment. Lightly armoured APC of average fighting power.

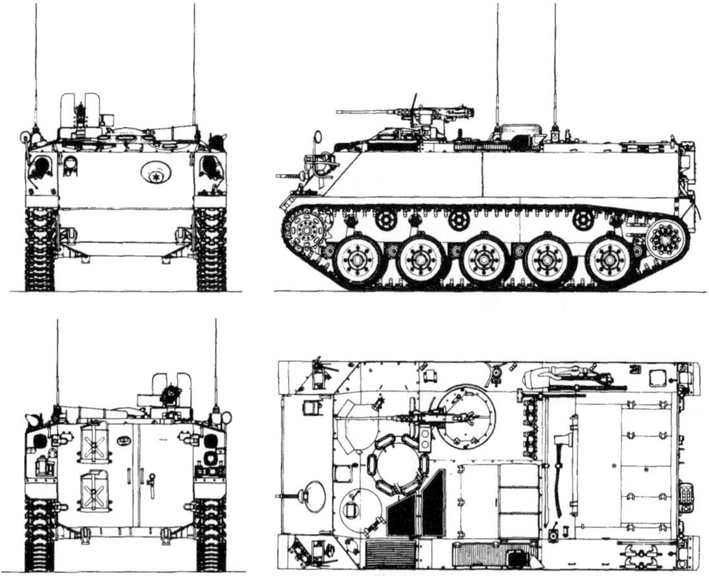

Type 60 APC

Type 60 AMC 107 mm

Type 60 ANBCRV

Type 60 armoured dozer

Type 73 armoured personnel carrier

Models.

Type 73 APC
Type 73 ACCC
Type 75 APC with ground-wind measuring installation
Type 87 armoured ammunition carrier
Type 96 mortar carrier

Development and production.

1967 start of development for a Type 60 APC successor.
1968 prototypes built, two by Mitsubishi (SUB-I) and two by Komatsu (SUB-II). Second SUB-I was adopted.
1973 start of series production running up to 1988.
1999 production of Type 96 mortar carriers starts.

Distinguishing features. Side. Five roadwheels, front sprocket. Shallow, low-silhouette hull with flaps at side which fold down to box in top run of track for swimming (not fitted to every vehicle). Sloped front with splashboard and hull MG. Driver front, commander central. External exhaust on left of hull roof. Vertical rear. Commander's cupola with external ring-mounted ADMG. Front. Large sponsoned hull. Bow MG in ball mounting front left. Commander's cupola right.

Other details. 12.7 mm MG can be fired from inside, elevation from +60 to -10 degrees. The section cannot fight from under armour, but there are two large roof hatches. Large double doors at rear. The APC is equipped with a NBC system and has aluminium armour. Infra-red equipment only for the driver.
ACCC with raised rear compartment.
Vehicle with wind-measuring device has an erectable mast (12.5 m).
Ammunition carrier with high box-like cargo compartment.
(TABLE)
Type 96 AMC with rear pointing 120 mm mortar in a raised superstructure. The hull has an additional roadwheel station and a total lenght of 6.7 m. The 23.5 t vehicle has a height of 2.95 m and carries 50 mortar bombs. It is powered by a 302 kW engine.

Role. In service with the JGSDF (337 APC, 2 with dozers, 15 Type 75, 15 Type 87).

Assessment. Low vehicle with good configuration. Good mobility, insufficient armament, lightly armoured.

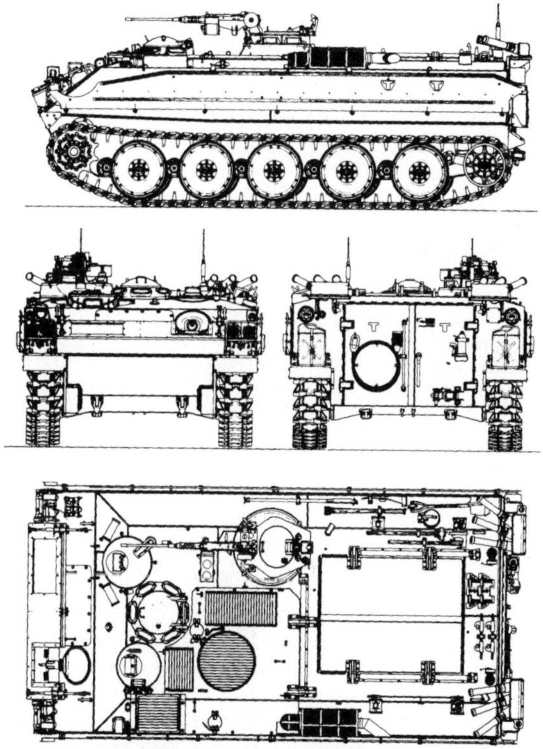

Type 73 APC

Type 73 ACCC

Type 75 APC with wind sensor

Type 87 AAC

Type 96 armoured general-purpose personnel carrier

Wt 14.7 t L 6.84 W 2.48 H 1.84 224 kW 100 km/h Crew 10

Models

Type 96 APC
Type 96 AMC

Development and production.

1992 requirement for a wheeled APC for UN missions.
1994 competing offers by Komatsu and Mitsubishi.
1996 first prototype issued.
1997 APC and mortar carrier ordered.

Distinguishing features. Side. Four mid-size wheels evenly gapped on boat-like hull with pointed bow. Cooling fan grill and exhaust outlet on left side. Two firing ports on both rear sides. A bank of four smoke dischargers on the top hull side towards the rear. Front. Box-shaped hull designed with inverted mine-deflector hull bottom. Sloped glacis and bow, driver's position on the right with three vision blocks and NBC air-intake grill on the left upper glacis. Rear ramp with single door integrated. APC with four roof hatches above the infantry compartment resp. two for the mortar carrier.

Other details. The APC is armed with a 40 mm Mk.19 grenade MG in an open mount; a cal. 50 MG is optional. The two front wheels are steerable, the two aft ones are fixed to swing arms with hydraulic bumpers. The vehicle has a central tyre-pressure regulation system. A laser warning device is mounted on the front left roof corner. The infantry squad is sitting on two benches facing each other. Driver's and gunner's compartments are accessible from behind.
AMC with rear pointed 120 mm mortar.

Role. In production for the JSDF, 29 APCs and 6 mortar carriers ordered.

Assessment. Conventional wheeled APC layout, only suitable for low-intensity conflicts. Unfavourable thermal signature due to exhaust position.

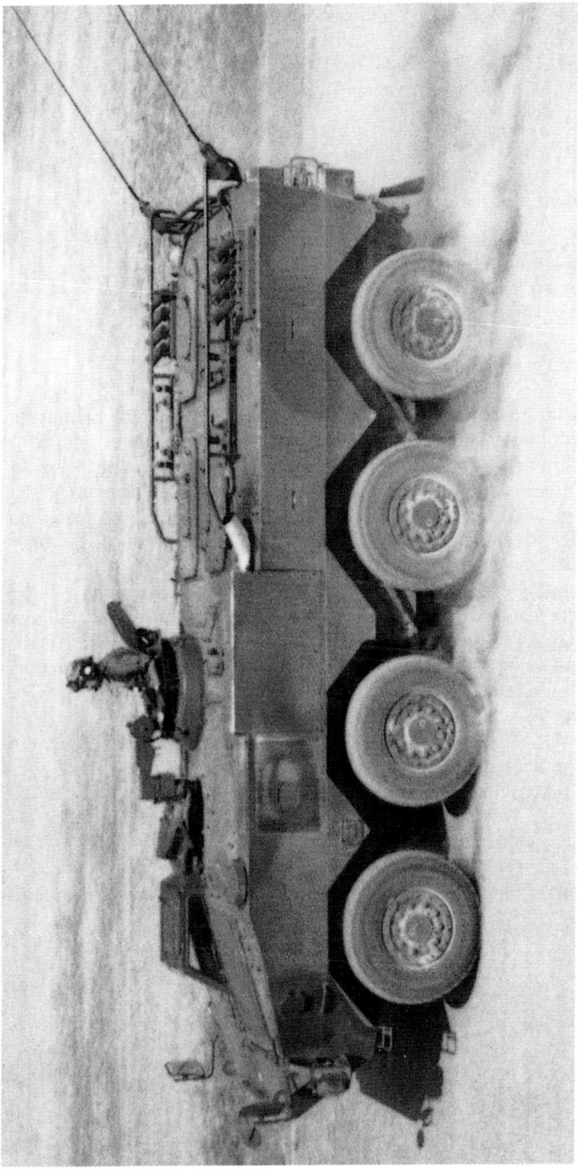

Type 96 APC

Type 75 155 mm armoured artillery howitzer

Wt 25.3 t L 7.79 W 3.09 H 2.55 331 kW 47 km/h 300 km Crew 6

Related vehicle.

Type 74 105 mm AAH

Development and production.

1969 start of development to meet requirements of JGSDF.
1971 completion of two prototypes (different loading system).
1975 adoption by the Japanese Army.
1978 start of series production running until 1988.
1983 project for a new SP howitzer with 155 mm FH-70 ordnance (Type 75 Kai) is suspended.

Distinguishing features. Side. Six large road wheels, no idler, front sprocket. Vertical front and rear, horizontal engine deck at front. Slightly sloped forward part of hull. Large angular turret, upper part sloped but rear vertical. Gun with internal drum mantlet and twin projecting recuperators, fume extractor half-way along barrel and double muzzle brake. Front. Hull fully sponsoned with vertical sides. Turret sides slightly sloped.

Other details. Engine left front, driver right front with own compartment behind front plate. Large ammunition-loading hatches in vertical hull rear. Main armament is a 155 mm 30-calibre weapon firing Japanese made HE rounds at a maximum range of 19 km. On the turret roof there is an external mounting for the 12.7 mm ADMG (1000 rounds). NBC system fitted.
(TABLE)
Type 74 AAH is a much smaller vehicle with five roadwheels and rear idler. Its weight is 16.5 t, length 5.9, width 2.9 and height 2,39 metres. The AAH is powered with a Mitsubishi 4ZF two-stroke V-4 diesel engine with 221 kW.

Role. In service with Japan (201 and 19 Type 74).

Assessment. Conventional design solution which leans heavily on the American M109 AAH. Comparable performance data.

Type 74 AAH

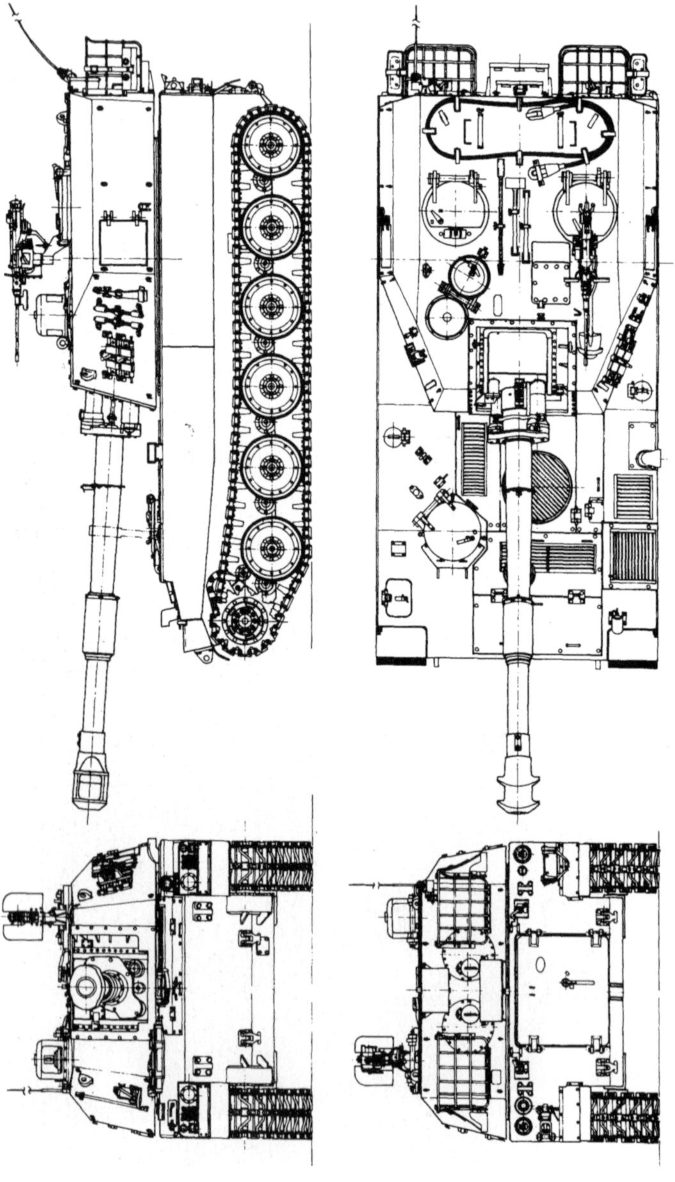

Type 75 AAH

Type 75 AAH

Type 75 130 mm armoured artillery rocket launcher

Wt 16.5 t L 5.78 W 2.8 H 2.67 221 kW 53 km/h 300 km Crew 3

Variant.

Type 76 artillery location radar vehicle

Development and production.

1965 start of development.
1969 issue of component prototype (15 rails).
1973 two prototypes built by Komatsu (chassis) and Aerospace Division (rocket).
1975 start of series production running until 1985.

Distinguishing features. Side. Suspension identical to Type 73 APC. Fighting compartment to rear flattened off with large box-like rack for 30 rockets in four rows on top of it. Front. Commander at right front with ADMG on external ring-mounting. Driver front left.

Other details. The launcher rack can be traversed 50 degrees to each side. It can be elevated up to 50 degrees. The rocket has a calibre of 131.5 mm, a length of 1.856 m, its weight is 43 kg (warhead 15 kg). Battle range is 15 km maximum.

Role. In service with the JGSDF (66).

Assessment. Effective use of the Type 73 APC vehicle. High survivability due to armour protection for the crew.

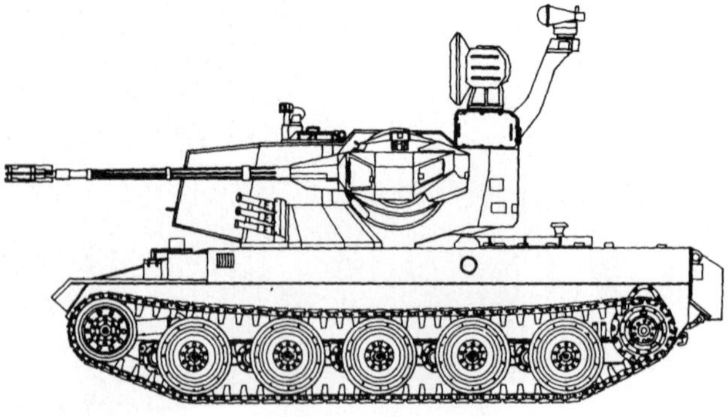

Type 75 AARL

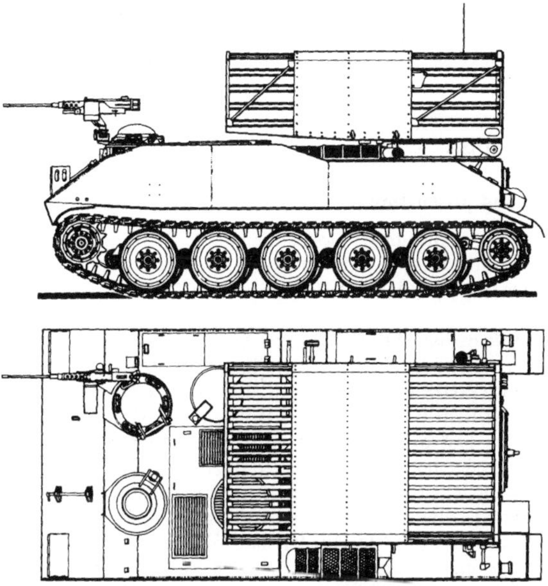

Type 75 AARL

Type 87 armoured air defence gun

Wt 36 t L 7.99 W 3.18 H 4.4 529 kW 53 km/h 500 km Crew 3

Related vehicle.

Type 91 AVLB

Development and production.

1981 production of a trials turret model.
1984 first complete prototype (AW-X) issued.
1990 first order (12) placed, first AVLB produced.

Distinguishing features. The chassis of the Type 74 MBT is used with slightly raised hull sides which are now vertical. Sloped glacis with driver on the left side and auxiliary power unit on the right side. Armament of Type 87 AADG comprise two Oerlikon-Bührle 35 mm KDA mounted externally on each side of the high turret. The layout of the turret is similar to the German GEPARD with the exception that both the surveillance and the tracking radar antennas are mounted on the top rear. Commander and gunner are placed in the forward part of the turret. A periscopic sight is in front of the gunner, both are provided with a vision block each. Externally on the left side of the tracking radar is a flat box with a laser rangefinder, optical tracker and a LLLTV camera.
AVLB has a MLC 60 bridge of the same laying principle of the German BIBER spanning up to 20 m. The total length of the vehicle is 10.9 m, width is 4.0 m, height 3.8 m and its weight is 41.8 tons. The tank is operated by a crew of two.

Role. In low-scale production for the JSDF, total requirement is 180 AADG; 2 AVLB issued, low-scale production.

Assessment. AD system with capabilities compared to the GEPARD AADG. AVLB comparable to German BIBER.

Type 87 AADG

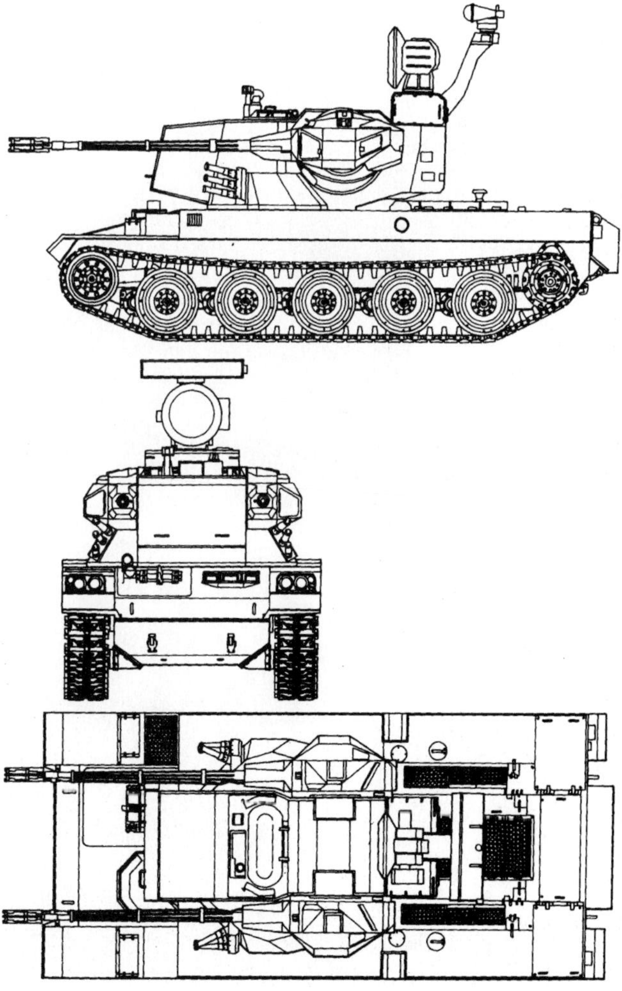

Type 87 AADG

Type 91 AVLB

Type 67 armoured vehicle launched bridge

Wt 36.7 t L 7.27 W 3.5 H 3.5 478 kW 45 km/h 200 km Crew 3

Development and production.

1967 start of series production.

Distinguishing features. Side. Suspension of Type 61 MBT. Scissors' type bridge mounted on hull top and launched over front of vehicle. Front. Rectangular launching pad in line with end of bridge while moving.

Other details. Bridge can be retrieved from both ends. Length of bridge is 12 m spanning a ditch of 10 m. It can carry vehicles up to 40 t. Launching time is 5 minutes.

Role. In service with the JGSDF.

Assessment. Old fashioned bridge type. Limited load capacity.

Type 67 AVLB

Type 70 armoured recovery vehicle

Wt 35 t L 8.4 W 2.95 H 3.1 478 kW 46 km/h 200 km Crew 4

Related vehicle.

Type 67 armoured engineer vehicle

Development and production.

1970 start of production.

Distinguishing features. Side. Suspension identical to Type 61 MBT. On top of hull in centre is a large superstructure with vertical rear. Dozer blade at front with A-frame attached on each side of superstructure. Large stowage box on hull rear. Front. Box-like superstructure instead of turret with vertical sides and slightly sloped, sharp nosed front. Cdr's cupola on top with external 12.7 mm ADMG mounting.

Other details. Winch exit on front glacis. Hydraulic winch is mounted in turret bustle. Armament consists of an 81 mm mortar which can be mounted on the glacis plate.
Type 67 AEV with a front dozer blade and a light crane. Weight is 35 t, crew of four.

Role. In service with the JGSDF.

Assessment. Limited lifting capability. No up-to-date performance data.

Type 70 ARV

Type 78 armoured recovery vehicle

Wt 38 t L 7.95 W 3.38 H 2.4 529 kW 53 km/h 300 km Crew 4

Development and production.

1974 completion of first prototype by Mitsubishi.
1978 start of production at slow rate (one per year).

Distinguishing features. Side. Type 74 MBT suspension. Hull with raised compartment at front left dropping down to engine compartment at rear. Sloped front with dozer blade. Crane pivoted at front of hull on right side has telescopic jib, laying to rear when not in use. Front. Driver in forward part of heightened superstructure, commander's cupola behind.

Other details. 3.5 m jib, can be extended to 5 m, 270 degrees traverse, maximum load 20 t. Two-speed winch in front of the tank with maximum pull of 38 t and 60 m of cable (32 mm diameter). The only armament is a roof-mounted 12.7 mm ADMG. Fuel capacity 950 l.

Role. In service with JGSDF (50).

Assessment. Comparable with German BERGEPANZER 2.

Type 78 ARV

Multi-purpose light armoured vehicle chassis

Wt 20 t L 7.0 W 3.2 H 2.6 (provisional data)

Development and production.

1985 first public appearance of an AVR and a 152 mm AAH.
1991 122 mm self-propelled gun spotted.
1992 identification of 130 mm SP gun, AARL and 30 mm ADG.

Distinguishing features. Side. Six large roadwheels, no return rollers, sprocket rear. Either vertical or boat-shaped bow front, horizontal roof, sloped rear. Reconnaissance tank with flat rounded turret, all sides sloped. Short gun barrel with bore evacuator in front of centre point. Cdr's cupola on forward part of turret. Ventilator dome on the right rear turret side. Different weapon installations in the central and rear section. Front. Well sloped glacis with driver's hatch on the left side. Sponsoned hull with sloped sides. Reconnaisance tank with round turret with full width over hull roof, handrail on upper part. Small gun mantlet.

Other details. Derivative of a locally built copy of the Chinese Type 63 APC. Main armament of the ARV is an 85 mm gun. A 7.62 mm coaxial MG is mounted on the right side of the gun. Above the main armament there is a launching rail for an ATGM, possibly SAGGER AT-3.
Artillery systems feature different gun installations ranging from 122 mm up to 152 mm and multiple rocket launchers.

Role. In service with the North Korean Army (several hundred).

Assessment. Similar to Soviet PT-76 or Chinese Type 63 light tanks.

M 1985 AVR

M 1991 122 mm AAH

K1 main battle tank

Versions.

K1 MBT
K1A1
K1 AVLB
K1 ARV

Non-serial versions.

K1A1 MBT
K1M MBT export version for Malaysia

Development and production.

1979 development of an indigenous MBT of 45 t in cooperation with USA (project named ROKIT).
1980 production of two prototypes ordered at General Dynamics Land Division.
1983 automotive test rig and FCS test rig trials in USA.
1985 production starts at Hyundai.
1989 contract awarded for AVLB development.
1993 production order for ARV.
1996 prototype of K1A1 issued with 120 mm M256 gun.

Distinguishing features. Side. Hull similar to US M1 ABRAMS but front plate more vertical, six roadwheels, sprocket rear. Return rollers and upper part of suspension covered by armoured skirts extending to top of hull and engine compartment. Turret similar to M1 but shorter and lower. Stowage basket extends from turret side round to rear. Front. Steep front plate, almost horizontal glacis plate, driver's hatch in left side. Sides of turret front slope slightly to rear, sides and rear slope inwards. Two roof mounted MG. Periscopic sight on right side of turret roof. Box-shaped main sight in front of it.

Other details. Main armament is the M68A1 105 mm gun with 47 rounds stowage. One 7.62 mm MG is mounted coaxially, another 7.62 mm MG in front of the loader station (7200 rounds). The cdr is provided with a 12.7 mm ADMG (2000 rounds). The first 210 tanks are fitted with a Hughes Gunner's Primary Sight. Follow-on tanks are expected to have a Texas Instruments GPS with a tank thermal channel and a CO2 laser rangefinder. Fire control system consists of a digital ballistic computer. The commander uses a French SFIM two-axis independent stabilised periscope. The hybrid suspension has both hydro-pneumatic and torsion bar suspension. MTU powerpack and ZF LSG 3000 automatic transmission. (TABLE)
The K1A1 features the M2 56 120 mm main gun with 32 rounds combat load. The vehicle weight increases to 53 tons.
The K1 AVLB features a Vickers DS scissors-type MLC-66 bridge with a span of 22 m. The vehicle has a total weight of 49.6 t and is 12.5 m long in travelling configuration. The height is 4.0 m, the width 4.0 m.

The K1 ARV uses the Bergepanzer equipment from MaK including a front-mounted dozer blade, a main winch with a maximum capacity of 343 kN. Cable length is 150 m. The crane which can be traversed through 270° has a maximum lifting capacity of 25 t. The vehicle has a weight of 51 t, is armed with a 12.7 mm MG and has an auxiliary power unit.

Role. In production for the South Korean Army (1000+ MBT, 149 ARV, 56 AVLB).

Assessment. High-technology transfer from USA to Korean industry. Above average mobility, good protection and sufficient fire-power.

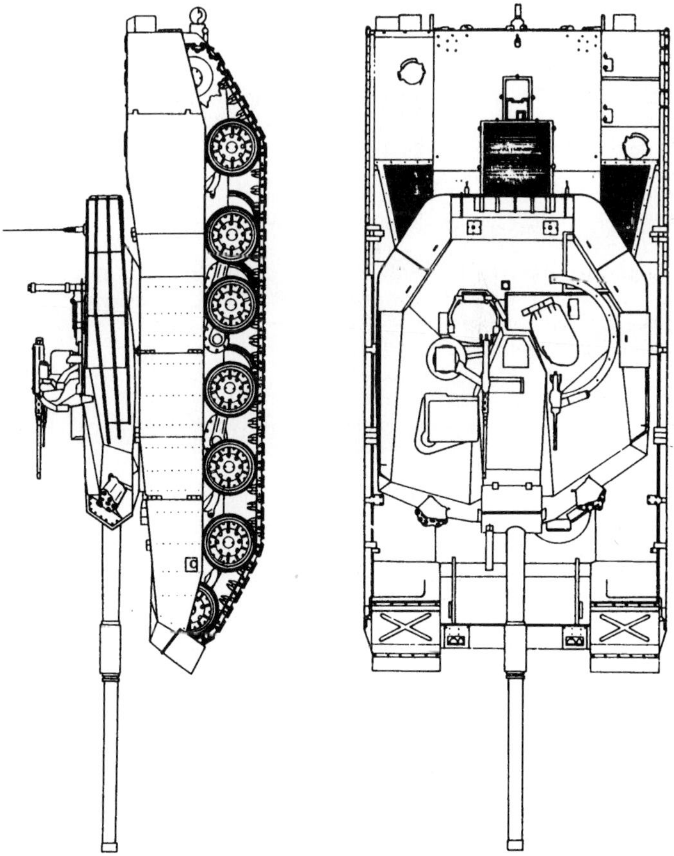

K1 MBT

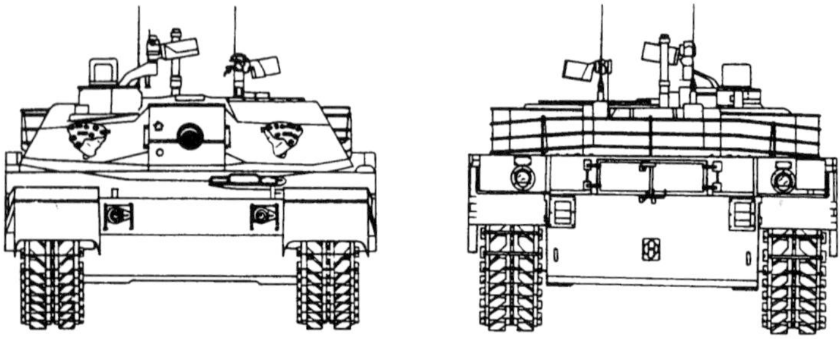

K 1 MBT

K 1 MBT

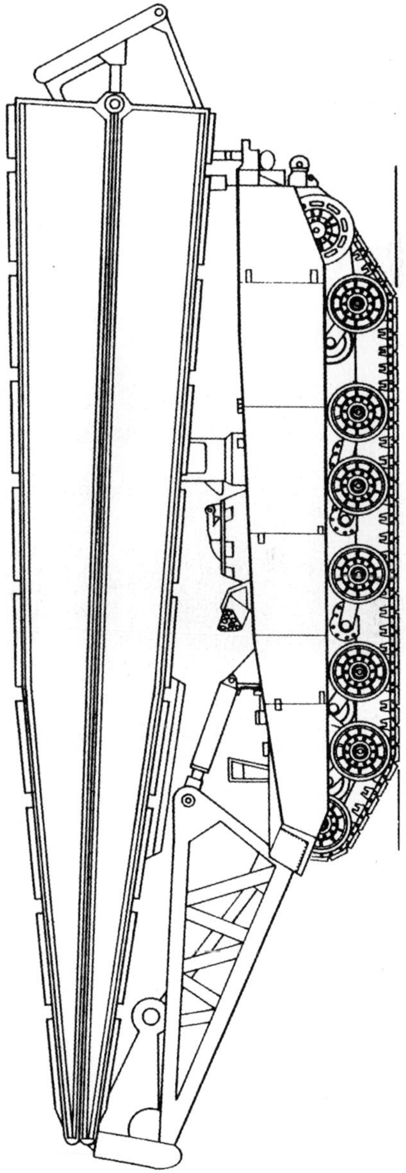

K 1 AVLB

K 1 AVLB

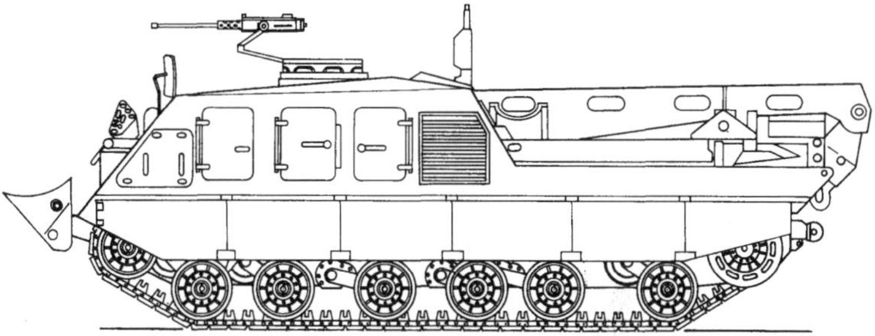

K 1 ARV

K 1 ARV

TR-580 main battle tank

Models.

TR-580 (TR-77) MBT
TR-85 MBT
TER-800 ARV

Non-serial version.

TR-800 MBT (TR-85 M1)

Development and production.

1977 first appearance. Locally refurbished T-55 with Israeli assistance using Centurion suspension parts.
1978 first appearance of the TR-85.

Distinguishing features. Turret identical to Soviet T-55 MBT, rear hull section slightly lengthened. Running gear with six road wheels with distinctive gaps behind first, third and fifth wheel stations. Three return rollers, track skirts. Five external ammunition boxes for the ADMG on the right turret rear.

Other details. Tank commander's and loader's cupolas welded to the turret roof (no use of bolts as on the T-55). The weight is increased to 42.5 t. Additional fuel tank on the left and stowage boxes on the right side of the track guards. TR-800 is a further development with a Chinese fire-control system with laser rangefinder.

Role. Equipment for several Romanian tank units (88 TR-580, 315 TR-85, some planned for TR-85M1 upgrade).

Assessment. The short living running gear of the Soviet version led to this local modification with better characteristics for this obsolete tank.

TR-580 MBT

TR-85 MBT

TR-85 M1 MBT

TAB armoured general purpose carrier

Models.

TAB-70 APC
TAB-71 APC
TAB-71M with 12.7 mm one-man turret
TAB-71A R1 (450, 451, 452) ACCC
TAB-77 APC
TAB-77A 451/M ACCC
TAB-77 PCOMA AACPV
TABC-79 AVR
TAB-79A PCOMA AACPV
TABRCH-84 ANBCRV
TAB-71AR AMC with 82 mm mortar
TAB-79AR AMC
TERA-71 ARV
TERA-71L ARV
TERA-77L ARV
TCG-80 ARV

Development and production.

1970 licence production of Soviet BTR-60P (TAB-70).
1972 adoption of TAB-71 (similar to BTR-60PB), sometimes designated TAB-72.
1984 licence production of BTR-70 (TAB-77).

Distinguishing features. Vehicles resemble closely their Soviet pattern, but differ in the turret design.
The TAB-71 has a circular turret with its forward roof curved.
TAB-77 has an enlarged circular turret with heavier MG armament (14.5 mm). One AT-3 SAGGER ATGM can be fitted on either side of the turret.
The TABC-79 is the Romanian version of the 4x4 BRDM-2 but with features of BTR-70 (front bow). It retains the original BRDM-2 MG turret. The vehicle lacks the central auxiliary wheels and has an entrance door instead.

Other details. TAB-70/71 are powered by two petrol-engines of 103 kW each providing the vehicle with a speed of 95 km/h. The fuel capacity is 290 litres with a maximum range of 500 km. The weight is 11 t.
The TAB-71 turret has an enlarged mantlet and a sight, protruding at left turret front for air defence aiming. The MG can be manually elevated up to 85 degrees.
TAB-73 lacks the MG turret and has large roof hatches which are swung open through 180 degrees either side.
TERA-77L ARV mounts a rotating crane on top of the rear hull.
The TAB-77 is powered by a single diesel engine.

Role. In service with Iran (100 TAB-77), Moldova (71 TAB-71), Romania (384 TABC-79, 474 TAB-70, 491 TAB-71, 167 TAB-77, 79 TAB-79AR)

Assessment. Indigenous deviations of former Soviet models with similar combat performance.

TAB-71 M APC

TAB-77 APC

TABC-79 AVR

TABC-79A PCOMA

TERA-71 ARV

BRDM armoured vehicle reconnaissance
Bronevaya Razvedyvatel'naya Dozornaya Mashina

Models.

BRDM-1
BRDM-2
BRDM-2 TDM
BRDM-2 RKh ANBCRV
BRDM-2 RKhb ANBCRV
BRDM-2U command vehicle
BRDM-2ZS psychological warfare carrier

Development and production.

1956 first prototype issued as a replacement for BTR-40.
1957 start of BRDM-1 production, still with open roof.
1958 BRDM-1 issued with roof closed with two hatches.
1960 development of an improved version, BRDM-2.
1962 production of BRDM-2 starts.

Distinguishing features. Side. Four-wheeled vehicle with two auxiliary wheels each centre side which are raised for road movement and can be lowered for cross-country operations (chain driven). High boatlike front with more vertical cdr's and driver's compartment. Horizontal hull roof sloping downwards towards rear. Vertical hull rear. Trapezoidal turret sloped in front. BRDM-1 with longer glacis (front engine) and without turret. Front. Pointed nose sloping upwards towards crew compartment. Lower part of hull has vertical sides and rear, upper part slopes inwards. Trim vane under nose.

Other details. Fully amphibious, propelled in water by rear hydrojet engine. Water speed 10 km/h.
BRDM-1 almost completely replaced by BRDM-2 has a weight of 5.6 t, a length of 5.7 m, a width of 2.25 m and a height of 1.9 m. Armament consists of an externally mounted 12.7 mm or/and one 7.62 mm MG. The GAZ-40P 6-cylinder petrol engine has 66 kW, vehicle's speed is 80 km/h. Its fuel capacity of 150 litres offers maximum road range of 500 km.
BRDM-2 is armed with a turret mounted 14.5 mm MG and a 7.62 mm MG with manual traverse and elevation. The only means of entry are two circular roof hatches in front of the MG turret. It has a rear mounted GAZ-41 V-8 petrol engine of 88 kW. NBC system is installed and a central tyre pressure system. Many vehicles use a land navigation system. AT versions with different ATGM systems. (TABLE)
BRDM-2 RKh is externally distinguishable by two rectangular racks containing poles with yellow pennants for marking the boundaries of contaminated areas. These two KZO-2 called devices carry 19 poles each planted using explosive charges. It has a GSA-12 automatic chemical indicator, a DP5V dosimeter and a DP-3B roentgenometer. It also carries a DK-4K vehicle decontamination kit.

The BRDM-2 RKhb is identical to the standard vehicle but is only fitted with a 7.62 mm MG in a reconfigurated mounting.
The BRDM-2U (Upravleniye) is lacking the MG turret and has a third roof hatch instead. It is equipped with two radio sets and a generator in the rear of the crew compartment. Besides the cdr and the driver up to three additional radio operators can be transported.

Role. In service with Afghanistan (250), Albania (13 BRDM-1), Algeria (50 BRDM-1, 80 BRDM-2), Angola (30 BRDM-1/2), Benin (12), Bulgaria (50 BRDM-2), Cape Verde (4), CAR (4 BRDM-2), Congo (10 BRDM-1, 15 BRDM-2), Croatia (5), Cuba (250), Egypt (240 + 30 ex-NVA RKh), Equatorial Guinea (6), Estonia (3), Ethopia (70 BRDM-1/2), Guinea (22), Guinea-Bissau (12), Iraq (400), Israel, Kazakhstan (140), Kyrgyztan (42), Latvia (2), Libya (242), Lithunia (10), Madagascar (26), Mali (20), Mongolia (130), Mozambique (28), Namibia, Nicaragua (70), North Korea (200), Palestina (40), Peru (15), Poland (510), Romania (121), Russia (withdrawn from front-line service), Sao TomÇ a. P. (8 BRDM-2), Seychelles (6), Slovenia (24), Somalia (?), Sudan (30), Syria (400 BRDM-1/2), Tanzania (24), Turkmenistan (14), Uganda, Ukraine (approx. 1000), Uzbekistan (35), Vietnam (80), Yemen (100), Yugoslavia (38), Zambia (76).

Assessment. Mobile amphibious reconnaissance vehicle, very lightly armoured, poor firepower.Versatile AT versions.

BRDM-2 AVR

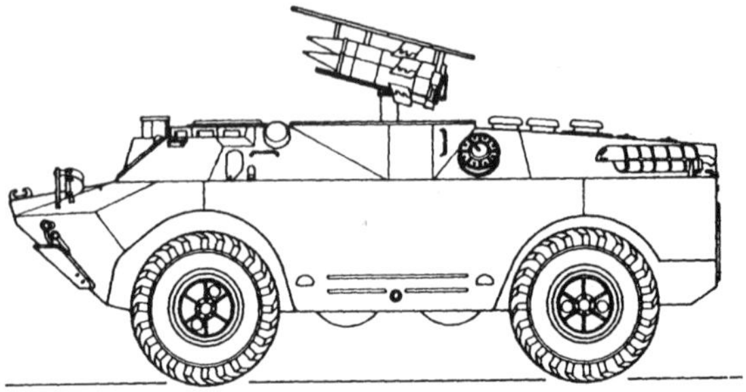

BRDM-2 TDM

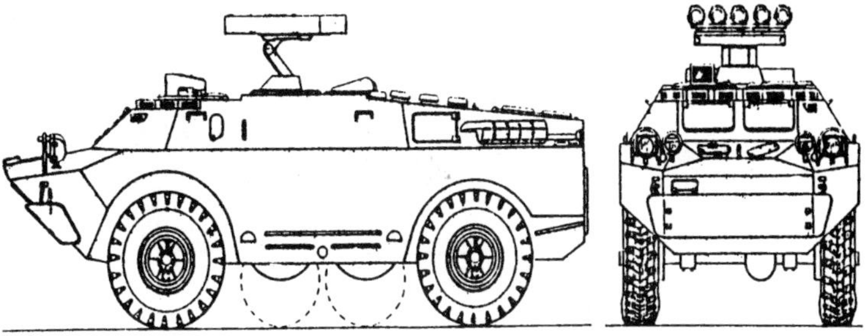

Steven Zaloga

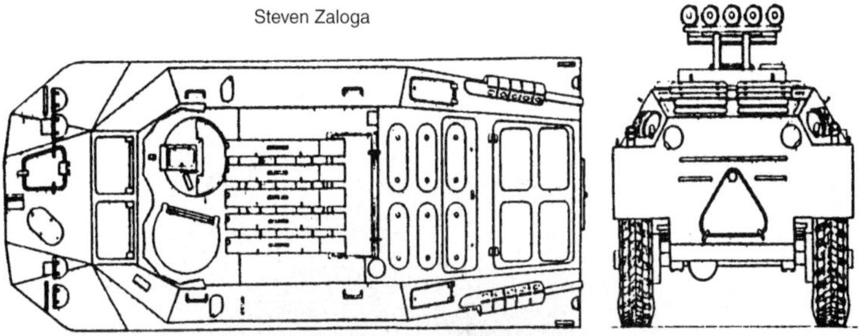

BRDM-2 TDM (AT-5)

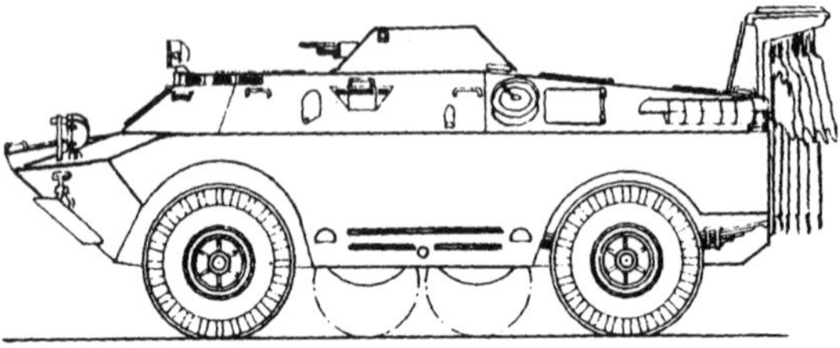

BRDM-2 RKh ANBCRV

BRDM-2 RKh ANBCRV

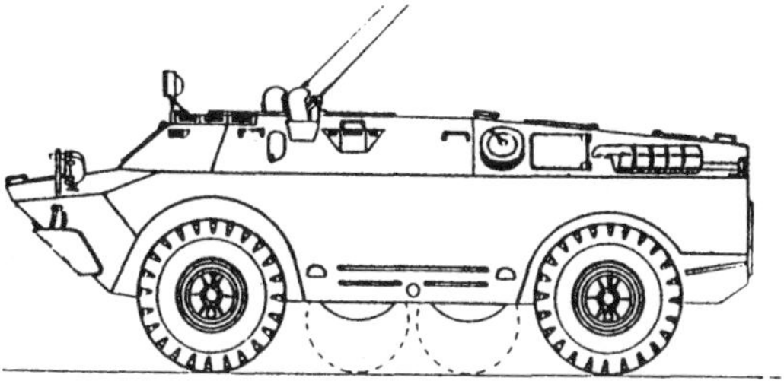

BRDM-2U ACCC

BRDM-2U ACCC

T-54/T-55 main battle tank

Models.

T-54A
T-54B
T-54AM
T-54M
T-55
T-55A
T-55M
T-55AM
T-55AMV with explosive reactive armour
T-55AD
TIRAN-67 MBT with 105 mm gun (Israel)

Non-serial versions.

T-55 APC with KLIVER turret
Numerous modernisation kits offered for the MBT.

Development and production.

1946 first T-54 prototype with T-44 hull and new round-shaped turret with improved 100 mm D-10 gun.
1953 final turret design appeared, distinctive rear overhang deleted.
1954 T-54A issued with D-10TG gun and one-axis stabilisation, initially with counterweight on muzzle, later with bore evacuator.
1955 T-54B entered production with D-10T2S gun with two-axis stabilisation, electric power traverse and IR night-fighting system. Latter device was gradually retro-fitted to older T-54. These vehicles are called T-54M resp. T-54AM.
1957 licence production of the T-54 started in the CSSR an in Poland, two years later in China (T-59).
1958 issue of T-55, originally fitted with IR-search-light directly clamped to the gun-tube.
1963 T-55A issued with smooth cdr's cupola base, NBC system, anti-radiation liner, and raised loader's and driver's hatch covers and other modifications.
1969 T-55 gradually fitted with new loader's cupolas with 12.7 mm ADMG (designation T-55A (M)).
1979 production ceased with about 95.000 T-54/T-55/T-59 built.
1983 T-55M and T-55AM issued.

Distinguishing features. Side. Suspension with five roadwheels (larger distance between first and second stations), no return rollers and rear sprocket. Sloped glacis and lower front, vertical rear. Horizontal hull top, engine compartment rear with exhaust pipe on left track guard above fourth/fifth wheel. Flat fuel/oil/stowage boxes above tracks. Semi-circular turret, well sloped all round with long gun. Front. Sloped glacis plate with driver's hatch front left. No hull sponsons. Mushroom turret with flattened top extends part away over suspension. Small gun mantlet.

Other details. Splashboard on glacis. Cdr's cupola with lid either bolted or welded on turret roof. Right side loader's cupola with external mounted ADMG. T-54 with fan-dome in front of loader's hatch. Longish telescopic sight slit to left of gun mantlet. 7.62 mm coaxial MG right of gun tube. IR spotlight recessed into rotating cdr's cupola. IR gunnery searchlight, linked to gun on bracket forward of right offside turret front. IR periscopic sight in turret roof in front of cdr's cupola. Snorkels for deep-wading attached on turret rear. Different arrangement of external stowage boxes on turret sides possible. Smoke laying system by injecting diesel fuel into exhaust outlet. The tank can also be fitted with front dozer blade or various types of mine-clearing devices. Many tanks have been recently fitted with box-shaped laser range-finder on top of gun mantlet.
(TABLE)
T-54M and T-54AM with V-55U engine developing 426 kW.
T-54M with external laser range finder in a box above the gun mount.
Lately modified tanks with smoke launchers, horse-shoe add-on armour on the front turret sides and additional armour plate on the hull front.
Command versions (Tanks with Suffix "K") have additional radio equipment.
The TIRAN-67 is an Israeli conversion of captured T-54/55 armed with the M68 105 mm main gun, US radio equipment and US MG armament. Additional stowage boxes on the turret and on top of the track guards were fitted. A 60 mm mortar is fitted on the left turret side for close defence.
T-55M and T-55AM with external additional armour shield on the turret front and external laser range finder. T-55AM capable to fire BASTION tube launched missile. Motorized by the V-55U engine.
Most of the NSWP armies (excl. Romania) modernized parts of their T-55 inventory to the T-55AM standard (Czech T-55AM1 and T-55AM2). These tanks are equipped with a Czech fire-control system (Kladivo), a laser rangefinder, cross-wind sensor, laser detector, thermal sleeve for the gun tube, add-on turret armour, increased hull front armour and smoke dischargers. The Polish tanks have a hammer-shaped cross-wind sensor on top of the front turret roof (instead of a sensor mast on the rear) and the laser rangefinder integrated into the main sight.
The T-55AD is the Soviet version comparable to the T-55AM2B but fitted with DROZD anti-missile launcher. Its fire-control system is called VOLNA, the missile BASTION (NATO designation AT-10).

Role. In service with Afghanistan (-400), Albania (15), Algeria (315), Angola (100), Azerbaijan (200), Bangladesh (31), Belarus (381), Benin (4), Bosnia-H. (30), Bulgaria (1142), Cambodia (60), CAR (4), Chad (45 T-55 ex-Libya), Congo (25), Croatia (140), Cuba (600), Czech Rep. (469), Egypt (850), Ethiopia (250), Finland (35 T-54, 150 T-55), Georgia (40), Guinea (8), Guinea-Bissau (24), Hungary (400), India (600), Iran (330), Iraq (several hundred), Israel (250 TI-67), Laos (30), Lebanon (100, Libya (-1600), Mali (25), Mauritania (35), Mongolia (100), Morocco (100), Mozambique (76), Nicaragua (130), Nigeria (54 T-55), North Korea (1500), Pakistan (51), Peru (280), Poland (400 and 492 upgraded), Romania (822), Russia (approx. 4000), Slovakia (600), Slovenia (40), Somalia (100), Sri Lanka (25 ex-Czech), Sudan (130), Syria (2100), Togo (2), Uganda (4), Ukraine (182), Uruguay (15 ex-Israeli T-55), Uzbekistan (80), Vietnam (1000), Yemen (600), Yugoslavia (390), Zambia (8), Zimbabwe (40).

Assessment. First post-war generation MBT of good mobility, average firepower and well sloped thick armour. Limited combat endurance. Now obsolete when compared with latest generation MBT even if modernised. Still in operation with many third world countries.

T-54B MBT

T-55A MBT

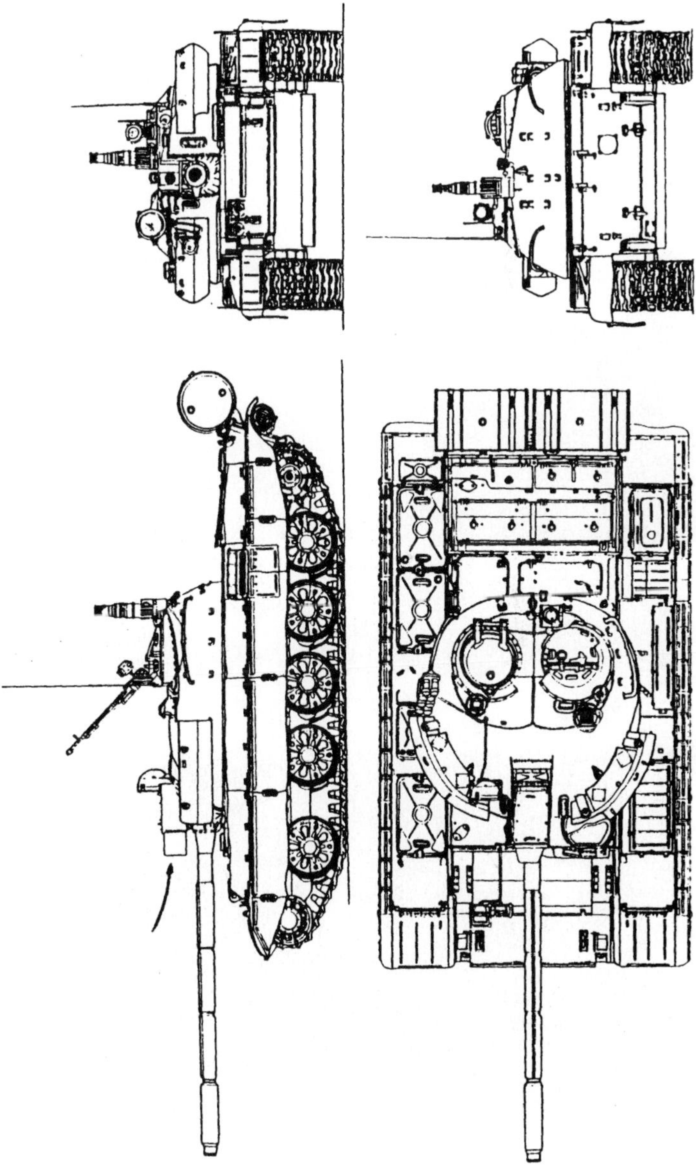

Steven Zaloga

T-55 AM MBT

T-55 AM MBT

T-55 AMV MBT

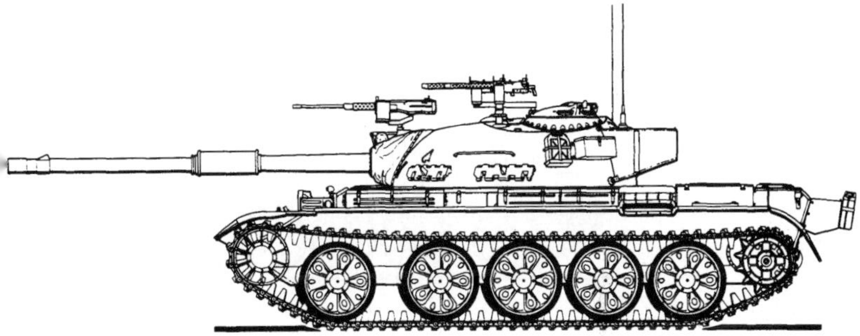

TIRAN-67 MBT

TIRAN-67 MBT

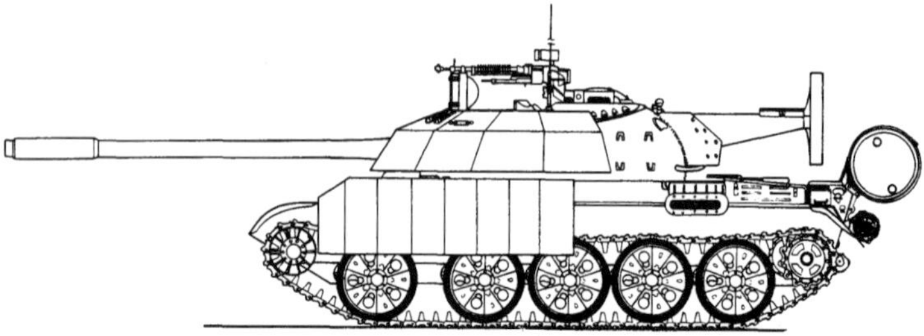

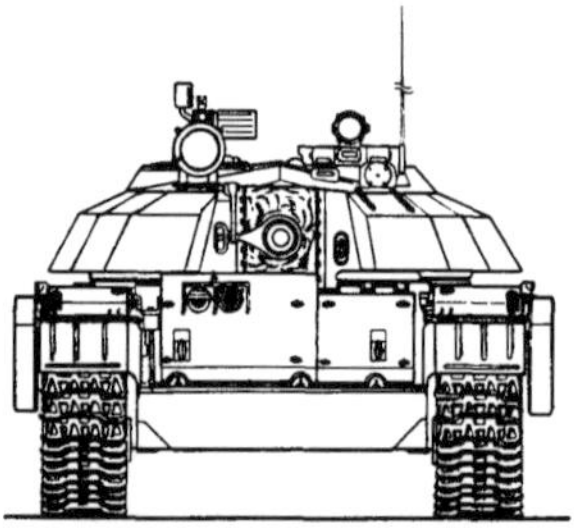

T-55 MBT with add-on armour (Iraq)

T-62 main battle tank

Models.

T-62
T-62M
T-62D
T-62MV

Non-serial versions.

Several modernisation kits offered.

Development and production.

1958 first prototype of a completely modernised T-55 appeared armed with new 115 mm smooth-bore gun.
1960 T-62 series production (without ADMG) starts.
1965 T-62 with modified engine deck produced.
1968 appearance of the T-62M fitted with external ADMG.
1973 licence production in the CSSR starts until 1978.
1983 appearance of T-62M in Afghanistan.
1985 version with explosive reactice armour (T-62-MV) issued.

Distinguishing features. Side. Similar, slightly longer hull appearance like T-55. Larger intervals between third and fourth and fourth and fifth roadwheel station. Egg-shaped turret with long gun tube with fume extractor half-way along. Front. Like T-55.

Other details. Automatic spent case ejection system through a hatch on the turret rear. Same standard as T-55B according to NBC protection, deep wading, stabilisation system IR equipment and mine-clearing devices.
Later versions have the VOLNA fire-control system and an external laser rangefinder in a box above the gun mantlet. (TABLE)
T-62M with horse-shoe add-on armour on turret front and additional front armour.
T-62D with Drozd anti-missile charge launchers on both turret sides.
T-62MV has ERA tiles on turret front sides and the hull sides. Capable to fire the 9K116-1 SHEKSA (NATO: AT-10) tube launched missile.
Command tanks (suffix "K") have additional radio equipment and gyroscopic compass.
Later versions feature the V-55 engine with 456 kW.

Role. In service with Afghanistan (200), Albania (35), Algeria (300), Angola (18), Belarus (142), Bulgaria (130), Chad (12 ex-Libya), Congo (14), Cuba (400), Egypt (500), Ethiopia (75), Iran (100), Iraq (1600 delivered), Israel (70), Kazakhstan (624), Libya (280), Mongolia (250), North Korea (1200), Sudan (24), Syria (1020), Uzbekistan (179), Vietnam (220) and Yemen (120).

Assessment. Noteworthy increase of firepower. Outstanding configuration of turret but cramped interior. Obsolete in comparison with new generation MBTs.

T-62 MBT

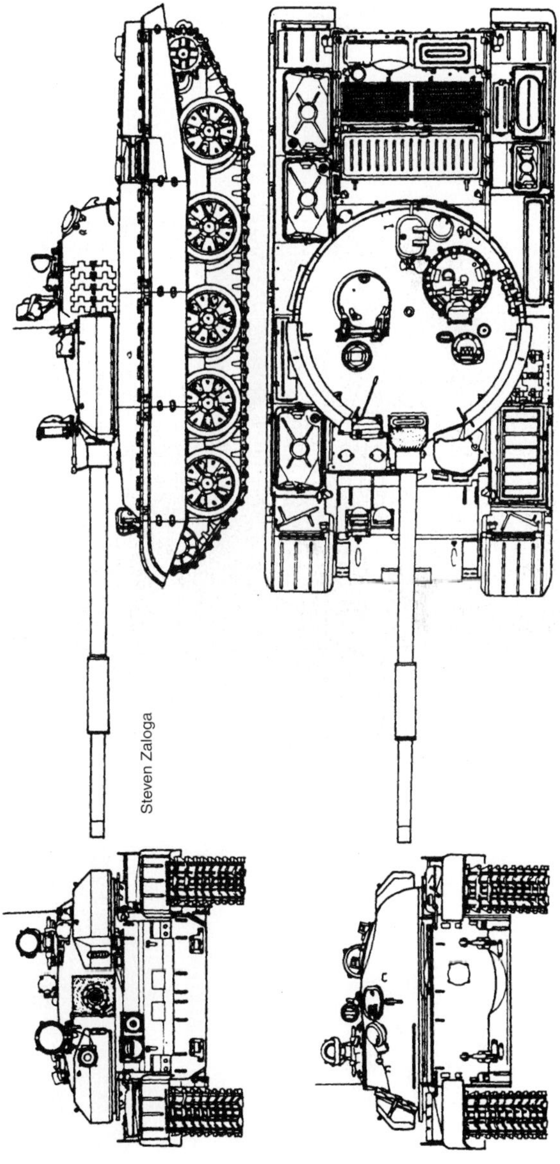

T-62M MBT

T-62M MBT

T-72 main battle tank

Models.

T-72
T-72A
T-72B
T-72BM
T-72B1
T-72M
T-72M1
T-72S
T-72CZ
T-90
M-84 (Yugoslavia)
M-84A (Yugoslavia, Croatia)
PT-91 TWARDY (Poland)
TR-125 (Romania)
T-72AG (Ukraine) with T-80UD add-on armour and engine

Non-serial version.

T-72-M2 MODERNA (Slovakia)
T-72MP (Czech Rep./Ukrainia)
M-95 Degmann (Croatia) with exlosive reactice armour

Many upgrade kits offered by different countries.

Development and prodcuction.

1970 first appearance of simplified T-64 MBT prototype.
1971 start of pre-series production.
1975 series production starts.
1978 licence production in CSFR and Poland commences.
1979 T-72A issued.
1980 first public appearance of T-72M.
1982 production of T-72M1 starts.
1984 series production in Yugoslavia, one year later in India.
1985 start of ERA (retro)fitting. T-72B and T-72BM issued.
1987 first appearance of T-72S.
1993 introduction of T-90.
1994 production start of PT-91.

Distinguishing features. Side. Six medium rubber rimmed roadwheels, three return rollers, sprocket rear. Upper part covered by skirt. Very low, extended glacis, horizontal hull deck, stowage boxes resp. fuel tanks on track guards. Circular turret centre of hull with cupola and external ADMG. Long extending main gun with thermal sleeve and fume extractor. Front. Well sloped glacis plate with

driver top centre, distinctive V-shaped splash plate in front of driver. Small dozer blade underneath front bow. IR searchlight right near gun. The slightly angled mushroom cast turret has the cdr's cupola on the right with ADMG and the gunner's hatch on the left with optical heads in front of it.
Late type T-72 with different arrangement of explosive reactive armour boxes.

Other details. Pre-series tanks had the search-light mounted on the left side and hand rails on the front turret sides. It was fitted with an optical coincidence rangefinder with two optical ports on both sides of the front turret roof. Later T-72 tanks had the searchlight moved to the right and the hand rails lowered, later totally deleted. Initially the T-72 was equipped with four hinged skirting fold-out plates on the front track guard. Later Soviet T-72 had slightly thickened front turret armour. The armour plates were later removed by side skirts, and the T-72 were gradually fitted with smoke dischargers.
The T-72A dispensed with the optical rangefinder and received a laser rangefinder housed in the gunner's sight. Later T-72A received a layer of anti-radiation cladding on the roof. The T-72A has a thickened turret front and an additional glacis armour plate with a rhomboidal cut-out around the driver's periscope. The engine received an improved cooling system. Later the Soviet T-72A were fitted with an external anti-radiation cladding on the turret top and the front hull section. The export and license produced T-72A are called T-72M1. Earlier licence versions lack thickened frontal turret armour and are called T-72M.
The T-72B has an even more extended turret front, the stronger V-84-1 engine (618 kW) and is capable to fire the 9M119 REFLEKS tube-launched missile. The smoke dischargers are arranged in a single group of eight on the left turret side thus preparing the tank for the adaption of external reactive armour blocks on the turret and the hull front. The T-72B1 is down-scaled not to fire the missile. The export versions of the T-72B/B1 are the T-72S/S1 with 155 ERA tiles, T-72A tanks have 175, T-72B 195, sometimes up to 227.
(TABLE)
The T-72BM features a different arrangement of second-generation ERA and also three panels of KONTAKT 5 on the forward side skirts.
The T-90 has the new fire control system 1A45T with the 1V528 fire control computer comparable to the T-80U and the 2E42-4 stabilisation system. The gunner has the 1A43 day sight with 1G46 laser rangefinder and the TO1-KO1 night sight with IR sight and thermal channel. The commander is equipped with the PNK-4S day/night sight. The export version is called T-90S.
Tanks retrofitted with explosive reactive armour have the suffix "V".
All T-72 version are also built in a command ("K") version with one additional radio set and an additional antenna pod behind the gunner's hatch.
Part of the Czech fleet is upgraded to T-72CZ featuring the new Officine Galileo TURMS fire control system with a cdr's periscopic stabilized day/thermal sight, a stabilized gunner's day/thermal sight, ERA protection package developed by VOP 025 and a new NIMDA CV-12 746 kW power pack coupled to an Allison XT 411-6 transmission.

The M-84 is similar to the T-72M but incorporates an indigenously-developed fire control system (SUV-84) with a large DNNS-2 gunner's day/night sight in front of the gunner's hatch. The cdr is provided with a new DNKS-2 sight. Externally distinguishable is the M-84 also by the meteorological sensor mounted on a small pylon on the centre turret roof. M-84A has a 735 kW engine and improved armour protection.
The M-84A has the improved SUV-M-84 FCS and a PK-59 gyrocompass. The engine power is increased to 735 kW by the installation of two turbo-compressors instead of one mechanical one and alterations on the cooling system. The tanks delivered to Kuwait (M-84AB) have light intensifier, RACAL radio sets, TELDIX FOA 40 havigation equipment and a 1 kW auxiliary power unit for the command version. From 1996 onwards some of the Croatian tanks have been upgraded to the M-84AB engine status and the OMEGA-84 fire-control system including a laser warning sensor (M-84A1/A3 or A4).

The Polish PT-91 TWARDY features a number of improvements including the stronger S-12U 625 engine, ERAWA-1/2 add-on armour blocks, a new FCS, new light intensification night sights and OBRA laser warning system. It is offered for export as T-72M1Z with different protection level.
The TR-125 is a conversion of the T-72 with a longer hull (seven roadwheels) and a heavier weight (48 t). The tank is powered by a stronger engine of 647 kW. A single-piece side skirt is fitted and a bank of eight smoke dischargers on the left turret side.
Main armament of all T-72 versions is the 125 mm smooth-bore gun fitted also to the T-64 and the T-80. The tank carries 44 rounds, 24 of them in an automatic loader carousel in a cassette system with charge above and projectile below loaded in two cycles. The stub of the combustable cartridge is ejected via a small hatch in the turret rear. Internal lining of synthetic material against neutron radiation, PAZ NBC detection system, FVU filtration system. Deep wading system with one snorkel mounted on the gunner's hatch. Attachment points for KMT-4 and KMT-6 mine clearing device.
Numerous modernization kits are offered for upgrading the T-72.

Role. In service with Algeria (285), Angola (50), Armenia (246), Azerbaijan (314), Belarus (1607), Bosnia (40 M-84), Bulgaria (475), CAR (4), Congo (25), Croatia (30 M-84), Cuba (75), Czech Rep. (531), Ethopia (40), Finland (162), Georgia (251), Hungary (224), India (1700), Iran (250), Iraq (400 delivered), Kazakhstan (600), Kuwait (150 M-84 + 15 command version + 15 ARV delivered), Kyrgyzstan (204), Libya (145), Peru (150), Poland (667 plus 150 PT-91, slowly increasing), Romania (40), Russia (9000+), Sierra Leone (2), Slovakia (272), Slovenia (50 M-84), Syria (1500), Tajikistan (40), Turkmenistan (530), Ukraine (1305), Uzbekistan (100) and Yugoslavia (283 M-84).

Assessment. Better mobility than T-64. Firepower and protection (T-72M1 /M2) below late generation MBTs.

T-72B

T-72 M MBT

T-72 BM MBT

T-72AV MBT

T-90 MBT

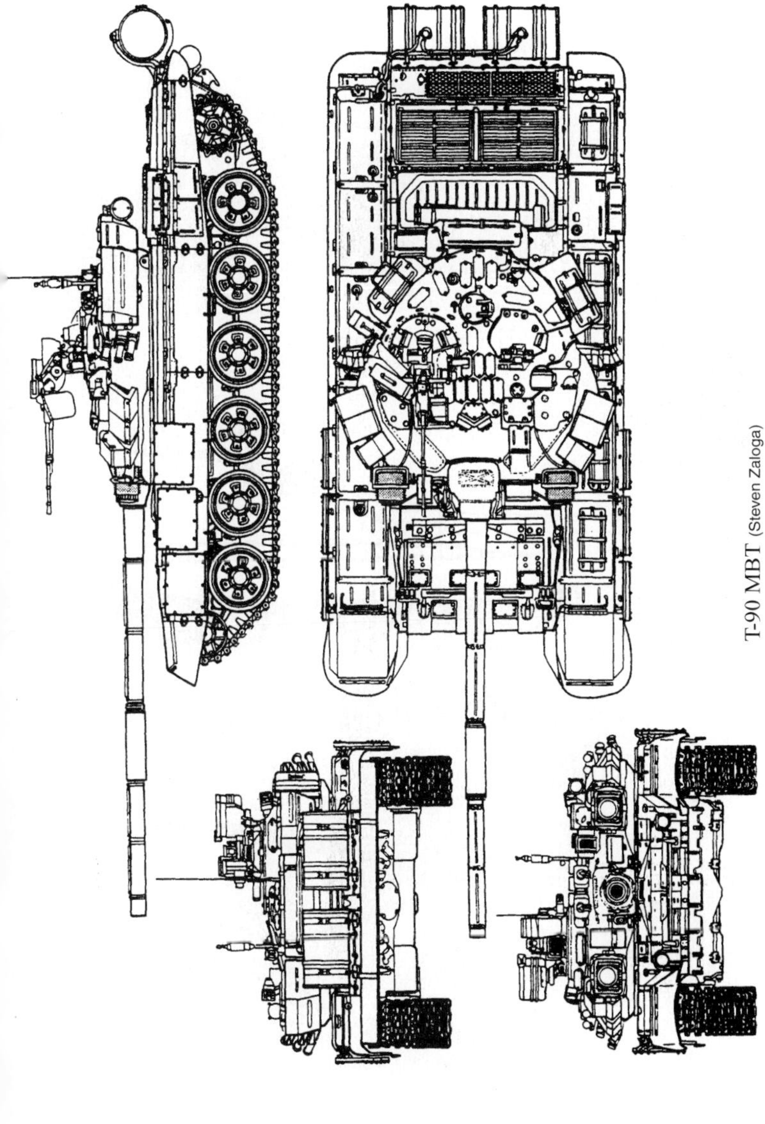

T-90 MBT (Steven Zaloga)

T-72 CZ MBT

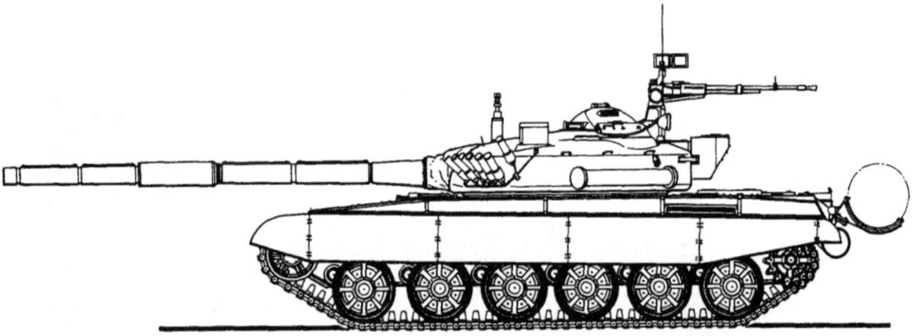

M-84 MBT

M-84A4 MBT

PT-91 MBT

TR-125 MBT

T-80 main battle tank

Models.

T-80
T-80A
T-80B
T-80BV
T-80U
T-80D
T-80UD
T-84 (Ukraine)
T-80UM

Non-serial version

T-80UM-1 with thermal sight

Development and production.

1976 production start of T-80 using T-64 components.
1978 T-80B is fielded.
1982 T-80A is issued.
1985 T-80B is (retro)fitted with explosive reactive armour.
Service enter of T-80U.
1987 diesel version of T-80 starts production.
1992 T-80UM is fielded.
1995 production start of T-84.

Distinguishing features. Side. Slightly lengthened hull with six rubber-rimmed roadwheels arranged in pairs with even narrower distance between third and fourth wheel, covered by skirts. Hull arrangement similar T-64B with protruding exhaust grill box at rear plate. Turret with larger socket of cdr's cupola. Installation rails for a third external fuel drum on top of engine deck. Front. Large optical periscope sight on left turret roof similar to T-64B.

Other details. Most T-80 versions are propelled by a gas-turbine coupled to a manual transmission with five forward and one reverse gear. and uses a new type of track with double steel pins. The T-80 reverted to the torsion bar suspension contrary to the T-64 hydro-pneumatic type.
While the T-80 (weight 42 t) is propelled by the initial gas turbine version GTD-1000T developing 735 kW, the later versions have the GTD-1000TF version with 809 kW. The T-80A (45,2 t weight) is a pre-step of the T-80U already provided with increased ammunition stowage of 45 rounds (still 28 in the loader carousel) compared to formerly 38 and has the GTD-1000M turbine with 882 kW. The three early versions have a fuel capacity of 1800 l (of which 700 are externally on top of the mud-guards) with a range of approx. 400 km.
Main armament is the 125 mm 2A46M-1 gun. The fire-control system is desig-

nated 1A45 with the 2Eh42 stabiliser and the 1V528 computer. The 1G46 gunner's day/night sight has a magnification from 3.6 to 12 times and a laser rangefinder The commander has the TKN-4S stabilised day/night sight. The T-80U can fire the 9K119 SVIR tube-launched missile (NATO: AT-11 SNIPER) laser beam-rider missile of range of up to 5 km. The T-80B versions fired the 9K112-1 (AT-8 SONGSTER).
Beginning in 1985 the T-80 was fitted with explosive reactive add-on armour (T-80BV).
T-80U has improved armour protection and second-generation explosive add-on armour all around resp. on top of the turret and on the front hull, too. It has the GTD-1000TF turbine.
(TABLE)
The T-80UM is a re-engined T-80U with the GTD-1250 with 919 kW.
The T-80UD has a 6TD-1 6 cylinder opposed-piston 735 kW diesel engine. The fuel capacity is 1300 l (560 externally) and the range is 560 km.
The Ukrainian T-84 has the 6TD-2 diesel with 883 kW.
All command versions have the additional suffix "K". All T-80 tanks can be fitted with three 215 l extra fuel barrels at the end of the engine deck. Late versions may be equipped with the IR interference counter-missile system SHTORA-1, a missile hard-kill system ARENA with radar sensors and laser warning systems.

Role. In service with China (200), Cyprus (81), Pakistan (320 on delivery), Russia (4200), South Korea (40), Ukraine (320).

Assessment. Successor of the T-64 overcoming the weakness of the power pack and modernising the running gear. Identical fire-power and improved protection.

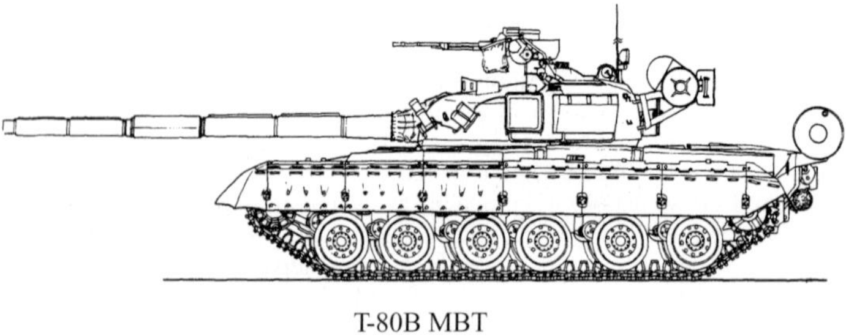

T-80B MBT

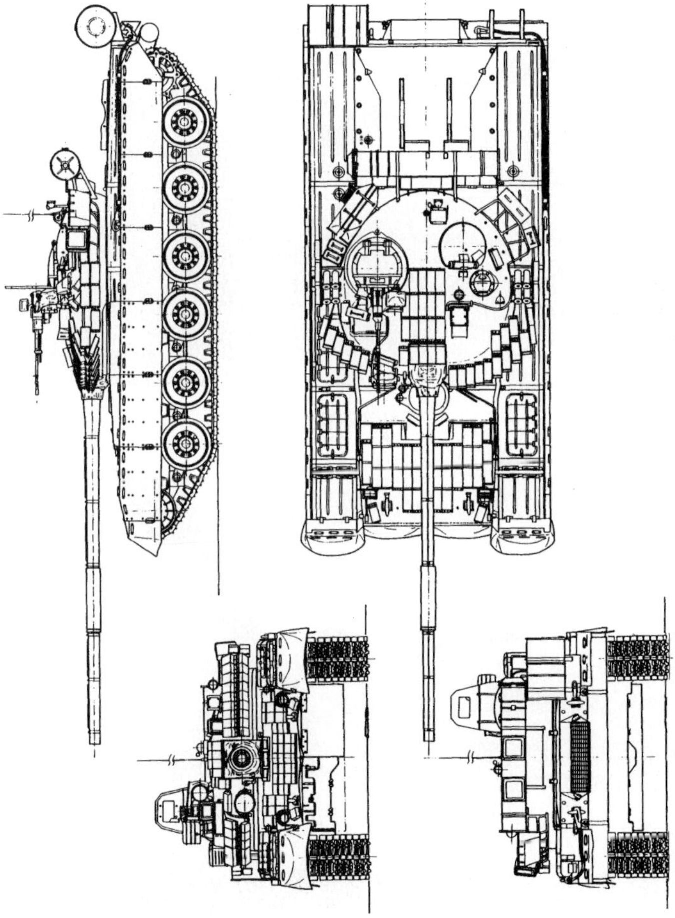

T-80BV MBT

T-80BV MBT

T-80U MBT

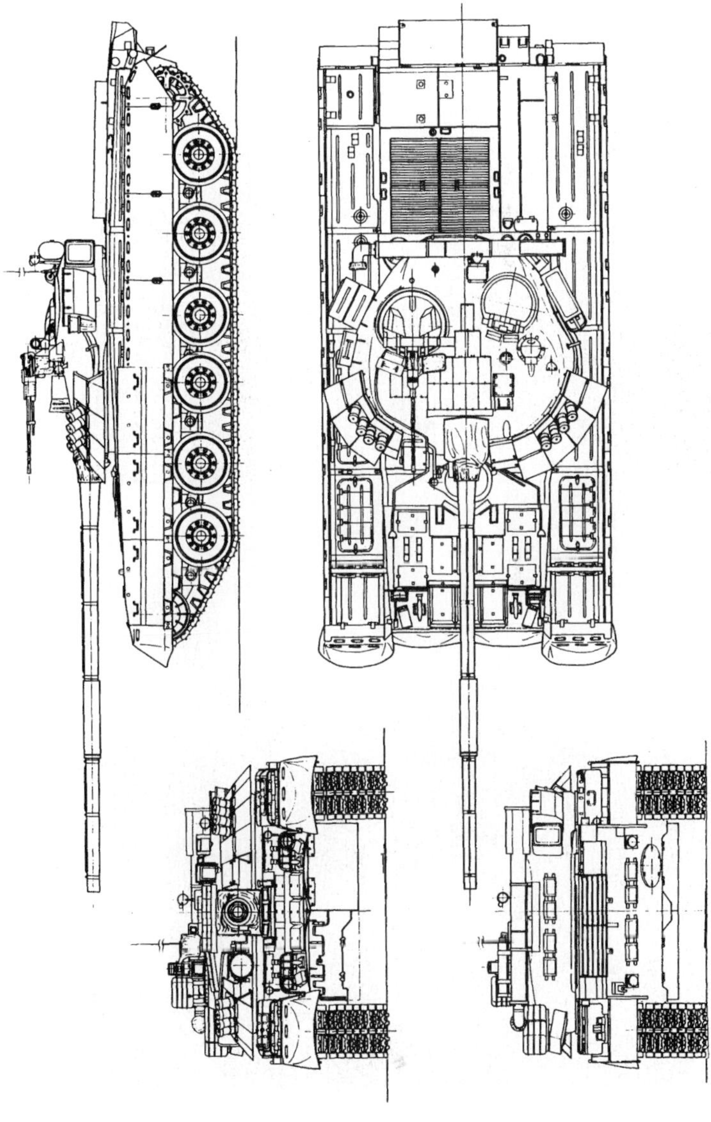

T-80UD MBT

T-80UD MBT

BMP armoured infantry fighting vehicle
Boyevaya Mashina Pekhoty

Models.

BMP-1
BMP-1D
BMP-1P
BMP-Sh ACCC
BMP-1KSh ACCC
BRM-1 AVR
BRM-1K ARRV
PRP-3 AARV (BMP-SON)
PRP-4 AARV
BMP-2
BMP-2D
IPR AEV
BREM-2 ARV
BREM-4 ARV (VPV Czech/Slovak. Rep.)
BPzV AVR (Czech/Slovak. Rep.)
SNEZKA AVR (Czech Rep.)
AMB-S AAV (Czech/Slovak. Rep.)
ShM-120 AMC (Czech Rep. formerly PRAM-S)
SMU mine layer vehicle (Czech/Slovak. Rep.)
SVO mine clearing tank (Czech/Slovak. Rep.)
OT-90 APC (Czech/Slovak. Rep.)
VP-90 AVR (Czech/Slovak. Rep.)
ZV-90 transport carrier (Czech Rep.)
DP-90 forward observer vehicle (Czech/Slovak. Rep.)
DTP-90 ARMV (Czech Rep.)
MP-31 ACCC (Czech Rep./Poland)
SARATH AAV (India)
MLI-84 AIFV (Romania)
MLI-84M (Romania)
MLVM AIFV (Romania)
ABAL ACC (Romania) turretless MLVM
M1989 AAH (Romania)
MVLM-AR AMC with 120 mm mortar (Romania)

Non-serial versions.

WZ501A AIFV (China)
WZ503 APC (China)
WZ504 TDM (China)
WZ505 AAV (China)
WZ506 ACCC (China)
OTR-5 ACCC (Czech Rep.)

PBK TDM with KONKURS ATGM
TRISHUL with AKASH tactical missile launcher (India)
TRISHUL radar carrier (India)
STROP AADG (Czech Rep.)
BWP-40 with 40 mm Bofors turret (Poland)

Related vehicle.

BORAGH APC (Iran)

Development and production.

1959 start of development of project 765 Korshun (Kite).
1965 troop trials and limited scale production of the initial version BMP beginning in
1966 followed by models with several modifications.
1970 appearance of BMP-1 as the major production version.
1974 final version BMP-1M issued.
1976 production start of BMP-2.
1982 issue of BREM-2.
1985 first appearance of BMP-2 with appliqué turret armour.
1987 Indian license production starts.

Distinguishing features. Side. Six roadwheels, three return rollers, sprocket front. Upper part of track covered by skirts. Pointed nose with slightly sloped ribbed glacis plate, horizontal hull roof with different hatch arrangements. Vertical hull rear with two bulged doors. Circular turret moved slightly to rear of vehicle. Front. High lower front, driver front left. Engine compartment louvres in roof to right. Sponsoned hull. Flat turret with angled sides.

Other details. The BMP-1P has the 9M14M missile loading hatch deleted and is fitted with a mounting post either for the 9M11 FAGOT or 9M113 KONKURS ATGM.
The BMP-1K lacks the three rear firing ports on the right side and has an additional antenna pod on the right hull rear.
The Czech resp. Slovakian versions are designated BVP-1/2, Polish versions BWP-1/2.
The BMP-Sh has a small turret with a 7.62 mm PKT MG, a raised rear deck and additional radio equipment.
The BMP-1KSh (Komandno-Shabnaya) has a fixed turret with its gun replaced by a telescoping communications antenna which can be folded up and down. It has two rear roof hatches and an externally transported generator.
The BRM-1 (Boevaya Razvedyvatelnaya Mashina) vehicles are equipped with a two-men turret and dispose of only two roof hatches on top of the rear compartment. BRM-1K (Kavalerskiy) is equipped with a small battlefield surveillance radar 1RL133 (NATO: TALL MIKE) fitted on the upper rear of the turret. The BPzV (Bojové Pruzkumné Vozidlo) is a Czech further improvement of

the BRM-1 with a better suspension (additional bumpers), two laser sensors, a KLADIVO laser rangefinder, a TNA-3 navigation system and a PSNR-5 radio locator. Vehicle's weight is 13.8 tons. The 1RL133 radar is mounted externally.
The PRP-3 (Podvizhnij Razvedyvatelnaniy Punkt) is equipped with a surveillance radar system (NATO: SMALL FRED) with a folding rectangular antenna mounted at the rear of a large two-men turret armed with a 7.62 mm PKT MG. It operates on the J-band, has a detection range of 20 km and a 7 km tracking range. The crew consists of five soldiers.
PRP-4 has a 1PN-59 thermal sight in the left part of the turret, rotating commander's periscope, a 1DM11M-1 laser range-finder and a 1PN-61 light-intensification sight. On the turret rear the 1RL133-1 surveillance radar (NATO: TALL MIKE) can be erected with a detecting range of approximately 10 km. The advanced PRP-4M system has improved observation systems.
While the BMP-1 is fitted with the 73 mm model 2A28 smoothbore low pressure, short-recoil gun fed from a 40-rounds magazine the BMP-2 has a 2A42 30 mm cannon mounted in a two-men turret. The BMP-2 carries six infantry soldiers in the rear compartment, the BMP-1 eight.
(TABLE)
Late BMP-2 vehicles can be equipped with KMT-8 or KMT-10 plough-type mine clearing system.
BMP-1 and BMP-2 (D-types) with add-on armour plates on the hull sides and in place of the track skirts.
Command versions have the suffix »K« and feature additional radio equipment and lack side firing ports.
All BMP can swim propelled in water by their tracks, trim vane must be erected.
OT-90 (Obrnemy Transporter) is equipped with an one-man turret with a KPVT 14.5 mm MG. It does not fall under the CFE AIFV restrictions. OT-90 M-1 and OT-90 M-2 are upgraded versions, the latter one additionally with add-on armour.
The BPzV (Slovakian BPsV) carries the PSNR-5 ground target radar with a range of 8 - 10 km (3 - 4 km for single persons), the TNA-3 navigation system, the NNP-21 night sight, the HLD-5 laser rangefinder, the MRP-4 radiolocative sensor, the SDIO nuclear detecting unit and the W 4P mine-finder.
The SNEZKA has a longer hull with seven roadwheels and carries a 14 m elevatable mast. The observation platform is equipped with a detection and direction finding camera JP-J-O-78 (range 4.5 km), a daylight camera JPJ-I-200 (5 km), a JPJ-II-200 light intensification camera (1.6 km), a laser rangefinder (20 km) and a radar system. The 17.4 t vehicle also has a GPS navigation system with digital maps.
The Czech command and ambulance versions are 2.11 m high and have a weight of 13.6 t. The AMB-S carries either 4 stretcher or 8 sitting patients or 4 sitting and 2 stretcher patients.
The DP-90 has a AST periscope, a PAB-2A compass and a LPR-1 laser rangefinder.
ShM-120 (Samohybny minomet) carries a vz.85 120 mm mortar on elongated

hull (7.47 m) at a vehicle's weight of 15.97, height is 2.25 m. The loading carousel carries 80 rounds, the rate of fire is up to 20 rpm, the range is 8000 m. The levation range is +40 to +80°, azimuth is 30°.

The SVO (Samohybny Vybusny Odminovac) is a 13.83 t mine-clearing version fitted with an array of 24 rocket-propellant charges (calibre 246.6 mm, weight 41.5 kg, explosive 26.6 kg, length 1457 mm) in an open compartment. A path of 5 m width and 100 m length can be cleared by one volley at a range between 250 - 540 m. Vehicle's height in battle position is 2.7 m.

The SMU (Samohybny Vybusny Odminovac) has a weight of 12.9 t and is able to place mines in intervals of 4, 6 or 8 metres at an operating speed of 3 to 15 km/h. The capacity is 110 mines. The vehicle is 6.74 m long and has a height of 2.07 m. It has a crew of four.

The MLI-84 is a slightly larger Romanian BMP-1 copy (weight 16.6 t, length 7.32 m and width 3.15 m) with a 12.7 mm DShKM ADMG mounted over the rear compartment. It is powered by a V8 1240 DT-S supercharged diesel engine of 265 kW.

Part of the fleet is now being upgraded with a Rafael OWS-25R (over-head weapon station) housing the Oerlikon KBA 25 mm cannon in an oscillating mount with a x8 day sight and a x7.5 light intensification night sight. A PKT 7.62 mm MG and 2 Malyutka M2T missiles are carried coaxially. 200 resp. 250 rounds are ready to use, 450 resp. 2000 more are stowed in the vehicle. The vehicle weight rises to 17.08 tons. Overall height is 2.942 m.

The MLVM has a similar layout like the BMP-1 but has a lower weight (9.8 t) and is armed with a flat 14.5 mm MG turret as used on the TAB-72 APC. Vehicle length is 5.85 m, its width is 2.72 m, height with/without turret 1.95/1.55 m. Ground clearance is 0.38 m, fuel capacity 480 litres, road range up to 740 km. A derivative is the ABAL armoured combat supply vehicle carrying up to 2 tons payload, armed with an external 7.62 mm MG.

The M1989 is an AAH on MLI-84 chassis resembling the 2S1. The amphibious vehicle with seven roadwheels is about 2.1 m high, 7.5 m long and has a width of about 3.15 m. Its weight is only 17 t.

The BREM (Bronirovannaya Remontno-Evakuatsionaya Mashina) ARV feature either a light crane in the forward part of the hull (BREM-2) or on the rear part (BREM-4). They are identical with BMP-1/2, turret removed. Rotating hydraulic 6.5 t crane installed on vehicle's rear (boom reach 2.8 m). A spade used as a counter-weight is folded up in the travelling position and lowered by means of two hydraulic cylinders. The BREM-4 has a winch on the middle of the front bow with a tractive force of 125 kW at 120 m cable length, to be increas-ed to 240 kW/60 m or 345 kW/40 m by means of one resp. two return pulleys. The cable drum is positioned instead of the AIFV turret. Fuel capacity has been increased to 480 litres. The 14.3 t Czech version is called VPV (Vyprostovaci Pásové Vozidlo).

The Iranian BORAGH is a vehicle mainly basing on the BMP-1 with minor modifications and a locally produced V-8 turbo-charged diesel engine developing 243 kW. Combat weight is about 13 tons; it is fully amphibious and is armed with a roof-mounted 12.7 mm MG with side and rear protection. Iran claims to have developed a complete vehicle family inluding a self-propelled 122 mm artillery system called THUNDER-1.

Role. BMP-1 in service with Afghanistan (206 1/2), Algeria (684), Angola (50+), Azerbaijan (119), Belarus (388), China (designation WZ-501), Cuba (400), Czech Rep. (623 designated BVP-1, 413 OT-90, 129 BPzv, 15 BRM-1), Egypt (200 with French Poyaud 228 kW diesel engine), Ethiopia (20), Finland (163 design. Pbv 501), Greece (500 ex-German), Hungary (502), Iran (200 + 300 WZ501, BORAGH), Iraq (1000 delivered), Kazakhstan (300), Kyrgyzstan (349), Libya (740), Mongolia (400), North Korea (200), Poland (1600 designated BWP-1/2, 38 BRM-1), Romania (178 MLI-84, 88 MLVM), Russia (22000), Slovakia (383 + 207 OT-90), Sri Lanka (210 1/2), Sweden (350 Pbv 501), Syria (2300), Turmenistan (250), Ukraine (1563), Uruguay (6) and Vietnam (300). The BMP-2 is used in Algeria (278), Angola (100), Armenia (164 BMP-1/2), Azerbaijan (197), Belarus (1265), Czech Rep. (187), Finland (110), Georgia (65 BMP-1/2), India (1000), Iran (approx. 100), Iraq (approx. 800), Jordan (35), Kazakhstan (300), Kyrgyzstan (15), Kuwait (46), Russia (3250), Sierra Leone (2), Slovakia (93), Sudan (6), Syria (100), Tajikistan (50 BMP-1/2), Turkmenistan (288), Ukraine (1403), Uruguay (10 ex-Czech), Uzbekistan (172) and Yemen (150 BMP-1/2).

Assessment. Small, mobile vehicle with very cramped interior. BMP-1 gun is in fact a rocket launcher with a poor hit capability and range. Fire-power and armour protection has been improved considerably with the BMP-2.

BMP-1 AIFV

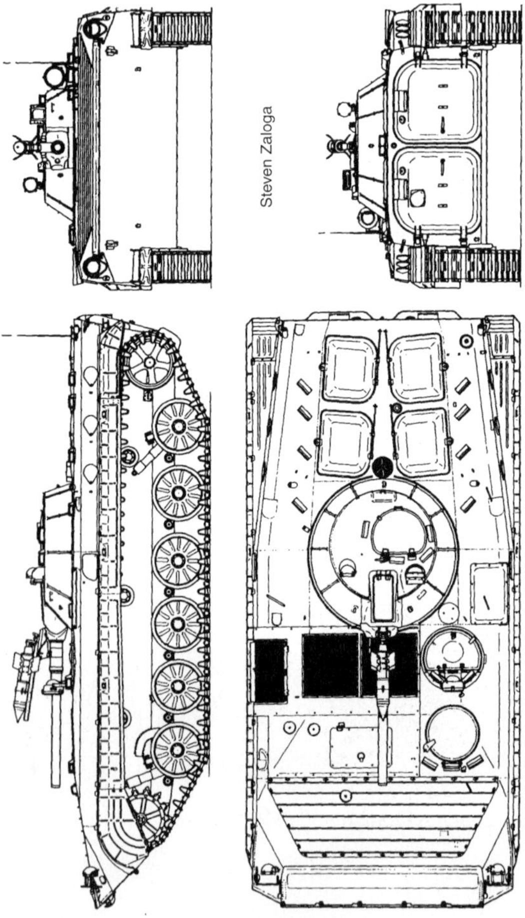

BMP-1 AIFV

BMP-1P AIFV

Steven Zaloga

BMP-1KSh AICV

BMP-1KSh AICV

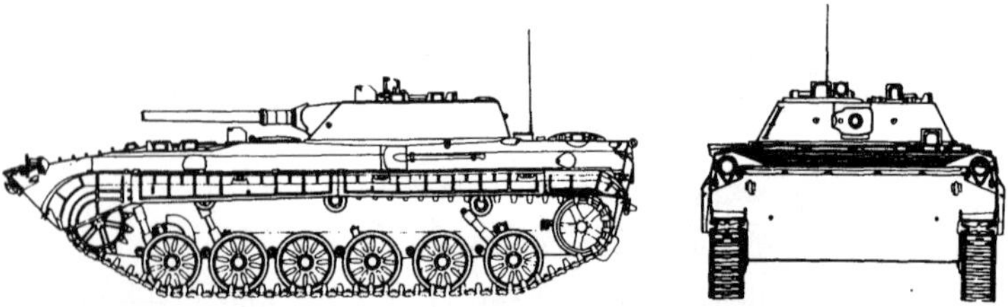

Steven Zaloga

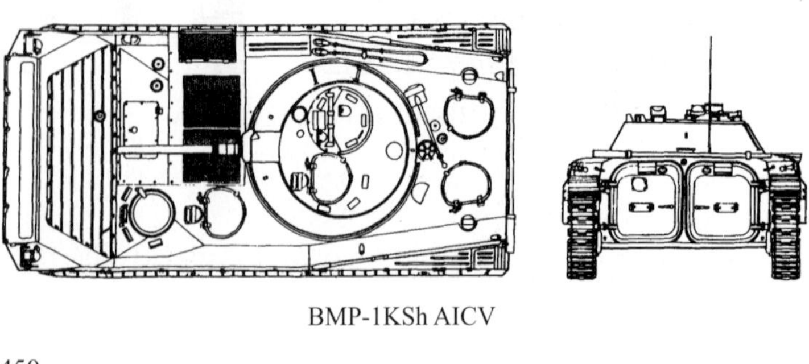

BMP-1KSh AICV

PRP-4 AARV

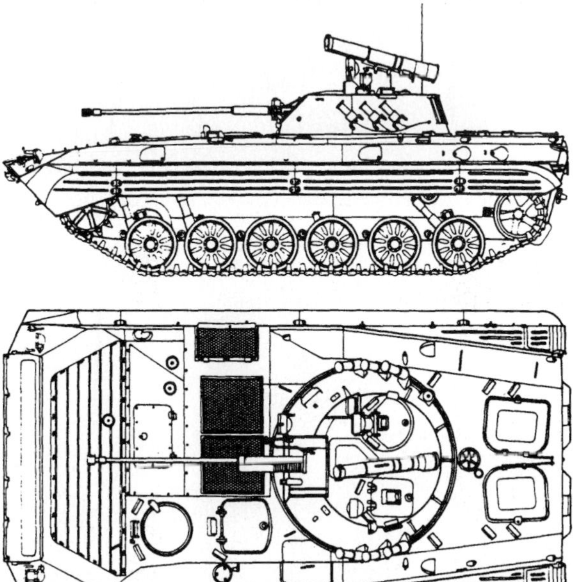

BMP-2 AIFV Steven Zaloga

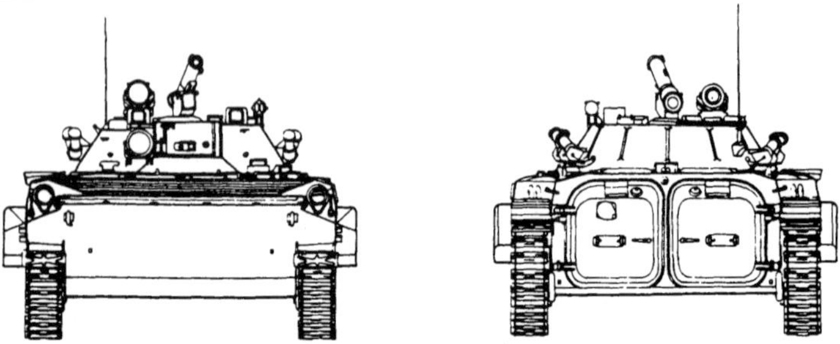

BMP-2 AIFV Steven Zaloga

BMP-2 AIFV

BPzV AVR

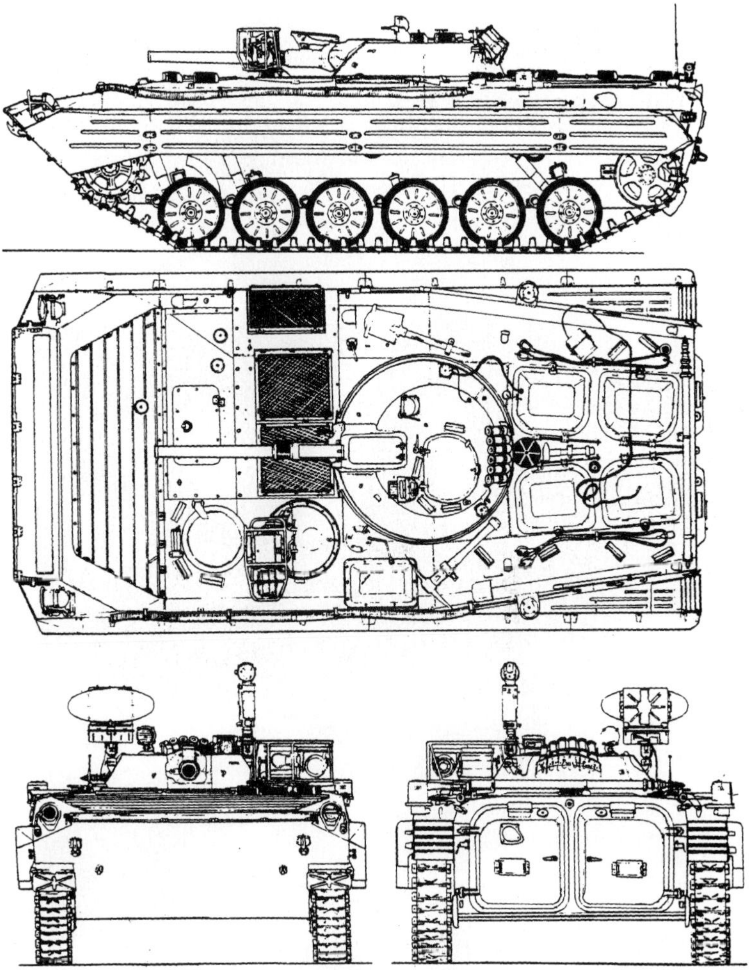

BPzV AVR

SNEZKA AVR

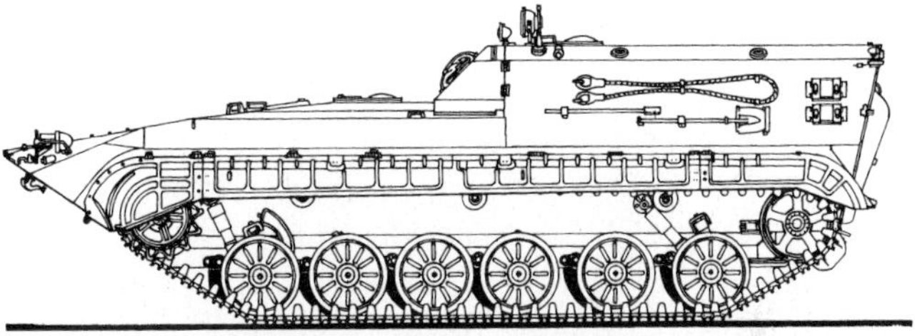

AMB-S AAV

SUM-120 AMC

SMU AMLV

SVO AMCV

OT-90 APC

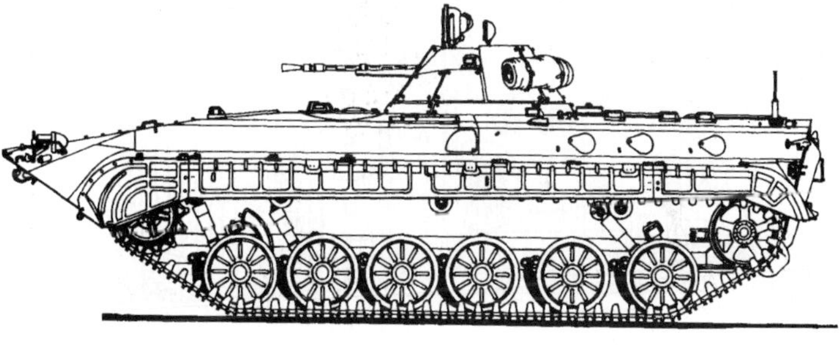

OT-90 M2 APC

VP-90 AVR

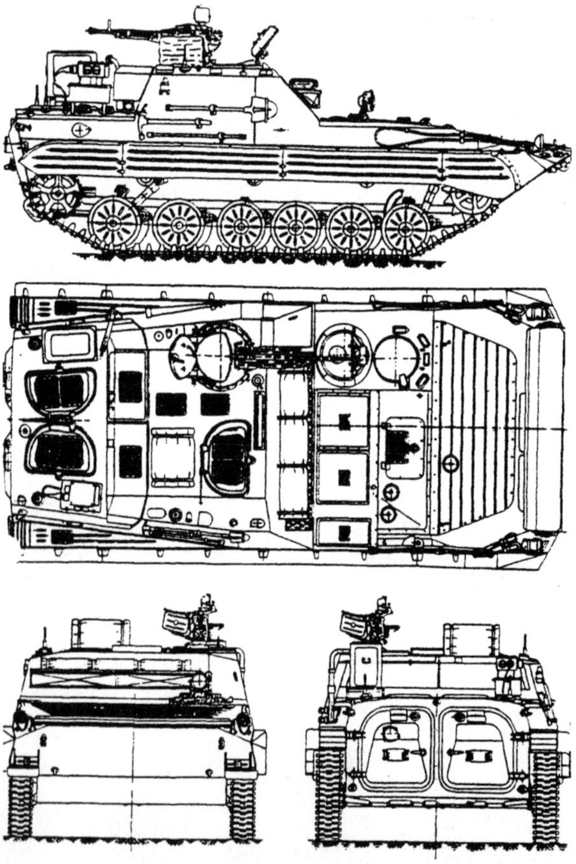

DP-90 AAOP

DP-90 AAOP

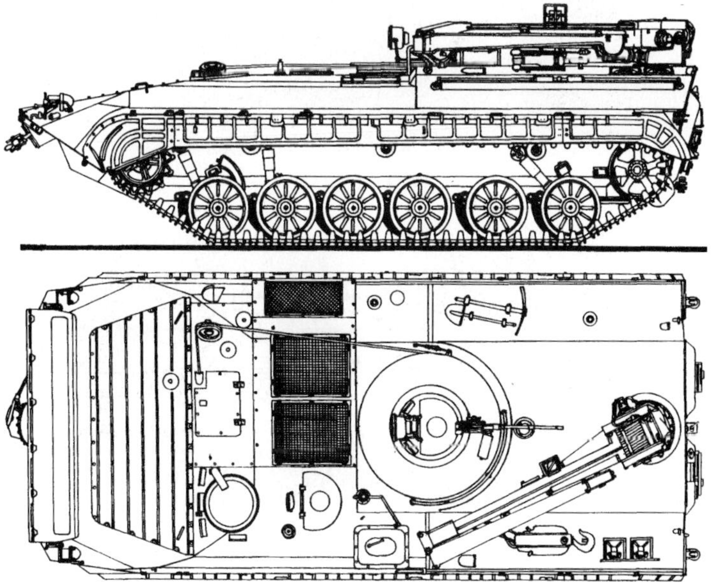

VPV ARV

VPV ARV

DTP-90 ARMV

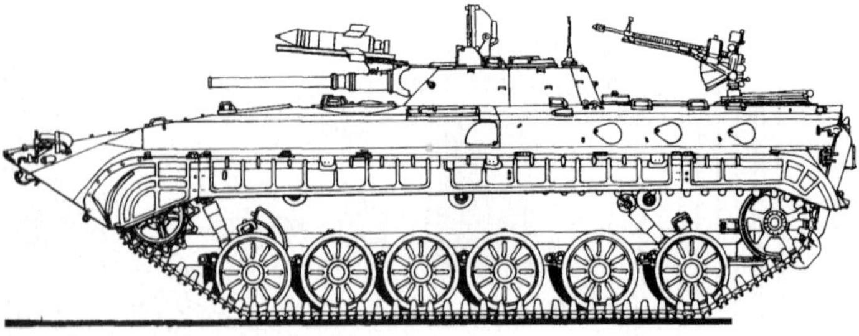

MLI-84 AIFV

MLI-84 AIFV

MLI-84M AIFV

MLVM AIFV

M1989 AAH

MVLM-AR AMC

BORAGH APC

BMP-3 armoured infantry fighting vehicle

Versions.

BMP-3 AIFV
BMP-3 AIFV with ATHOS thermal imager
BMP-3 KORNET-E TDM
BRM-3 RYS AVR
PRP-S AICV

Non-serial versions.

BMP-3 with 40/45 mm cannon
BMP-3 KHRIZANTEMA TDM
2S31 VENA 120 mm SP gun mortar system
BREM-L ARV

Development and production.

1990 service enter.
1992 first export deliveries to the UAE.
1993 BRM-3 fielded.

Distinguishing features. Side. Six roadwheels with slightly larger gap between second and third wheel station. Three support rollers, small track skirt. High box-like upper hull, pointed bow. Flat circular turret similar to BMP-2 with additional armour plates. Long gun tube. Front. High bow with trim vane, sloped glacis, driver's vision blocks centrally. Flat turret with commander's cupola on the right side and big main sight on the left.

Other details. Main armament consists of two guns arranged coaxially. Centrally mounted is a rifled 100 mm gun and on its right side a 30 mm cannon. On its left there is a coaxial 7.62 mm MG. The 2A70 stabilised main gun fires 3UOF17 and 3OF32 HE fragmentation rounds fed from an 22 rounds automatic loader or the manually loaded 9M117 (NATO: AT-10 STABBER) laser-guided missile with a range of 4 km. Two bow machine guns are placed in the sponsons. Two doors in the rear has to be used for (dis)mounting in conjunction with two large roof doors in order to cross the rear engine decking. Two firing ports on either side of the hull. The vehicle is amphibious by two rear mounted hydrojets.
The first production vehicles were powered by the 331 kW UTD-29 engine, the later produced ones have the stronger UTD-29M version with 368 kW. (TABLE)
Export vehicles are equipped with the French ATHOS periscopic thermal imager which has two magnifications x3 and x10. It is mounted on the left turret rear. Its vision is shown on a small monitor beside the gunner.
The BRM-3 lack the 100 mm gun, the two bow MG ports and has only one firing port on both hull sides. Main armament is the dual-feed 30 mm 2A72 cannon. The 7.62 mm coax MG is retained. Special equipment is the 1RL-133-1

battlefield surveillance radar which is mounted on an elevatable mast. It detects vehicles out to a range of 8 km and personnel up to 4 km. Two thermal night observation systems (1PN71 and 1PN61) are mounted on the left resp. right turret side. The 1PN61 has a magnification of x7 and a laser illuminator. The 1PN71 has magnifications of either x3.7 or x11.5. On the roof top there is a 1D14 periscope with integrated laser range finder; it has magnifications of x7.3 and x18. The vehicle is equipped with the TNA-4-6 navigation system and a 4 kW auxiliary generator. It has a combat weight of 19 t and is operated by a crew of six. Ammunition stowage is 600 rounds 30 mm resp. 2000 for the MG.
The tank destroyer version is equipped with the KORNET-E 9P162 anti-tank guided missile system. The twin-round launcher can be retracted into the hull for automatic loading. 12 missiles are carried internally in two revolving drums. An additional four missiles are stowed in the lower part of the hull. In addition the standard tripod launcher is carried for external employment. The gunner sits on the left side of the driver. He has a periscopic sight which includes thermal imaging. The missile range is up to 5.5 km. The commander sits on the right side of the driver.
Command versions have additional radio systems and optional navigation devices.

Role. In production for Russia (700), in service with Azerbaijan (3), Cyprus (43), Kuwait (76), South Korea (40), Ukraine (4) and UAE (300).

Assessment. Agile and versatile AFV family usable for different application. In the AIFV version with powerful armament but sub-standard armour protection. Very low ground pressure.

BMP-3 AIFV

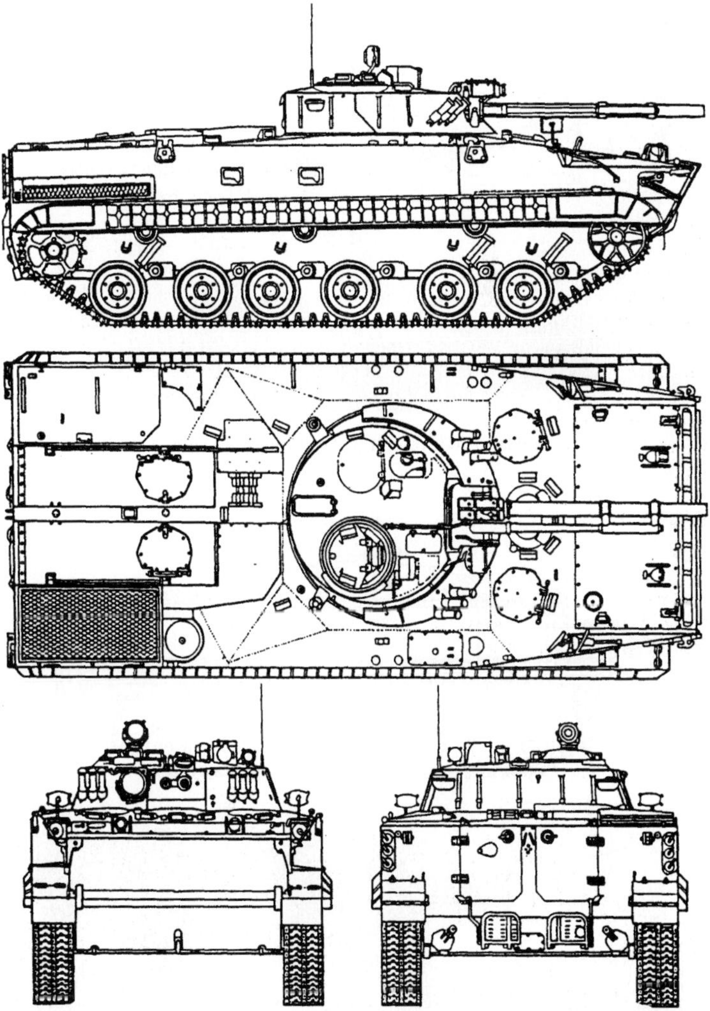

BMP-3 AIFV Steven Zaloga

BMP-3 AIFV with ATHOS thermal sight

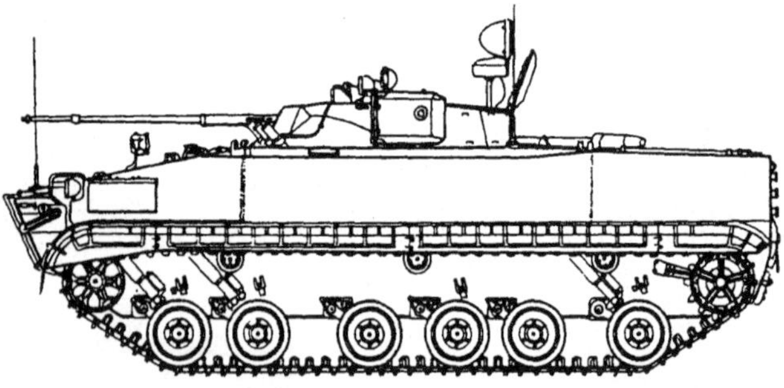

BRM-3 RYS AVR

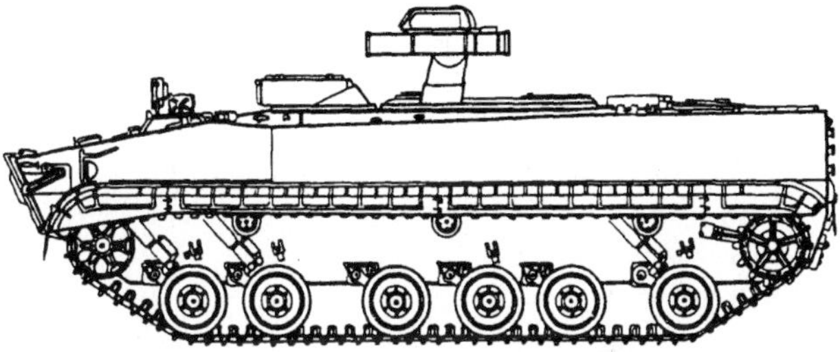

BMP-3 KORNET-E TDM

BMP-3 KORNET-E TDM

BMD airborne armoured general purpose carrier
Bronevaya Mashina Desantnaya

Models.

BMD
BMD-1
BMD-1M
BMD-1P
BMD-1R ACCC
BTR-D
BTR-RD AADFDV
BTR-R-440-ODB SATCOM vehicle
BMD-2
BMD-2 KShM
BMD-3
1V118 AACPV
1V119 AACPV
BMD launching vehicle for Shmel-1 RPV
BREM-D ARV
BMD-1ZD with ZU-23-2 AA guns
2S31 AMC

Development and production.

1970 introduction of BMD.
1973 issue of BMD-1.
1979 service enter of BTR-D versions.
1984 introduction of BREM-D.
1985 BMD-2 enters service.
1991 CFE notification of the BMD-3.

Distinguishing features. Side. Five small roadwheels, four return rollers, sprocket rear. High pointed front with heavily sloped glacis plate. Horizontal hull roof. Low truncated cone turret mounted well forward. Front. High prow front with trim vane on glacis plate, hull sides vertical with slight chamfer along two-thirds of hull side.
BTR-D without turret, six roadwheels.

Other details. Suitable for air dropping, adjustable suspension. The turret is similar to the one fitted to the BMP AIFV armed with the same 73 mm 2A28 gun and a 7.62 mm coaxial MG. Elevation and traverse mechanism is electric. Two forward firing bow MG on either side of the vehicle's front. Mounted over the main gun there is a launch rail for a SAGGER-C ATGM later replaced by the AT-4 SPIGOT or AT-5 SPANDREL system (BMD-1P). The crew consists of seven soldiers. The driver sits in a front central position with the commander left and the bow machine gunner on the right. Behind the one-man turret three

infantry soldiers are carried. The gunner is provided with a dual mode 1PN22M1 monocular periscopic sight mounted on the left side of the turret roof. In the day mode it has a magnification of x6 and a field of view of 15 degrees In the night combat mode the magnification is x6.7 resp. 6 degrees field of view.
The BMD was something like a pre-production model with a square NBC filter cover on the right hull roof.
The BMD-1 has a more prominent circular vent cover and later a single side weapon port.
The BMD-1M has a new internal ventilation system. It is distinguishable by a vent grill on the hull bow.
The BTR-D has two crescent shaped hatches on the roof to the rear of the cdr's place.
The BMD-2KSh is a command post tank with a folding rail antenna around the superstructure with the front firing ports deleted and without side firing ports.
The BMD-2 is armed with the 30 mm 2A42 cannon. The right front bow MG port and the front right hatch cover are omitted.
BMD-3 has a slightly larger hull and features the BMP-2 turret armed with the 30 mm 2A42 cannon. In the front bow one AGS-17 30 mm grenade machinegun is mounted on the left and one 5.45 mm RPKS-74 MG on the right side.
(TABLE)
All BMD versions are fully amphibious propelled by two hydrojets.
2S31 "VENA" has a lengthened BMD-3 chassis with 7 roadwheels and armed with a long-barreled 120 mm gun mortar with a firing range of ca. 12 km.
The BREM-D ARV has also a longer chassis with six roadwheels and has weight of 8 t. Vehicle's length is 5.89 m, the height is 1.82 m. It is operated by a crew of three and is armed with one 7.62 mm MG.

Role. In service with Armenia (10), Azerbaijan (79), Belarus (115), Moldova (54), Russia (2000), Ukraine (41 BMD-1, 78 BMD-2) and Uzbekistan (110).

Assessment. With this multi-purpose armoured vehicle a highly effective airborne tank has been created capable of providing airborne troops with a wide range of support and substantially enhancing their operational capability. BMD-3 with significantly increased mobility.

BMD-1 APC

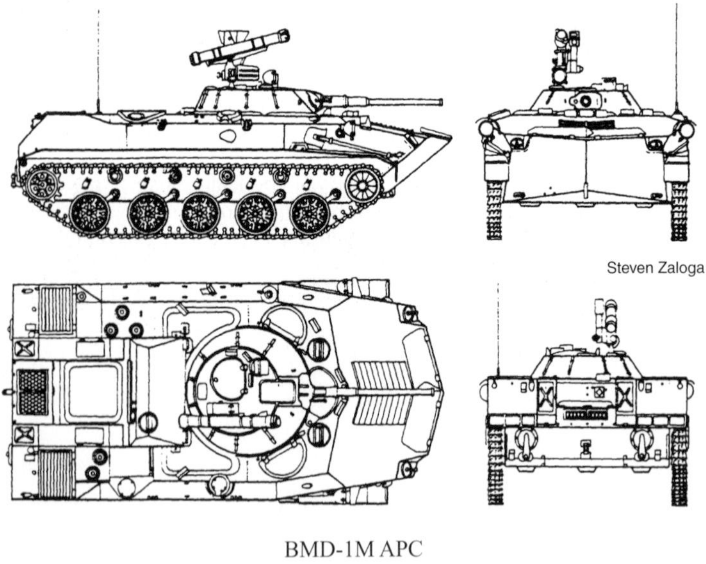

Steven Zaloga

BMD-1M APC

BMD-1M APC

BTR-D APC

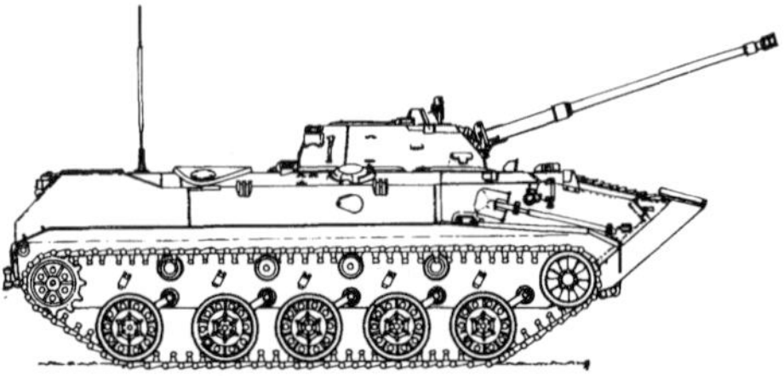

BMD-2 APC

BMD-2 APC

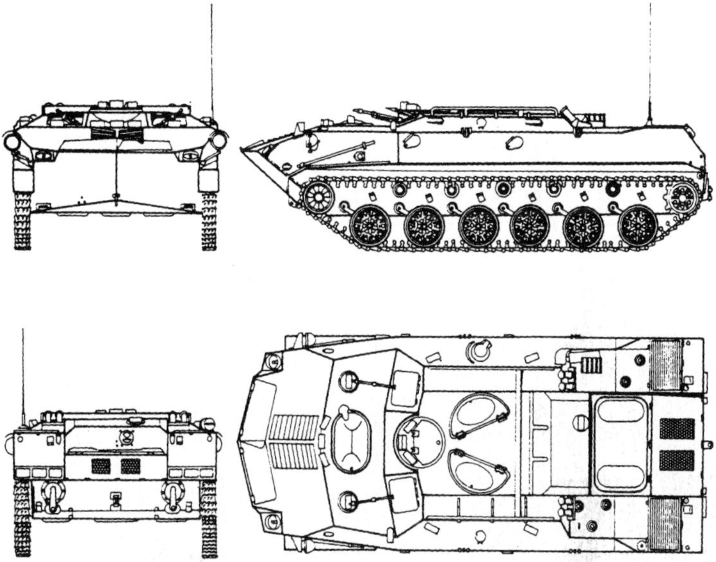

BDM-KShM ACCC Steven Zaloga

BDM-KShM ACCC

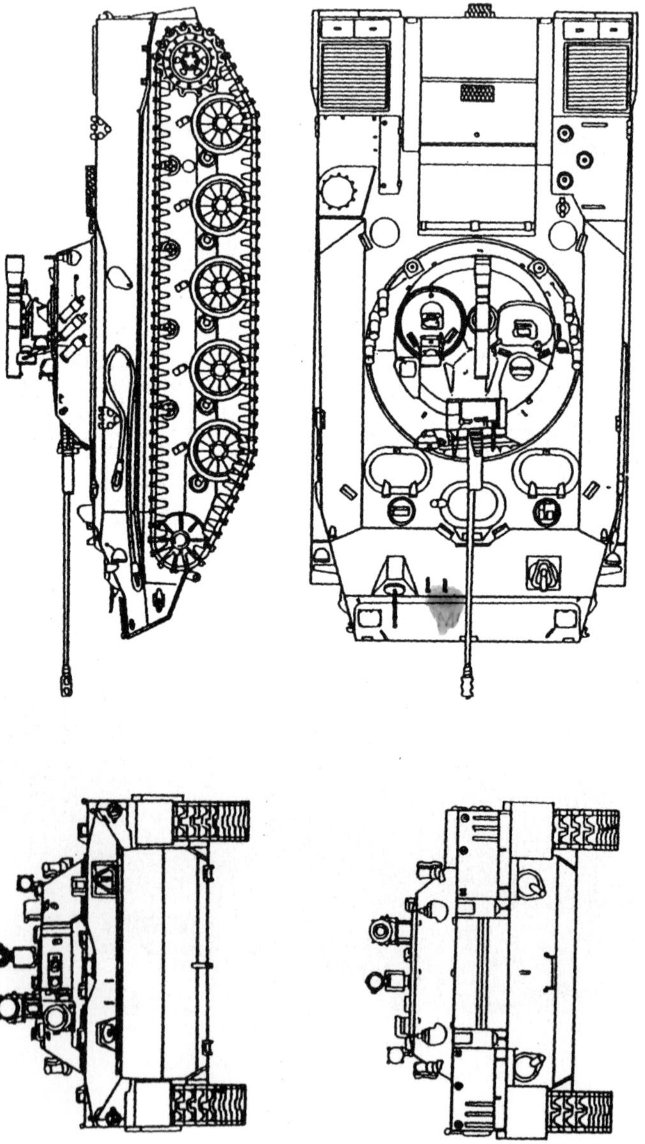

BDM-3 AGPC Steven Zaloga

BDM-3 APC

1V119 AACPV

BDM with Shmel-1 RPV

BDM-1ZD AADG

2S31 AMC

BTR-50P armoured general purpose carrier
Bronetransporter

Wt 14.2 t L 7.08 W 3.14 H 1.97 176 kW 44 km/h 400 km Crew 2+20

Models.

BTR-50P APC
BTR-50PA APC
BTR-50PK APC
BTR-50PU ACCC
MTP AMV
Type 77-1 APC (China)
Type 77-2 APC (China)
HQ-2J missile carrier (China)
TOPAS-2AP APC (Poland)
OT-62R-2 / R-2M battalion ACCC (TOPAS-WD R-2AM)
OT-62R-3MT regiment ACCC (TOPAS-WD R-3M)
OT-62R-4MT division ACCC
DTP-62 AMC (Czech Rep./Slovakia)
TOPAS-WPT AMV

Development and production.

1951 start of development of a tracked APC using the PT-76 chassis.
1954 service enter of BTR-50P.
1959 BTR-50PK enters service use.
1963 start of Czechoslovakian licence production OT-62.
1974 first appearance of Chinese licence version.

Distinguishing features. Side. Six 'waffle'-type roadwheels, no return rollers, sprocket rear. Boat-like sharply angled front, high front troop compartment, low rear engine deck and vertical rear. Front. Almost horizontal glacis plate, driver's hatch central with periscope below. Hull with sponsons and vertical sides. Dome-like cdr's hatch on the front left with slightly raised hatch.
The OT-62/TOPAS is identical with the BTR-50P except in the following details: two crew stations with raised hatches in the glacis, collective NBC protection/ventilation system.

Other details. The BTR-50P versions are either armed with one/two 7.62 mm MG or one 12.7 mm/14.5 mm MG on pintle mounting. Originally built with open troop compartment from BTR-50PK onwards it had a fully enclosed one. TOPAS-2AP is a Polish OT-62 version with turret-mounted 14.5 mm and 7.62 mm MG using the turret of the SKOT-2AP. Side doors have been added for quicker dismounting of the 12 infantrymen carried. The TOPAS weighs 16.4 t and has a permanent crew of three.
The command versions are equipped with additional radio equipment.

The BTR-50PK (B) is optimized for amphibious recovery operations with additional towing ropes and attachments. Its weight is 14 t.
The MTP (Mashina Tekhnicheskoy Pomoshchi) has a heightened crew compartment (2.5 m) and is provided with a pulley of 16 t and a jib expounder (2.85 m) with a capacity of 1.5 t.
The DTP-62 (Dilna TechnickÇ Pomoci) and the TOPAS-WPT are without turret but with a MG on protected ring-mounting. The special equipment includes a mechanical winch with a 2.5 t pull and 60 m of cable, a hand-operated crane with 1 t capacity, welding eqipment and a pontoon.
The Chinese versions are slightly larger (length 7.4 m, width 3.2 m) and are powered by a Type 12150L-2A diesel engine of 294 kW. Fuel capacity is 416 litres. Weight is 15.5 t. The driver is positioned on the left front, the cdr's dome is nearly in the centre, armed with an external 12.7 mm MG. The Type 77-1 has three rear loading ramps to enable the transport of an 85 mm Type 56 gun (30 rounds stowage) or a 122 mm Type 54 howitzer (20 rounds) on top of the vehicle's rear. The HQ-2J carrier has a lengthened chassis with seven roadwheels.

Role. In service with Afghanistan, Albania, Algeria (75), China (Type 77), Congo (18), Egypt, Guinea, uinea Bissau (20), Iran (50 Type 77, 2 HQ-2J), Iraq, Libya (300), North Korea, Slovenia (12), Somalia, Sudan, Syria and Vietnam.

Assessment. Simple APC with big troop capacity. No provision for fighting mounted under armour.

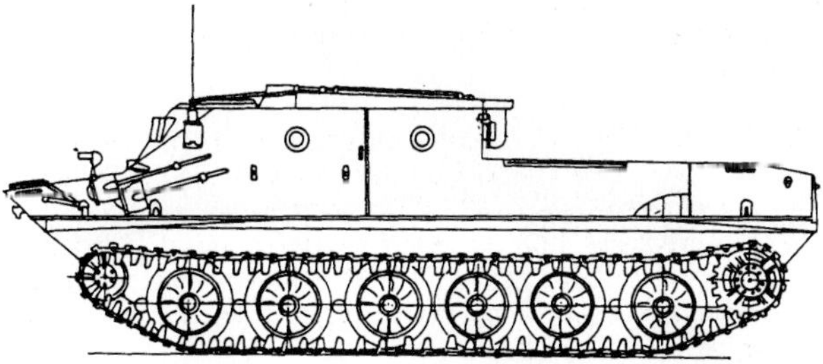

BTR-50 PK APC

BTR-50 PK APC

BTR-50 PU ACCC

Type 77-1 APC

HQ-27 missile carrier

TOPAS-2AP APC

TOPAS-WPT AMV

BTR-60 amoured personnel carrier
Bronetransporter

Models.

BTR-60P
BTR-60PA
BTR-60PB
BTR-60P-..../R-.... ACCC (different configurations)
BTR-60PU / PU-12 / PU-12M ACCC
1V18 KLEN-1 AACPV
1V19 KLEN-2 AAFDV
BTR-60E-351BR
BTR-60KShM ACCC
BTR-60MBP AACPV
BTR-60MS ACCC
BTR-60PB forward air control vehicle
MTP-2 AMV
BTR-60PAU (BUL) AACCC
BTR-60 with twin 30 mm AD cannons in open mounting (Cuba)

Development and production.

1957 start of development for an amphibious replacement vehicle for the BTR-152 called project 49 with two competing teams.
1960 selection of the Dedkov version as BTR-60P (Plavayushchiy = amphibious). This version still had an open roof.
1963 introduction of BTR-60PA with closed roof, troop size reduced from 16 to 12.
1965 issue of BTR-60PB with 14.5 mm MG turret.
1979 first appearance of the evolutionary BTR-70, BTR-60 remained in production for export.

Distinguishing features. Eight-wheeled vehicles with larger space between two centre axles. Boat-shaped front and slightly sloped glacis with two armoured flaps. Long hull with engine compartment at rear. Trapezoidal turret sloped in front behind driver compartment (PB version, B for Vyshka = turret).
(TABLE)
Company command versions bear the additional letter "K".
Command and control versions mostly lack MG turret and are provided with additional radio equipment showing extra aerial arrangement and external power-generator. These command versions have the suffix according to their communication equipment.
BTR-60P versions: P-219, switchboard P-238BT, P-239BT, radio/wire integration switchboard P-240, P-240BR, P-240BT, P-240TBR, field telephone system P-241-BT, P-241-TBR, P-251, P-286, P-537;
BTR-60R versions: VHF set R-137, R-137B, R-137BR, HF set R-140, R-

140BM, VHF/HF set R-145, R-145 RM, R-145K, R-156, R-156B, R-156BTR, R-396, R-409, radio relay station R-409 BM, R-409BR, R-419B, R-419BTR, SAT Com R-440B, R-975, R-975M, R-975M1.
BTR-60PU with 10 m mast and rail antennas, P-12 and PU-12M is used by air defence unit.
1V18 (battery level) and 1V19 (division level) use a different turret ("Darth Vader" resemblance) with special night sights and observation devices under armoured covers including a laser rangefinder and the APPK artillery computer.
BTR-60E-351BR is a battery loading station.
BTR-60MS radio system with a telescopic antenna (NATO: "High Ball").
FAC vehicle is a BTR-60PB with armament removed and the resulting port covered by a plexiglas observation window.
MTP-2 has a raised tarpaulin cover oder the reconfigurated troop compartment that runs almost to the rear.
Beside that there exist numerous local configurations.

Role. In service with Afghanistan, Algeria (370), Angola (62), Armenia, Azerbaijan (25), Belarus (188), Botswana (24), Bulgaria (781), Cambodia (110), Chad (10?), Congo (28), Cuba (100), Djibouti (10), Estonia (10), Ethopia (80), Finland (110), Guinea (8), Guinea-Bissau (10), Iran (160), Iraq (approx. 500), Laos (22), Libya (260), Lithunia (10), Mali (10), Moldova (2), Mongolia (70), Mozambique (80), Nicaragua (19), North Korea, Peru (12), Russia (approx. 4000), Somalia, Syria (approx. 600), Tajikistan, Turkmenistan, Uganda (20), Ukraine (231), Vietnam (approx. 400), Yemen (approx. 300), Yugoslavia (150?) and Zambia (10).

Assessment. Mobile APC with poor main armament and light armour. Dual-engine transmission liable to frequent malfunction. Consequent evolutionary improvement of a cheap, versatile carrier family.

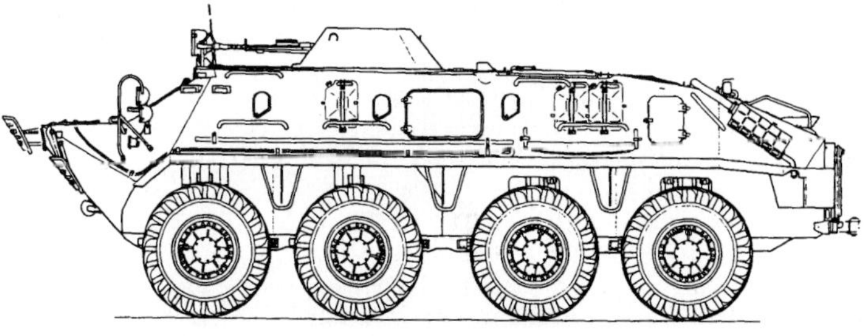

BTR-60PB APC

BTR-60PB APC

BTR-60PBK

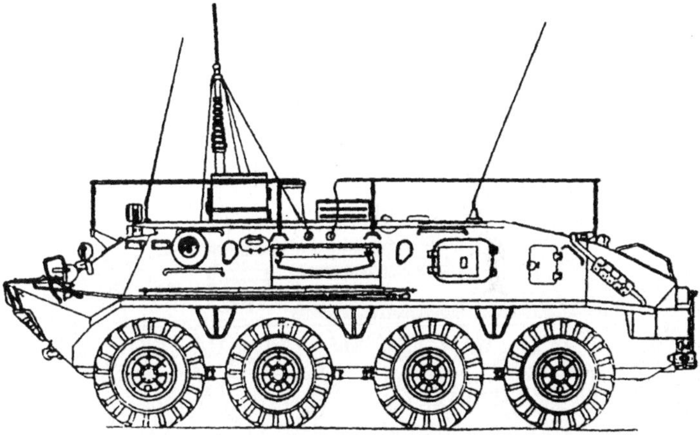

BTR-60PU ACCC

BTR 60PU ACCC

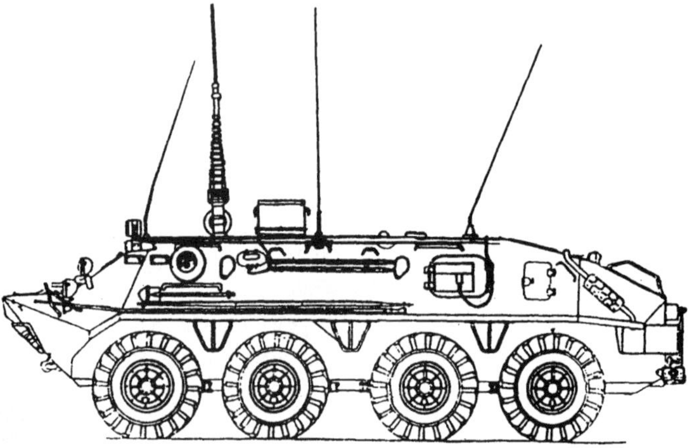

BTR-60PU-12M ACCC

BTR-60PU-12M ACCC

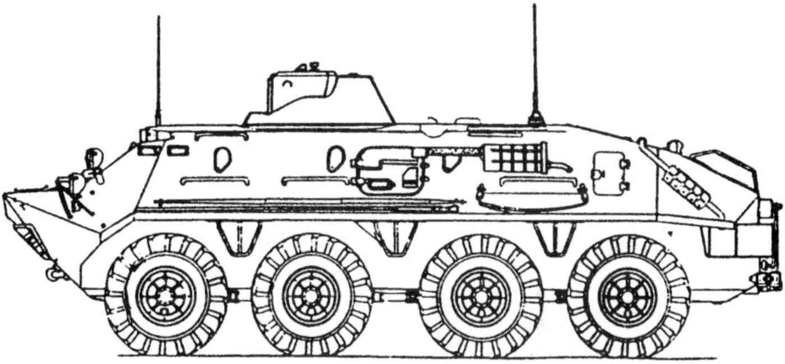

1V19 AAFDV

1V19 AAFDV

BTR-70 amoured personnel carrier (8x8)
Bronetransporter

Models.

BTR-70
BTR-70W
BTR-70Kh ANBCRV
BTR-70MS ACCC
BTR-70KShM ACCC
BTR-70MBP AACC
SPR-2 with jamming equipment
BTR-70ZS-88 psychological warfare vehicle
BTR-70 AAV
BREM AMV

Development and production.

1979 first appearance of the evolutionary BTR-70, BTR-60 remained in production for export.

Distinguishing features. Eight-wheeled vehicle with lengthened hull and larger engine compartment. Similar to BTR-60PB but with small door in lower part of hull between second and third wheel. Trim vane folds back onto glacis plate (BTR-60 has it beneath the nose).
(TABLE)

Other details. BTR-70W features the BPU-1 turret with higher elevation angle using the 1PZ-2 sight and the TNP 205 periscope for the gunner. This version can be equipped two 30 mm AGS-17 grenade launchers near the engine compartment partition wall above the third roadwheel station reducing the number of the infantry section from 10 to 8.
Command and control versions lack MG turret and are provided with additional radio equipment showing extra aerial arrangement and external power-generator.
The ambulance version can transport 4 wounded laying resp. 6 sitting.
BREM is a turretless BTR-70 with a bow-mounted jib crane and repair equipment.

Role. In service with Afghanistan (180), Armenia (83), Azerbaijan (88), Belarus (451), Estonia, Georgia, Hungary (150), Kazakhstan, Kyrgystan (102), Moldova (7), Nepal (135), Pakistan (120), Russia (3000), Tajikistan, Turkmenistan (199), Turkey (300 ex-NVA), Ukraine (1108) and Uzbekistan (11).

Assessment. Mobile APC with light main armament and light armour. Dual-engine transmission liable considerably improved compared to BTR-60.

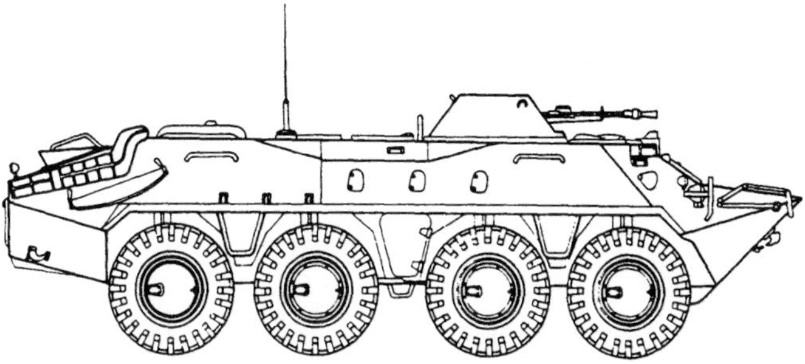

BTR-70 APC

BTR-70 APC

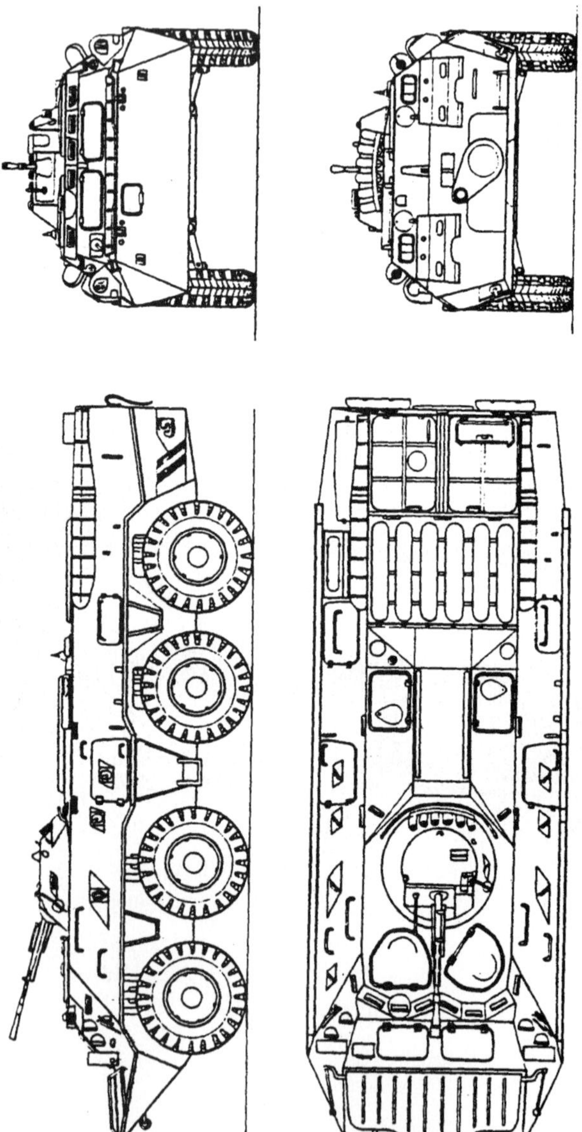

BTR-70W APC

BTR-70W APC

BTR-70 AAV

BTR-80 amoured personnel carrier
Bronetransporter

Models.

BTR-80
BTR-80A
BTR-80S
BTR-80 ACCC (different versions)
BTR-80K1SH1 ACCC
1V118 ACCC
BTR-80 radar carrier (NATO: "Clothes Horse")
RKhM-4 ANBCRV
BTR-80ZS-88 psychological warfare system
BMM AAC versions
2S23 NONA-SVK AMC
BTR-80 "Vetluga" rocket launcher
BREM-K ARV

Non-serial versions.

BTR-80 with KLIVER turret (30 mm cannon plus KORNET ATGM)
BTR-90 (GAZ-5923)

Development and production.

1979 first appearance of the evolutionary BTR-70, BTR-60 remained in production for export.
1985 adoption of further improved BTR-80.
1991 appearance of 2S23 AMC.
1994 issue of BTR-80A.

Distinguishing features. Eight-wheeled vehicles with larger space between two centre axles. Boat-shaped front and slightly sloped glacis with two armoured flaps. Long hull with four forward directed firing ports on each side and engine compartment at rear. Trapezoidal turret sloped in front behind driver compartment.
BTR-80 is almost identical to BTR-70 but has a large two-piece entrance door between second and third wheel. (TABLE)

Other details. BTR-80A (GAZ-59034) has an oscillating 30 mm 2A72 cannon (combat load of 300 rounds) increasing the elevation tp +70°. BTR-80S is identical having only a 14.5 mm MG. Command and control versions (except company command version BTR-80K) mostly lack MG turret and are provided with additional radio equipment showing extra aerial arrangement and external power-generator.
NBC reconnaissance version is easily recognizable by the rear marking poles emitter.

The ambulance version (GAZ-59039) has a weight of 12.6 t. There are three versions, the BMM-1 evacuation vehicle, BMM-2 battalion medical station and BMM-3 mobile dressing station.
"Vetluga" is a 22 rounds' multiple rocket launcher with a maximum range of 6 km.
2S23 is a BTR-80 hull with all fire-ports and the side doors deleted provided with a turret very similar to the one mounted on the SO-120 (2S9) armed with a 120 mm 2A60 mortar. The turret differs from that on the 2S9 having a raised cdr's cupola towards the turret rear on the left side which overhangs the side of the turret. Elevation ranges from -4 to +80 degrees. Traverse is 30 degrees to both sides. The mortar has a maximum range of 8.7 km resp. 7.1 km in the direct fire mode; RAP ammunition has a range of 12.8 km. The vehicle's weight is 14.5 t, the combat load 30 rounds. It is operated by a crew of four.
The BREM-K (GAZ-59033) retains the same turret but only with a 7.62 mm. The vehicle has towing bars and lifting appliance and a rigid crane jib to be attached on the bow plate. Just behind the turret there are two cable drums for towing purposes. The vehicle has a weight of 14 t, the main winch tows 60 kN, the lifting crane has a lifting capacity of 15 kN.
Several civilian applications.

Role. In service with Afghanistan, Armenia (83), Azerbaijan (14), Bangladesh (58), Belarus (195), Estonia (20), Finland (2), Georgia, Hungary (229, 540 more on delivery), Kazakhstan, Moldova (7), Russia (2000), Sierra Leone (12), Tajikistan, Turkey (141), Turkmenistan, Ukraine (449) and Uzbekistan (95).

Assessment. Mobile APC with light main armament and light armour. Dual-engine transmission is abandoned in favour of a single diesel engine concept. Consequent evolutionary improvement of a cheap, versatile carrier family.

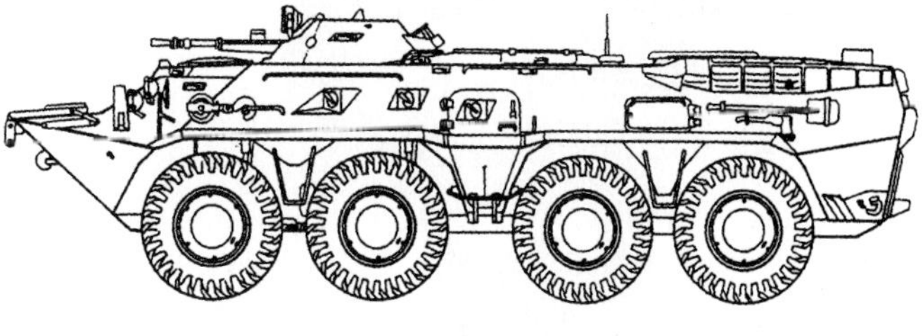

BTR-80 APC Steven Zaloga

BTR-80 APC

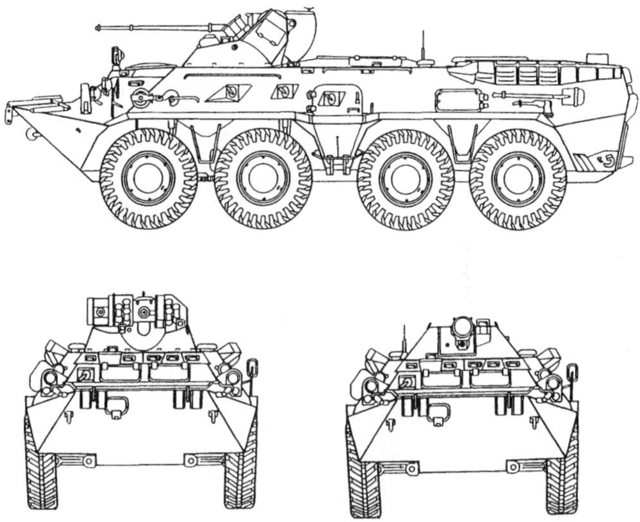

BTR-80A APC Steven Zaloga

BTR-80A APC

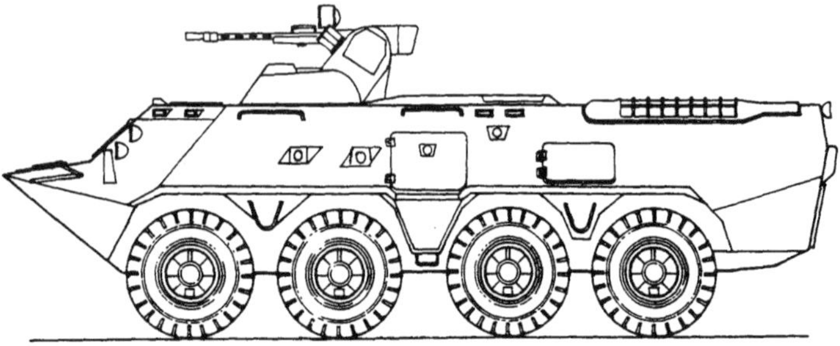

BTR-80S APC

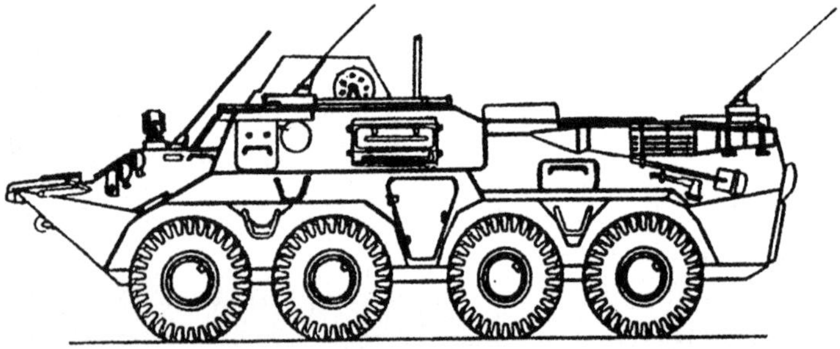

BTR-80 ACCC

1V118 ACCC

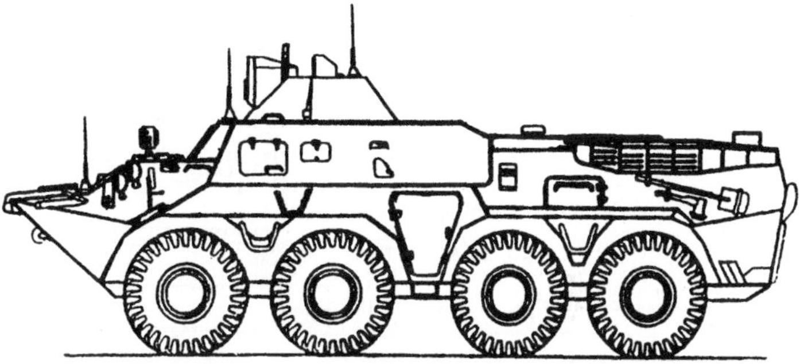

1V118 ACCC

BMM AAV

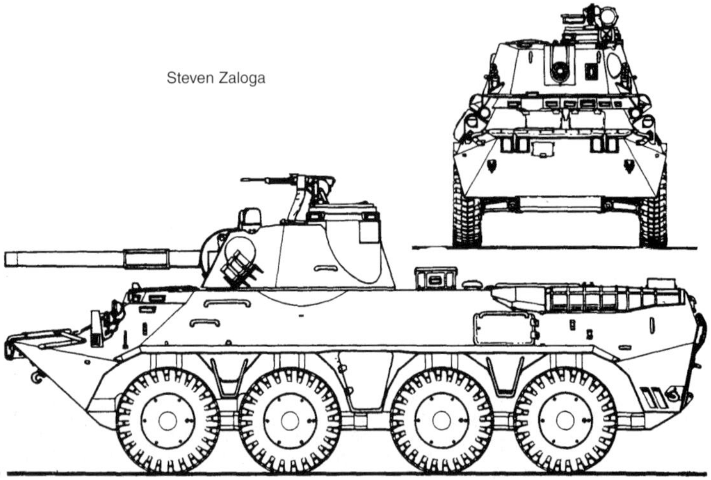

2S23 AMC

2S23 AMC

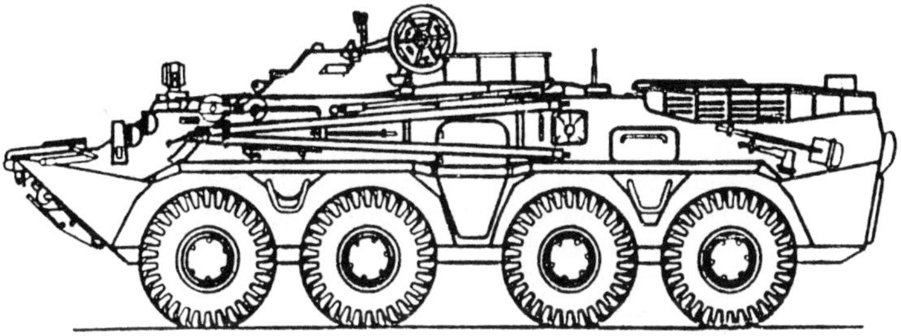

BREM-K ARV

BREM-K ARV

MT-LB armoured general purpose carrier
Mnogocelovyj Transporter Legko-Bronirovannyj

Models.

MT-LB APC (MT-LBV with wider tracks)
9P149 Shturm-S TDM
Pbv 401 TDM with Bill ATGM (Sweden)
K-611 nuclear recce vehicle
K-612 nuclear radiation detector vehicle
MR-HR ANBCRV (Bulgaria, Poland)
KShM R-80 AACPV (Bulgaria) with raised hull
KShM R-81 ACCC (Bulgaria)
R-137 and R-137T ACCC (Poland)
SOVA (Bulgaria) with ground surveillance radar
SNAR-10 AARV
TRI AERV (Poland)
MT-LB AAV (different configurations)
MT-LB AMC with 82 mm mortar (Bulgaria, PODNOS in Poland)
B1.10 TUNDZHA AMC with 120 mm mortar (Bulgaria)
MT-LB with 3x4 107 mm rocket launcher (Iran)
MT-LB with twin ZU-23 cannon (Iran)
MT-LB AEV
WPT ARV (Poland)
MTP-LB AMV

Non-serial versions.

Numerous conversions and prototypes.

Development and production.

1966 production start as a derivative of the unarmoured MT-L tractor.
1974 service enter of SNAR-10.

Distinguishing features. Side. Six equally spaced chrysanthemum patterned road wheels, no return rollers, sprocket front. High front, slightly sloped glacis. Crew compartment extends to the rear troop compartment. Almost vertical rear with twin doors. Small flat-topped front turret. Front. Small turret to right. Cdr's and driver's compartment to immediate rear of glacis plate. Sponsoned hull with almost vertical sides with chamfered top. For swimming, flow plates are fitted in front of and behind the tracks, and a protective casing is placed over the radiator cover.

Other details. Transmission and engine compartment left centre, access tunnel on right side. Load platform on rear deck. Used in different applications e.g. troop carrier and gun tractor. (TABLE)
The Shturm-S is equipped with an external ATGM launcher on top of the hull rear firing the AT-6 SPIRAL ATGM with a range of 400 - 5000 m. The missile

is 1830 mm long, 130 mm in diameter, has a weight of 35 kg (46,5 kg incl. the launch tube). A total of 12 missiles is carried. The main sight is situated on top of the right front hull roof.
The SNAR-10 carries a 1RL-127-1 mortar/artillery locating radar (NATO: BIG FRED) operating at 34-35 GHz with a range of 16 km. The vehicle's weight is 12.6 t, its height is 2.39 m with antenna folded down resp. 2.89 with erected antenna, and it is operated by a crew of four.
The engineer version has a plough blade behind the rear wall and a rectangular stowage box on the rear deck.
Ambulance version with medical equipment, later versions with a raised superstructure at the rear.
TRI is slightly longer (6.63 m), is powered by the SW680/167/1 six cylinder super-charged diesel engine with 180 kW and is equipped with NBC and mine detection equipment (weight 12.9 t).
The MTP-LB is used for field maintenance and has an A-frame at the front which can lift a maximum load of 1.5 t. It also has a winch with a capacity of 6.7 t with a cable length of 85 m (13 mm diameter) and a front dozer blade. It has a weight of 12.3 t, a length of 6.8 m, a height of 2.3 m, and is operated by a crew of two. It lacks the MG one-man turret.
WPT used as a mobile battlefield damage rescue and service stand-by vehicle featuring a roof-mounted crane.

Role. In service with Armenia (75), Azerbaijan (approx. 200), Belarus (71), Bulgaria (1113), Finland (220), Greece (500 ex-German), Hungary (30), Iran, Iraq (approx. 500), Kazakhstan (200), Moldova (62), Poland (185), Russia (4000), Sweden (550 ex-East German now designated Pbv 401 resp. Pbv 4012 ex-MTP-LB), Ukraine (2300) and Yugoslavia (50).

Assessment. A cheap versatile vehicle used as a multi-purpose carrier family. Particularly low ground pressure gives the vehicle good cross-country performance.

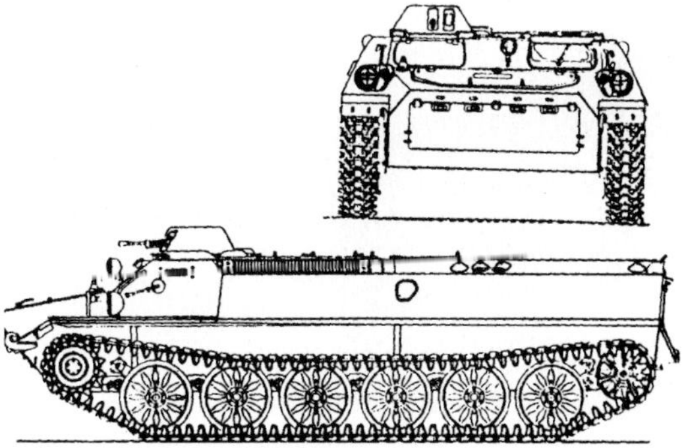

MT-LB APC

MT-LB APC

Shturm-S TDM

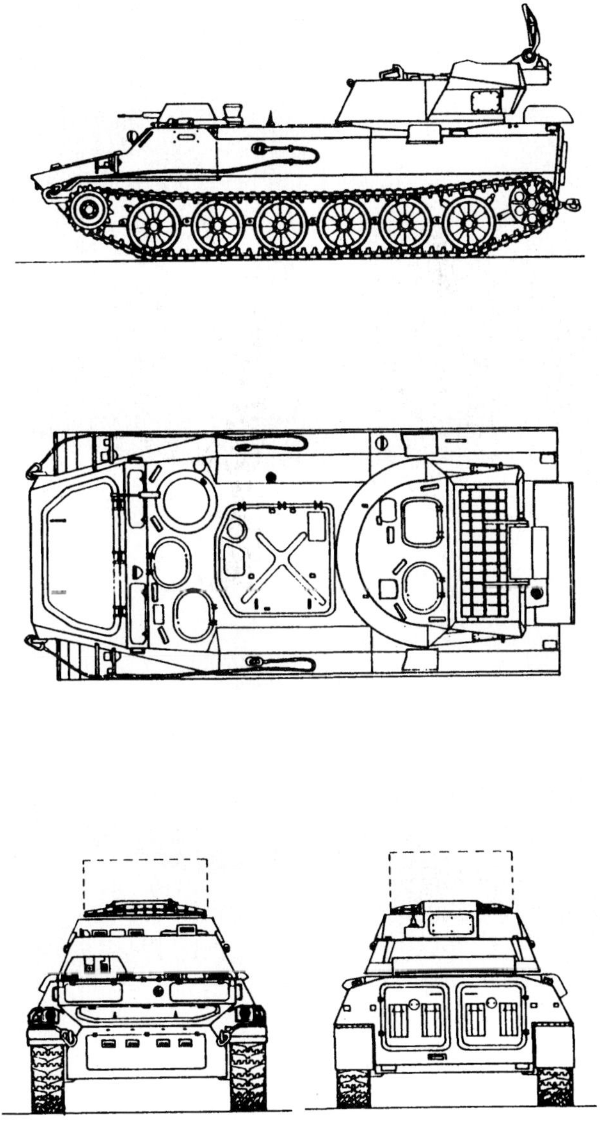

SNAR-10 AARV Steven Zaloga

SNAR-10 AARV

TRI AERV

MT-LB with 107 mm rocket launcher

MT-LB with ZU-23-2

MT-LB AEV

MTP-LB AMV

MT-LBu armoured general purpose carrier

Models.

MT-LBu ACCC
1V12 artillery forward observer vehicle
1V13 AACPV
1V14 AAOP
1V15 AACPV
1V16 AAFDV
1V21 staff command vehicle (different versions)
1V22 AADFDV
1V22-1 RANGIR air-defence battery CP vehicle
1V23 AADFDV
1V24 artillery divisional CP vehicle (diff. versions)
1V24R artillery command and control version
1V25/1V25R air-defence command and control carriers
1V95 divisional data-processing carrier
9S37 ACCC for TOR AADL units
BETA 3M (Polish 1V95 version)
KShM9S743 ACCC (Bulgaria)
R-161B divisional CP carrier
R-330B VHF jammer
R-330KB radio direction finding carrier
R-330P HF/VHF jammer
R-378B HF jammer
R-381T HF resp. VHF interceptor carriers (diff. versions)
R-409BM radio relais station
KDKhR-1N ANBCRV
1L219 ZOOPARK-1 AARV
PPRU-1 AADRV
MTP-1 AMV (Bulgaria)

Numerous conversions and non-serial versions exist.

Development and production.

1974 first public appearance of the 1V13-16 vehicle family.
1993 ZOOPARK enters production.

Distinguishing features. Side. Modified MT-LB chassis with seven road-wheels, sprocket front, no return rollers. High hull superstructure with pointed bow and vertical rear. "U" stands for upravleniye = command and control). Front. Protruding bow with inclined bottom and sloped glacis normally with two large vision flaps on upper part. Sponsoned hull with upper part sloped inwards.

Other details. 1V13 is a battery CP tank equipped with a land navigation system, a gyroscope to provide orientation, and a firing data processing com-

puter. On the forward roof of the turret there is a director mounted to pass orientation to the firing battery. This version has two antennas on the front and rear left hull roof.
1V14 is the battery cdr's OP vehicle. Besides a navigation system and a gyroscopic orientor it is provided with a laser rangefinder with a maximum range of 20 km. Besides the laserhead on the left front edge of the turret is a binocular angle measuring device. Attached on the right turret side is a searchlight.
The 1V15 battalion cdr's vehicle differs from the 1V14 externally being equipped with a 10 m telescopic antenna stowed on the upper left hull when not in use while the 1V14 only has whip antennas.
The 1V16 is the battalion's fire direction vehicle and contains an artillery computer and two additional antennas on the right hull roof. It is also equipped with the telescopic antenna. It differs externally from the other models lacking the periscopes on top of the turret and the lack of the box-like sponson bay on the right hull sides.
Vehicles with "M" suffix have modernized electronics and optronics and the inboard auxiliary power-unit replaced by a new one mounted externally on the right vehicle's rear.
1V16M is no longer fitted with a turret.
R-... designations correspond to the technical equipment model which is used in the certain role.
KDKhR-1N is used for stand-off chemical reconnaissance. It is equipped with lasers to scan at a chemical agent detection range from 1000 to approx. 3000 m. The different laser systems are integrated in a rotating rear-mounted turret. The vehicle also carries other radiological and chemical agent sensors processed by an on-board computer.
ZOOPARK-1 has the 1L259 3D monopulse phased-array artillery location radar system operating in the H band. The scanning sectors are 60 degrees in azimuth and 40 deg. in angle of sight. 12 targets can simultaneously be tracked. It is able to identify up to 20 artillery or mortar firing positions per minute within an area of 10 - 15 km resp. missile launch positions within an area of up to 45 km. It is operated by a crew of three.
PPRU-1 is an early warning surveillance and target acquisition radar (NATO: DOG EAR) with an acquisition range of 80 km and a tracking range of 35 km operating at 4930 MHz. This vehicle has seven roadwheels.
The MT-LBu jamming vehicle is provided with the VHF R-330 P jamming set. When in movement a frame antenna is erected on the hull roof centre with eleven antenna elements. In the stationary mode a telescopic antenna is used. An auxiliary power unit is installed externally on the right rear hull.
MTP-1 is a technical support vehicle with a 3 t crane in a small turret-shaped cupola, a rear-mounted entrenching blade and a 30 t winch.
Most models are amphibious but flow plates have to be fitted. NBC filter system.

Role. In service with the former WARPAC armies.

Assessment. Roomy, highly mobile vehicle with standardised hull. Versatile applications offer a wide range of employment.

1V13 AAPV

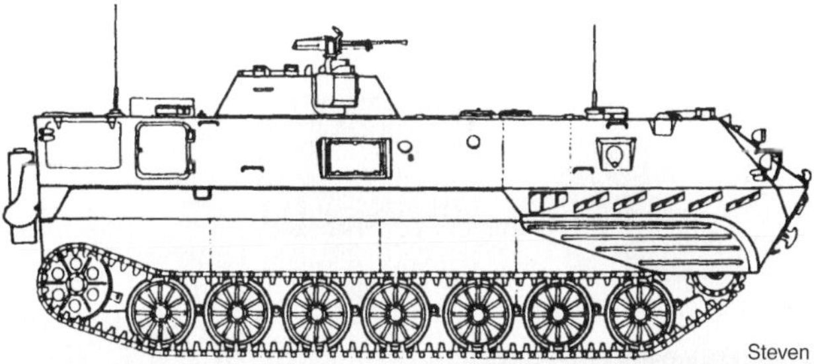

Steven Zaloga

1V14 AAOP

1V15 AACPV

1V16M AAFDV

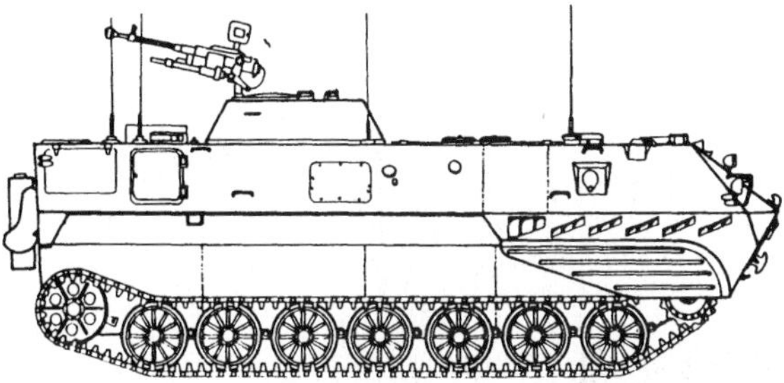

1V16 AAFDV

1V22-1 RANGIR ADCPV

9S37 ADCPV

R-330KB radio direction finder

R-330P HF/VHF jammer

ZOOPARK-1 AARV

PPRU-1 AADRV

2S1 GVOZDIKA (SO-122) armoured artillery howitzer

Models.

2S1 AAH
BRM-23 AVR (Bulgaria)
BMP-23 APC (Bulgaria)
BMP-23A APC (Bulgaria)
BMP-30 APC (Bulgaria)
RKhM ANBCRV
2S12 AMC

Development and production.

1974 first public appearance in Poland and USSR. Later export.

Distinguishing features. Many automotive components of MT-LB carrier are used including engine and suspension (seven instead of six roadwheels). Glacis very shallow, lower front plate slightly sloped. Boat-like front and large sponson over tracks. Low profile, rear-mounted turret at rear of long hull. Gun with muzzle brake and fume extractor half-way along barrel in a narrow drum mantlet, no gun barrel front overhang. Engine front right, driver front left.

Other details. The howitzer is armed with a modified version of the D-30 towed howitzer. A power rammer and extractor are fitted on a folding guard rail. NBC system installed.
(TABLE)
2S12 is a 2S1 chassis with turret removed and rear mounted 120 mm mortar. The mortar compartment can be closed by two large roof hatches hinged on both sides.
BRM-23 has a weight of 14.8 t and a crew of 5. Height is 2.53 m.
BMP-23 has a weight of 15.2 t and a crew of 10. Both vehicles are armed with a 23 mm cannon with 450 rounds on band and 150 spare rounds. Coaxially mounted is a 7.62 mm PKT MG with 2000 rounds. Both tanks are provided with externally mounted SAGGER ATGMS. Fuel capacity is 540 l.
BMP-23A has different roadwheels (non-ribbed) and AT-4 SPIGOT.
BMP-30 is armed with the BMP-2 turret.
RKhM has a raised centre superstructure with a 7.62 mm MG turret and a rear decking box on the right side with marking pennants.

Role. In service with Afghanistan (15), Algeria (145), Angola, Armenia (10), Azerbaijan (14), Belarus (239), Bulgaria (740 + 256 BMP-23), Croatia, Cuba (20), Czech Rep. (97), Ethopia, Finland (72), Hungary (151), Iran (60), Iraq (approx. 50), Kazakhstan (100), Libya (130), Poland (521), Romania (6), Russia (3000+), Slovakia (49), Slovenia (18), Syria (400), Ukraine (644), Uzbekistan (18), Yemen (approx. 50) and Yugoslavia (75).

Steven Zaloga

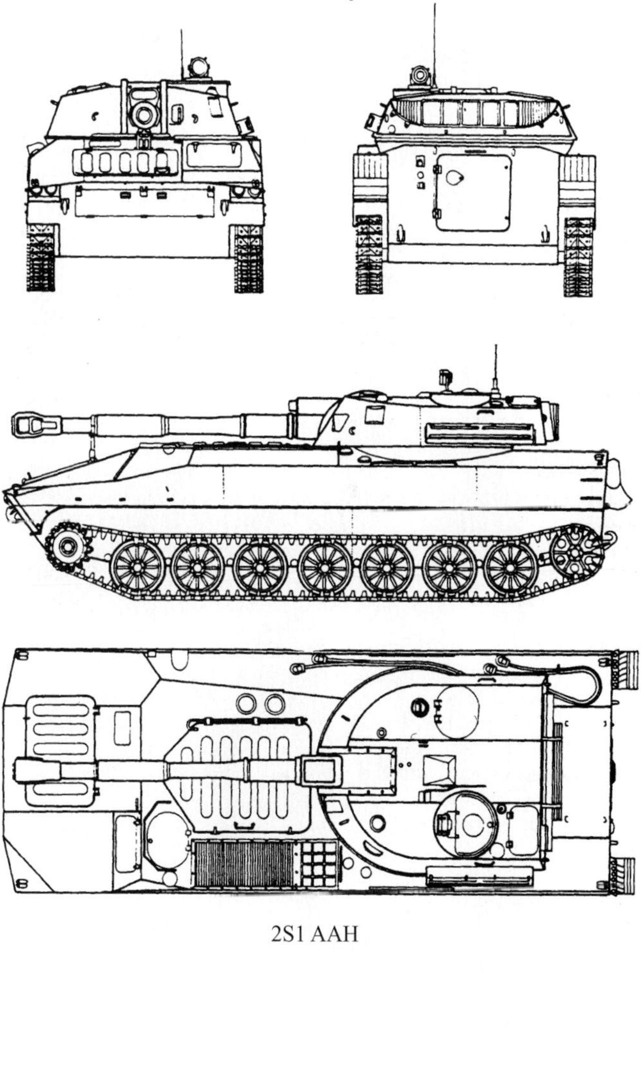

2S1 AAH

2S1 AAH

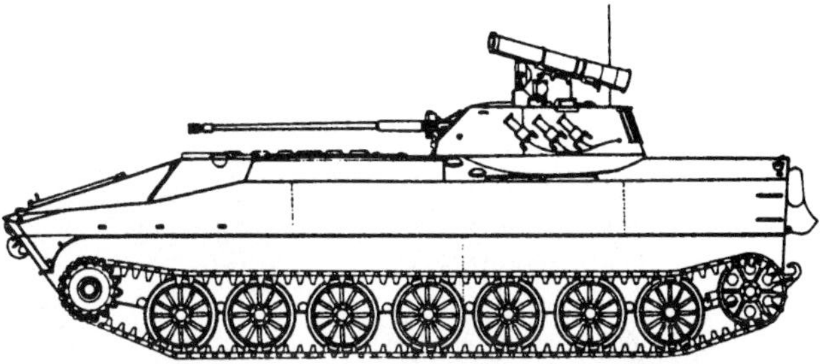

BRM-23 AVR

BMP-23 APC

BMP-30 APC

RKhM ANBCRV

2S3 AKATSIYA (SO-152) armoured artillery howitzer

Models.

2S3
2S3M
2S3M1

Development and production.

1967 start of development.
1971 service introduction into Soviet Army, later exported.

Distinguishing features. Side. The 2S3 has a similar chassis as used for the SA-4 GANEF AADML with six instead of seven roadwheels with a prominent gap between the first and second and second and third wheel station, four return rollers, front sprocket. Box-shaped turret with sloped front, cutaway rear and raised portion around cupola to left, on which ADMG and spotlight are mounted. Drum mantlet, squat barrel. Double baffle muzzle brake with four deflector apertures, cylindrical fume extractor behind it. The barrel protrudes over the hull front and is secured by a clamp behind the fume extractor during movement. Twin buffers protrude beyond the mantlet above the barrel. Front. Superstructure partly sponsoned. High truncated cone turret with raised part around cupola to left. Driver front left, power train right.

Other details. The howitzer is armed with a modified version of the towed D-20 152 mm gun/howitzer. It is also nuclear-capable (2 kt). Of the six-men crew two are transported by the ammunition truck and when in firing position they are deployed as ammunition handlers via two circular hatches in the lower rear wall. The AAH is non-amphibious and is provided with a NBC system.
(TABLE)
2S3M and 2S3M1 has an ammunition stowage of 46 (compared with 40 of the basic 2S3 version).
2S3M1 additionally has a fire control display.

Role. In service with Algeria (40), Armenia (28), Belarus (166), Cuba (20), Hungary (40), Iraq (approx. 50), Kazakhstan (approx. 50), Libya (55), Russia (2000 +), Syria (50), Turkmenistan (10), Ukraine (501), Uzbekistan (17) and Vietnam (30).

Assessment. The 2S3 is similar to the western M109 AAH. Lightly armoured, average fire-power.

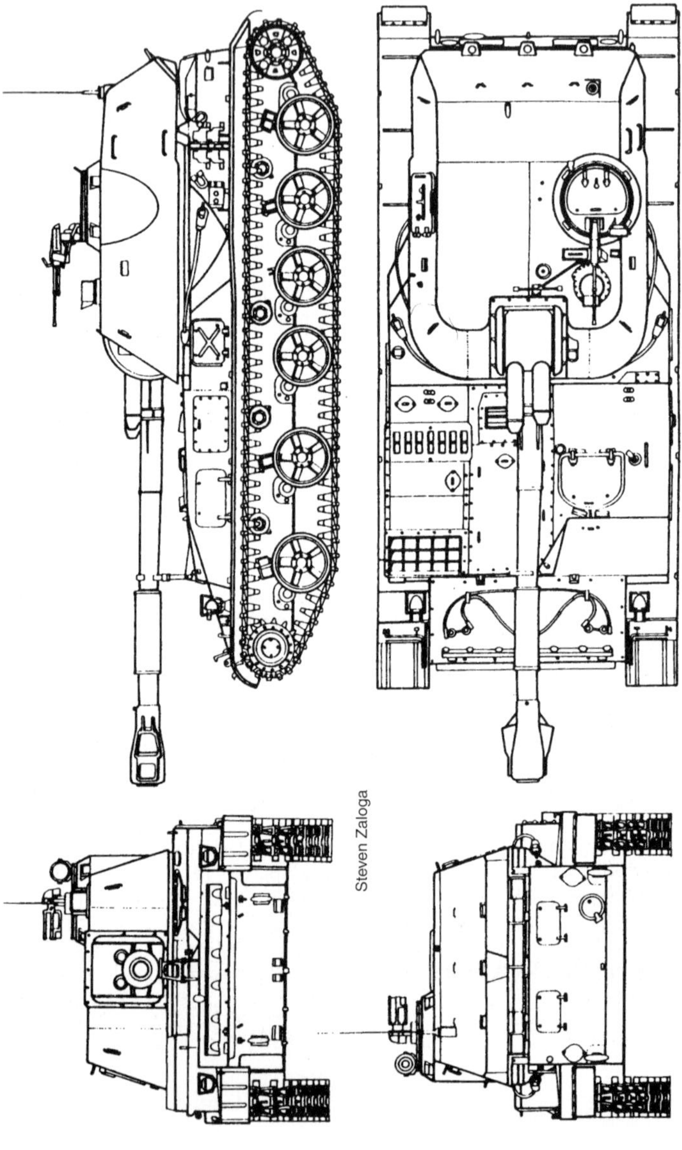

2S3 AAH

Steven Zaloga

2S3 AAH

2S3M AAH

2S4 TULPAN (SM-240) armoured mortar carrier

Wt 25 t L 7.5 W 3.25 H 3.2 382 kW 60 km/h 500 km Crew 5

Development and production.

1964 deployment starts.
1971 production start.

Distinguishing features. Chassis used similar to the GMZ tracked minelayer or SA-4 launch vehicle with six small roadwheels (distinctive gap between the three forward wheels), sprocket front and four return rollers. Small turret on left front part of the carrier. Engine in the front hull. The mortar is carried wholy on top of the vehicle. The superstructure allows the mortar to be embedded on its large baseplate at the rear of the vehicle when firing.

Other details. Armament consists of a 240 mm breech loaded mortar which can be elevated from +50 to +80 degrees with a traverse of 8 degrees to left and right. The traverse angle is 23 degrees to both sides. It fires a 136 kg mortar bomb at ranges of 800 m minimum to 9.51 km maximum with HE resp. 19.69 km with rocket boosted round. Types of ammunition are HE, HE-RAP, minelet round with 20 PFM-1 mines, a concrete-piercing round, chemical ammunition, and even nuclear ammunition. It can be remote-controlled via a small control box. The ammunition capacity is 20 bombs. One round per minute can be fired. Two of the crew members are transported by the ammunition replenishing vehicle. Emplacement time is 3-5 minutes. The carrier has a NBC system, it is non-amphibious.

Role. In service with the Czech Rep. (24), Russia (-400), Slovakia (12) and Ukraine.

Assessment. Simple but clumsy solution of a heavy self-propelled mortar carrier. When firing the crew has to operate outside the vehicle without armour protection.

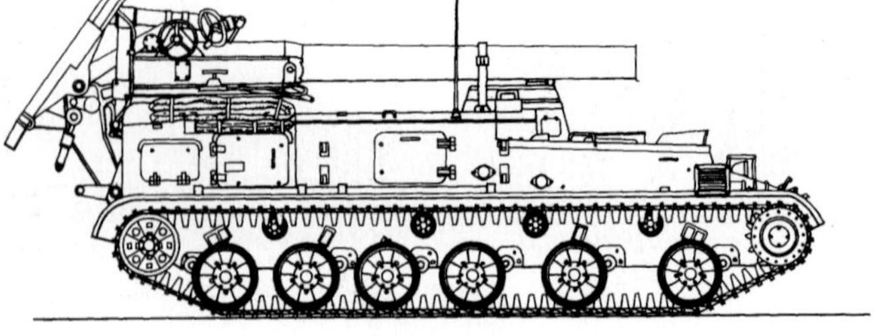

2S4 AMC

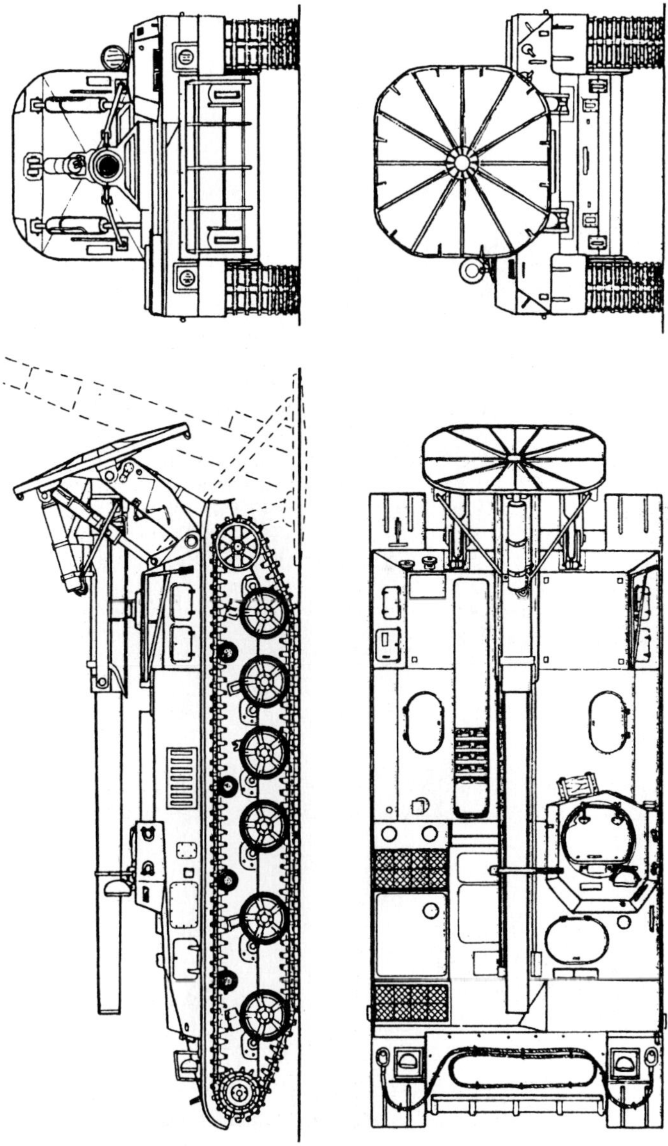

2S4 AMC

2S4 AMC

2S5 GIATSINT-S armoured artillery gun

Wt 28.2 t L 8.33 W 3.25 H 2.76 382 kW 60 km/h 500 km

Development and production.

1977 production start.

Distinguishing features. Side. Chassis as 2S3 AAH. Long barreled gun in unprotected mounting at rear, pepperpot muzzle brake. Large spade is lowered before firing. Front. Small turret located front left with vision blocks left front and left rear with search light and external ADMG. Driver's hatch to the right front of the cdr's turret, another hatch to the right rear.

Other details. Main armament is a 45 calibre 152 mm gun. Elevation is -3 to +65 degrees. It has a firing range of more than 27 km (RAP 37 km) and is nuclear capable. At the rear of the breech there is a hydraulically assisted loader and rammer, the rate of fire is four rounds per minute maximum. Four soldiers of the six-men crew travel in an ammunition carrier.
GIATSINT-B is the towed gun version.

Role. In service with Afghanistan (12), Finland (18) and Russia (approx. 2000).

Assessment. Similar capability as M110A2 AAH, crew unprotected, too.

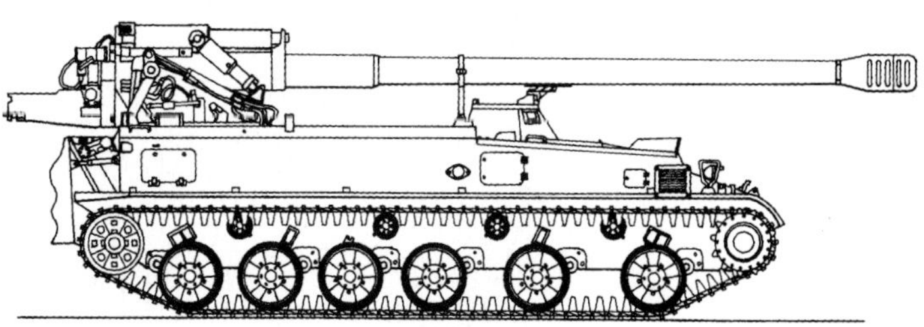

2S5 AAG

2S5 AAG

2S7 (SO-203) armoured artillery gun

Wt 40 t L 12.8 W 3.2 H 2.8 336 kW 50 km/h 500 km Crew 4

Development and production.

1975 first appearance.

Distinguishing features. Side. Chassis may be related to that of the SA-12 GLADIATOR ADG with seven roadwheels, five return rollers, front sprocket. Crew enclosed cab protruding in front above transmission. Very long gun in open mounting at the rear. Large spade at rear with ammunition handling system right. Front. Protruding front crew compartment, engine immediate rear.

Other details. The 2S7 is armed with a 203 mm gun with a power elevation of -3 to +65 degrees. After the barrel is unclamped the loader rotates the loading arm to the right. Then the rear idler wheels are lowered to the ground together with the rear spade. The loader operates the loading arm which clamps the projectile and positions it behind the vertically opened breech block. The projetile is rammed by a chain rammer. Maximum range of the gun is 30 km, highest rate of fire is two rounds per minute. The gun is nuclear capable. Four additional crew members are carried on assigned ammunition vehicle.

Role. In service with Belarus (48), Poland (8), Russia (80) and Ukraine (99).

Assessment. Big and clumsy chassis with crew and ordnance unprotected.

2S7 AAG

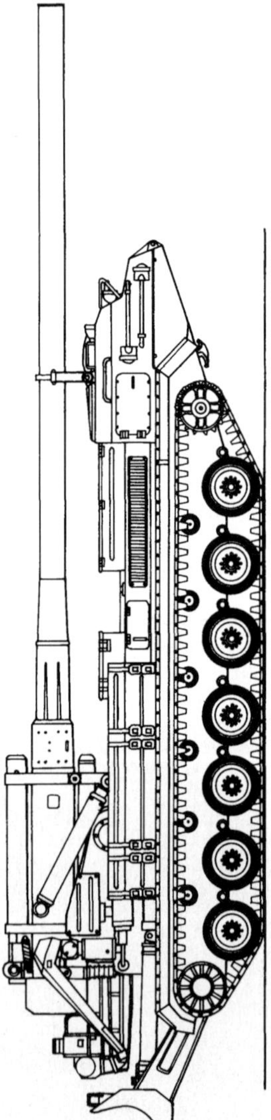

2S7 AAG

2S9 NONA-S (SO-120) armoured mortar carrier

Development and production.

1981 deployment of an airborne artillery assault vehicle.

Distinguishing features. Stretched version of the BMD airborne combat vehicle with six roadwheels and five return rollers, sprocket rear. It is lacking the front bow machine guns. A nearly all-circular two-man turret is located in the middle of the hull. It has a sloped front with a cylindrical mantlet. Rotating periscope on the left front turret roof. Driver front central as on BMD, commander left of him.

Other details. Main armament consists of a breech-loaded 120 mm mortar with an effective range of 1.7 to 8.8 km (12.8 km with rocket assisted bombs) at a rate of fire of 10 rounds per minute. It can be employed in the direct firing mode with HEAT projectiles (penetrating 600 mm of steel). When in the firing position the suspension is raised to provide a more stable platform. Gun elevation is -4 to +80 degrees, turret traverse is 35 degrees to each side.
(TABLE)

Role. In service with Russian airborne troops (700+) and in Afghanistan, Azerbaijan (29), Belarus (54), Kyrgystan (12), Moldova (9), Turkmenistan (12), Ukraine (64) and Uzbekistan (60).

Assessment. Consequential use of the BMD basic vehicle to provide the airborne troops with a light-weight artillery gun. Advanced design with crew operating under armour protection.

2S9 AMC

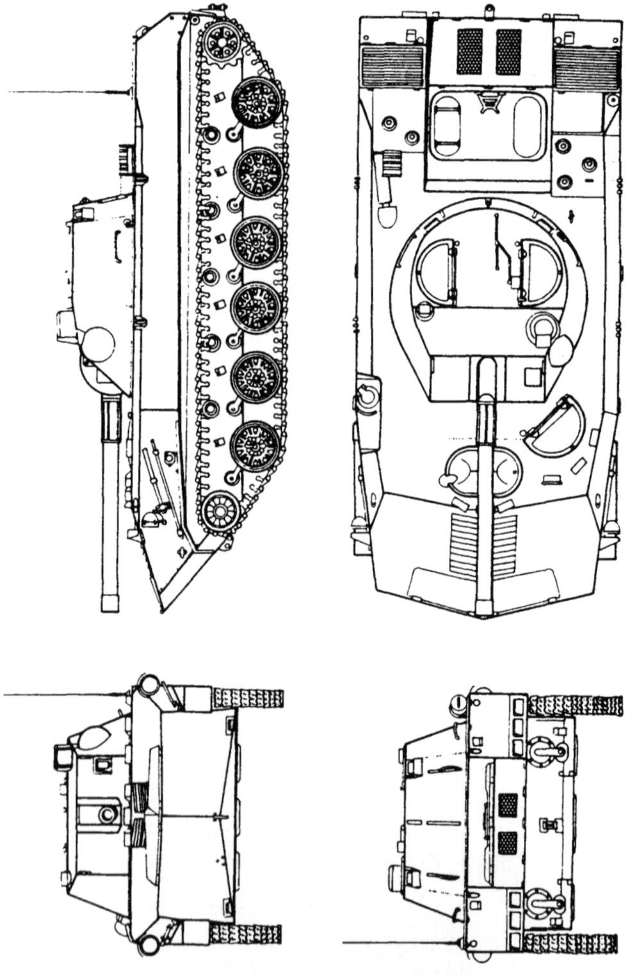

Steven Zaloga

2S9 AMC

2S19 (MSTA-S) armoured artillery howitzer

Development and production.

1989 first systems issued to the Soviet Army.

Distinguishing features. Hull and running gear are similar to the T-80 MBT having six roadwheels and five support rollers, drive sprocket rear. High box-like superstructure on the rear covering two thirds of with a long barrelled gun with a fume extractor and a muzzle brake.

Other details. Main armament is the 152 mm 2A65 MSTA-B howitzer with a range of 24.7 km (MV 810 m/sec) or about 36 km with RAP. The gun can be elevated from -3 to +68 degrees. Turret traverse is 360 degrees. Due to an automatic loading system a maximum rate of fire of 8 rpm can be achieved. 50 rounds are carried inside the vehicle. Laying-in elevation is automatic and laying-in traverse is semi-automatic. The howitzer can be reloaded during firing through the turret rear by means of a loading arm.
The light armoured turret provides protection against small arms fire and splinters. The gunner sits in the front left provided with a rotating surveillance periscope and the main periscope with its exit just left of the mantlet. The cdr on the front right position has a small cupola with 12.7 mm NSVT ADMG which can be remote-controlled from inside the turret. Two more roof hatches are situated behind the cdr's cupola.
A 16 kW autonomous gas turbine auxiliary power unit in the turret provides power when the main engine is shut off. An NBC system is fitted, too.
(TABLE)

Role. Introduced into Belarus (15) and Russian service (approx. 300) and also offered for export.

Assessment. Most advanced Soviet artillery system of high range and excellent manoeuvrability.

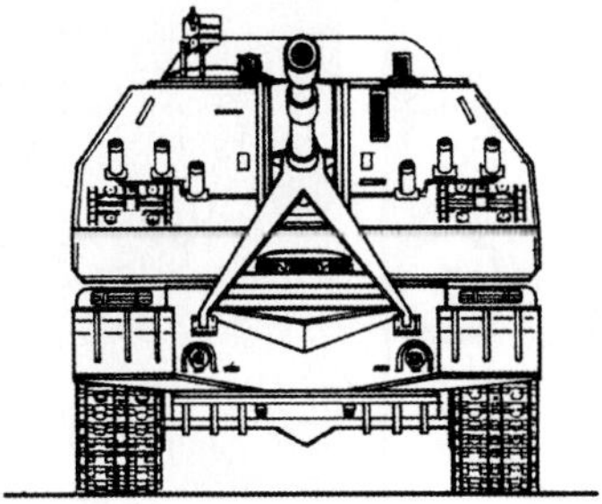

2S19 AAH

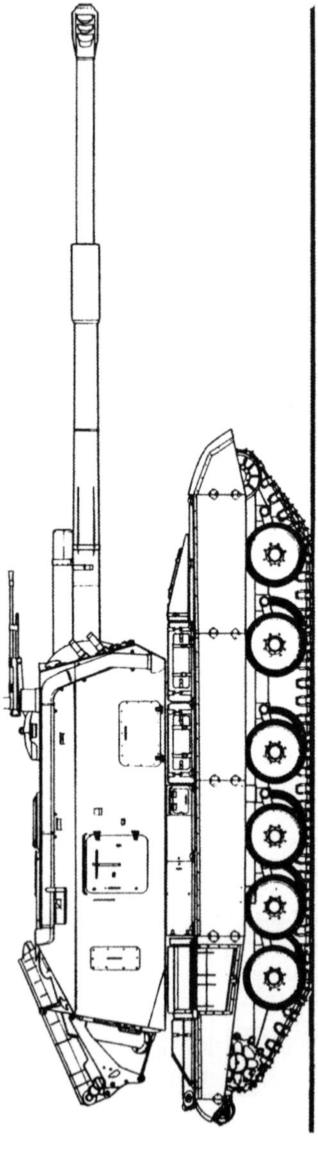

2S7 AAG

ZSU-23-4 SHILKA armoured air defence gun
Zenitnaya Samokhodnaya Ustanovka

Models.

ZSU-23-4
ZSU-23-4V
ZSU-23-4V1
ZSU-23-4M

Development and production.

In the late 1950s development started by the Astrov design bureau using the ASU-85 chassis.
1965 series production starts.
1968 ZSU-23-4V appears to overcome problems with the electronics.
1972 issue of the ZSU-23-4V1.
1977 appearance of ZSU-23-4M. Production ceased in 1983.

Distinguishing features. Side. Six roadwheels, no return rollers, sprocket rear. Near-vertical glacis, horizontal hull top, sloped rear. Large square turret quadruple cannon arrangement in forward part and large dish-shaped radar antenna over turret rear. Front. Box-shaped superstructure with sponsons over tracks. Large driver's vision flap on front left. Turret same width as hull superstructure. Internal mantlet for cannons. Radar antenna dish on V-shaped mounting. Small commander's sub-turret.

Other details. Each pair of barrels is carried on a central mounting and can be quickly changed. Short flash eliminators. The quadruple AZP-23 mounting has an elevation arc of -5 to +85 degrees. Battle range approximately 1500 m. Both API and HEI projetiles have a muzzle velocity of 970 m/sec. Radar and armament fully stabilised. The radar system (NATO: GUN DISH) operates in the J band and has a range of 20 km in the search mode and about 8 km in the target tracking mode. An optical sight enables the use of the system in an ECM environment. Vehicle navigation system, fire-surpression and NBC collective protection systems.
ZSU-23-4 had small bins on the turret sides.
ZSU-23-4V is provided with large full-length bins.
ZSU-23-4V1 with larger turret ammunition panniers and improved ventilation system.
ZSU-23-4M with three instead of two access doors on left hull side. Armoured cover for mantlet. Digital computer. (TABLE)

Role. In service with Afghanistan (20), Algeria (110), Angola (20+), Bulgaria (48), Chad (4 ex-Libya), Cuba (36), Czech Rep. (66), Egypt (350), Hungary (20), India (75), Iran (100+), Iraq (200+ delivered), Jordan (16), Mozambique (25), North Korea (100+), Peru (36), Poland (100), Russia (3000), Slovakia (34), Somalia (5), Syria (300) and Yemen (40).

Assessment. First AADG with radar-controlled automatic cannon to enter service. High rate and density of fire, low effective range.

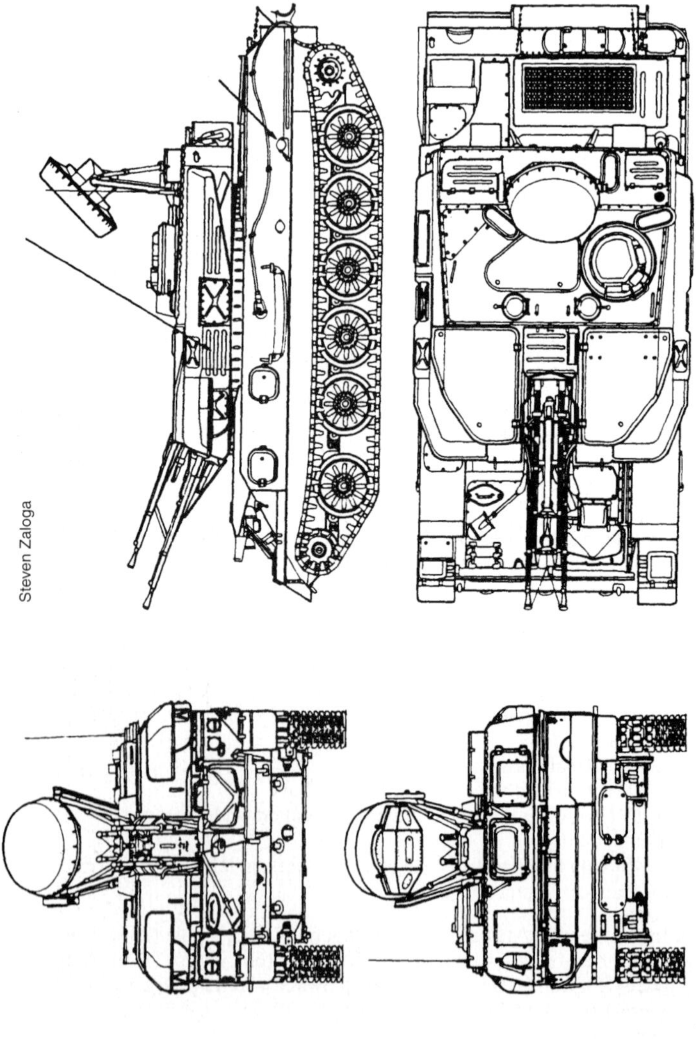

Steven Zaloga

ZSU-23-4V1 AADG

ZSU-23-4M AADG

2S6 TUNGUSKA armoured air defence gun

Models

2S6
2S6M

Wt 34 t L 7.84 W 3.47 H 3.89 382 kW 65 km/h 500 km Crew 4

Development and production.

1986 first identification.
1988 service enter in the Western Group of Soviet Forces.

Distinguishing features. Side. Six roadwheels, three return rollers, sprocket rear. Sloped front, inclined bottom, horizontal hull roof with sloped end, nearly vertical rear. Long high turret with twin cannon in external mounting on either side behind a launcher box of four missiles positioned in two pairs of cylindrical tubes. In azimuth and traverse movable tracking radar on turret front, rotating target acquisition radar above turret. Two periscopic sights on top of forward turret. Front. Sloped glacis with driver's hatch on the left side. High sponsoned hull. High small turret, spanning the whole vehicle's width due to missile launcher boxes on either side. Front mounted circular tracking radar, shell antenna of acquisition radar high above turret rear.

Other details. The 2S6M is armed with two twin 30 mm cannons on either side (ammunition stowage is about 1000 rounds). Eight 9M311 (SA-19 GRISON) missiles (four on both sides) have an effective range of max. 8 km at an altitude 15 - 3500 m. After the boost phase of the missile (900 m/sec) a pulsed light source in the missile's tail shows the flight position to the optical sight. The gunner has to keep the crosshairs of the optical sight on the target and the correction signals are transmitted by the tracking radar. The missile has an impact fuze and a laser proximity fuze with an activation range of 5 metres. The 9 kg warhead has a pre-fragmented casing. The effective range of the water-cooled and gas-operated 2A38M cannons is 4 km maximum at an altitude up to 3 km. The rate of fire can be varied from 1950 - 2500 rpm. The two radar systems (HOT SHOT) are either operated by the commander or the radar operator sitting on his left side consist of a J-band (10 - 20 HGz) tracking radar mounted in the front turret section and an E-band (2 - 3 HGz) search radar on the turret's rear. The surveillance range is 18 km and the tracking range 13 km. The commander is located in the forward right-hand turret section. He has the PUIM console above the search-radar screen to control the 1A26M computer and the OK1M console to operate the radar and the 1RL-138 IFF. The gunner is supplemented with the stabilized 1A29M optical sight, the PP operating console on his left for the missile engagement and the control of the NBC protection system and the PN console in front of him to select the operating modes of the optical sight and the firing if delegated by the commander. The chassis is derived from

the GT-S transporter, its suspension is of the hydro-pneumatic type which can be raised on movement and lowered in the firing position to provide a more stable platform. The tank is also equipped with an auxiliary engine of 50 kW.
(TABLE)
The initial version 2S6 had only four missiles.

Role. In service with the Russian Army and India (12).

Assessment. Thoroughgoing evaluation of western AADG led to this GEPARD copy with high abilities enhanced by the additional missile armament.

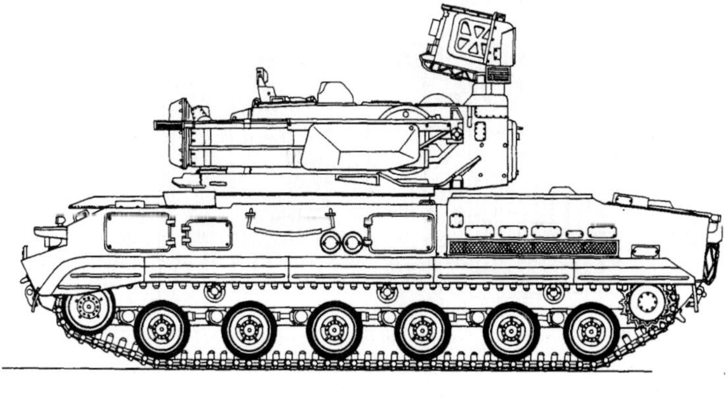

2SGM AADG/ML

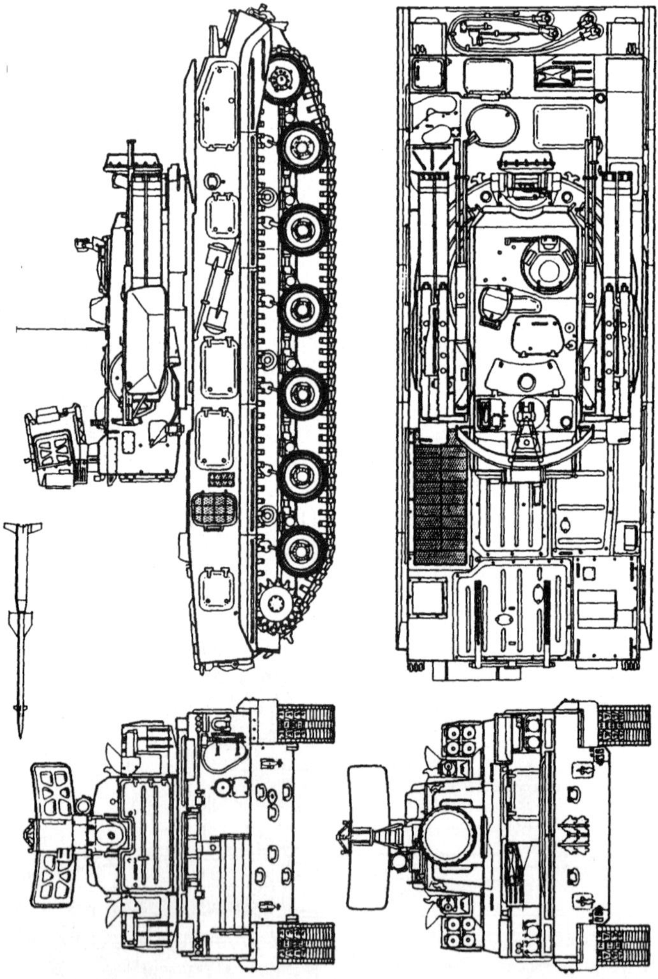

2S6M AADG/ML Steven Zaloga

2S6M AADG/ML

ZRK-SD 2K11 KRUG armoured air defence missile launcher
Zenitnyy Raketnyy Kompleks-Srednoye Deistviye
(NATO: SA-4 GANEF)

Models.

2P24 KRUG SPU ADML (SA-4A GANEF Mod 0)
2P24 KRUG SPU ADML (SA-4B GANEF Mod 1)
KRUG SSNR AADRV

Non-serial version.

2K22 TREUGOLNIK (NATO: SA-19 GRISON)

Development and production.

Development started in the mid 1950s on a medium/high altitude SAM system.
1964 first public appearance. First batch with sloped hull sides.
1965 series with vertical sides and reconfigurated front missile travel lock brace.
1973 improved system with 9M8M2 missiles.

Distinguishing features. Side. Chassis of heavy vehicle family with seven road wheels, four return rollers, front sprocket. Engine centre right. Traversing and elevating launcher system for two missiles mounted centrally on turret-ring type race. Sloped front, vertical rear. Front. Wide, rail-like manually operated travel lock. Small conical turret supporting launcher rail. Driver front left. Missile. Large cylindrical missile with four external jettisonable boosters. Conical warhead separated from motor by about one-third of missile body length. Cruciform tail unit with four large fins; four smaller stabilising fins mounted on body opposite front of motor at 45 degrees to tail fins.

Other details. Three launcher vehicles operate in conjunction with one radar carrier. Missile is guided to target by guidance beam of PAT HAND with semi-active terminal homing in final stage. Two missile configurations are operated side by side, the long-nosed 9M8, later the 9M8M or the 9M8M1 (SA-4a), length 8.8 m with a range from 9.3 to 72 km at maximum height of 27 km and the short-nosed 9M8M2 (SA-4b) with an improved close range performance from 1.1 km to 50 km at maximum height of 24 km, missile length 8.3 m. Fuselage diameter of both types is 0.9 m, the war-head weighs 135 kg, the missile weight is about 2.5 t.
(TABLE)
The radar carrier (NATO: PAT HAND) uses the same chassis. The continuous wave fire control and command-guidance radar works in the H-band (6.44 - 6.88 GHz) and has a range of 55 - 70 km.
Under development is reported to be a successor system called 2K22 TREUGOLNIK with the 9M311 missile. No further information is available at the moment.

Role. In service with Armenia, Azerbaijan, Bulgaria (27), Czech Rep. (27), Hungary (37) and Russia (500).

Assessment. Cumbersome, obsolete system. Vulnerable to ECM due to old tube technology and first generation guidance system.

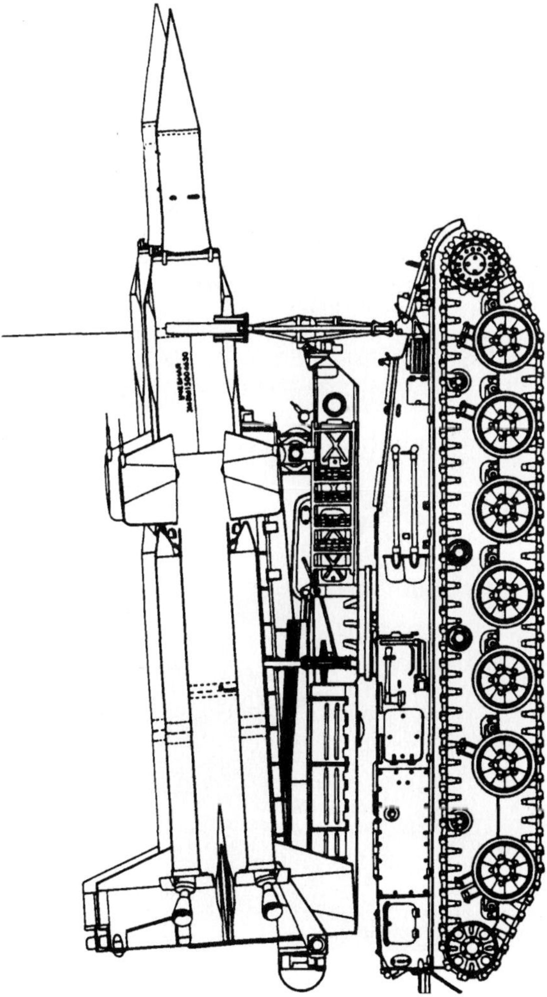

KRUG AADML (SA-4A)

KRUG AADML (SA-4A)

KRUG AADML (SA-4B)

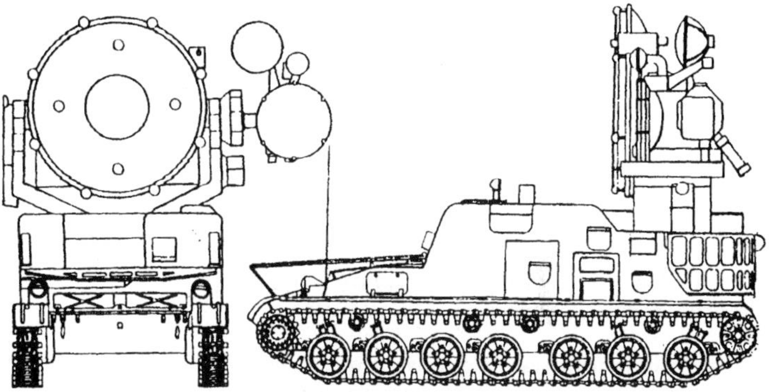

KRUG SSNR AADRV

KRUG SSNR AADRV

ZRK-SD KUB armoured air defence missile launcher
Zenitnyy Raketnyy Kompleks-Srednoye Deystviye
(NATO: SA-6 GAINFUL)

Models.

2P25 (KUB SPU) AADML
1S91 (KUB SSNR) AADRV

Development and production.

1959 development starts at Toropov's design bureau.
1967 first public appearance on Moscow Parade.

Distinguishing features. Side. Chassis of the light armoured vehicle family (PT-76) with six roadwheels, no return roller and rear sprocket. Armoured superstructure over whole length with vertical-sided sponsons with several hatches. The hull has a stepped roof, with a raised part containing pericopes above the driver and commander, then the lower platform for the launcher base, and then the higher engine deck. Front. Near-vertical glacis with two vision flaps. Large cylindrical dome at right front. Launcher system for three missiles mounted on rotating base occupies full width of vehicle.

Other details. Four launcher vehicles operate in conjunction with one radar vehicle. Up to three missiles are guided by monitoring the missile's G/H band (4 - 8 GHz) tail fin tracking beacons. The fire control computer in the radar vehicle generates course corrections (I-band) transmitted to the reference antenna on the lower left tail fin. The missile's terminal phase is flown in a semi-active homing mode of the reflected target's energy illuminated by the SSNR continuous wave setting.
The 9M9 missile has a length of 5.8 m and a body diameter of 0.335 m. The warhead weighs 80 kg, launch weight is 580 kg. The missile range is from 3.7 to 24 km at a height from 80 m to 12 km.
Starting in 1977 an improved missile 9M336 was used (NATO: SA-16B) on the chassis of the later SA-11 system with the Soviet designation KUB-M2.
The radar (NATO: STRAIGHT FLUSH) has a range of 60 to 90 km. The surveillance radar antenna and the higher circular tracking antenna are also mounted on a modified PT-76 chassis. Overall height is 5.16 m.

Role. In service with Algeria (40), Bulgaria (20), Cuba (12), Czech Rep. (80), Hungary (60), India (120), Iraq (180 delivered), Jordan (16), Libya (154), Poland (120), Romania (50), Russia (400), Somalia (12), Syria (216), Tanzania (12), Yemen (12), Yugoslavia (80).

Assessment. The effectiveness of this system depends much on the sophisticated fire control system which is not ECM proof. Success rate in conflicts were below 5 % hit probability at a 2 % kill rate.

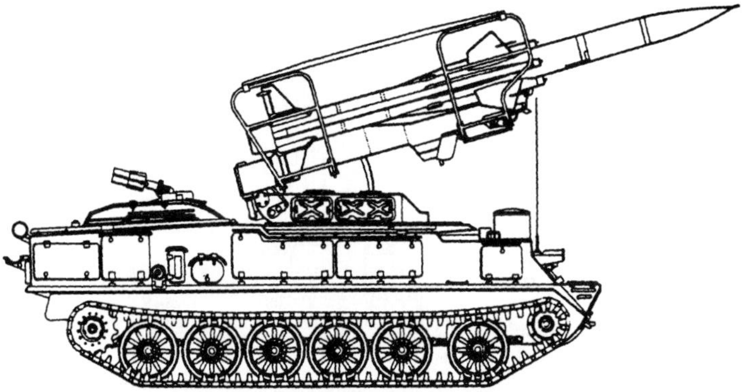

KUB SPU AADML Steven Zaloga

KUB SPU AADML

Steven Zaloga

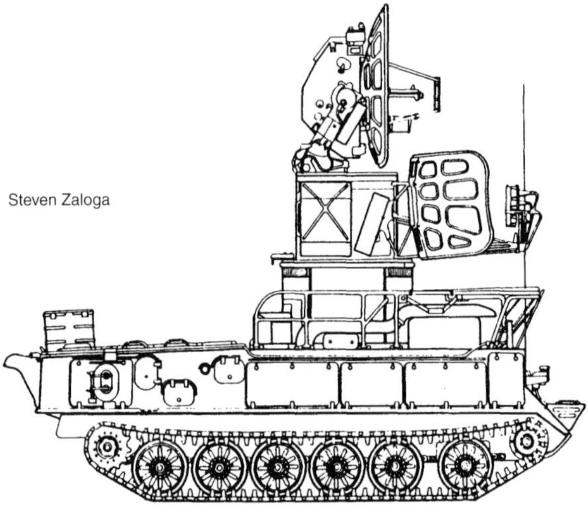

KUB SSNR AADRV

ZRK-BD STRELA-1 armoured air defence missile launcher (NATO: SA-9 GASKIN)

Models.

SA-9a (GASKIN Mod 0)
SA-9b (GASKIN Mod 1)
SA-9b with HAT BOX passive radar detection antenna

Development and production.

1968 system enters service.

Distinguishing features. Side. Hull of BRDM-2 AVR used (without chain-driven support wheels) with turret removed and replaced by new launcher turret carrying four missiles in long square container boxes in two pairs. The boxes are back-folded onto the hull rear during movement. Front. Controller mounted in the turret with a large window for optical target acquisition.

Other details. The original GASKIN version (SA-9a) used the 9M31 missile with uncooled IR seeker. It was followed by the 9M31M missile (SA-9b) with a cooled IR seeker to provide greater sensitivity. Normally a salvo of two missiles is fired per acquired target via an optical-mechanical sighting system.
The missile is 1.8 m long and has a diameter of 0.12 m. Its weight is 43 kg and it has a warhead of 7 kg with a proximity fuze. The minimum range of the 9M31 is 800 m the maximum range 6.5 km (560 m resp. 8 km with the 9M31M). Altitude limits are 15 m to 6.1 km.
(TABLE)
One vehicle in each battery is provided with HAT BOX, one on either side above the front wheel, one pointing forward under the left launching canister and one rearwards pointing above the engine deck.
Both variants have been spotted with four appendages located above the operator's window which are assessed to be IFF receivers.

Role. In service with Afghanistan, Algeria (46), Cuba (30), Czech Rep. (66), Hungary (50), India (200), Iraq (100+ delivered), Jordan, Libya (-60), Nicaragua (12), Poland (225), Russia (200 replaced by SA-13), Slovakia (34), Syria (350), Vietnam (100+), Yugoslavia (-100).

Assessment. A light, highly mobile low-level fair weather AD system. Heat-seeking missile susceptible to flares.

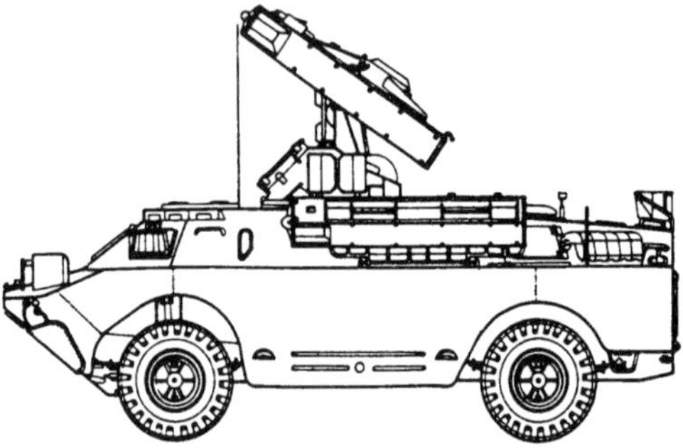

STRELA-1 AADML Steven Zaloga

STRELA-1 AADML

BUK-M1 GANG armoured air defence missile launcher (NATO: SA-11 GADFLY)

Models.

9A310M1 SOU launch vehicle
9A39M1 PZU launcher loader vehicle
9S18M1 SOT target acquisition radar vehicle
9S470M1 command and control vehicle

Non-serial version.

BUK-2M URAL (NATO: SA-17 GRIZZLY)

Development and production.

1977 stop-gap solution KUB-M2 (NATO: SA-6B) with 9M336 missiles. 1982 service enter to complement the SA-4 and SA-6 systems.

Distinguishing features. Side. Vehicle is based on the MT-S tractor with six small roadwheels, four return rollers, sprocket front. Snub-nosed front, horizontal hull top and vertical rear with bottom sloping inwards. Missile launcher at rear of large platform, large radar mounted in front of missiles. Front. Sloped glacis with two front hatches. Sponsoned hull with sloped sides. Prominent radar over full width of vehicle on top of hull.

Other details. One BUK-M1 system consists of one CP and one radar vehicle, six launchers and up to three launcher-loaders. Six targets can be engaged at a time.
The launch vehicle has four 9M38M1 missiles in the ready-to-launch position. It is capable to engage targets between the range limits of 3 to 35 km up to a maximum height of 22 km. The single-stage, solid-fuel Mach 3 semi-active homing missile has a launch weight of 690 kg with a warhead of 70 kg. It has a length of 5.55 m, a span of 1.2 and a diameter of 0.4 m. The vehicle carries an H/I-band engagement radar (NATO: FIRE DOME). It operates at 6900 MHz resp. 7700 MHz in the tracking mode. It is cued to incoming targets by the CP vehicle. The effective guidance range is 3 - 32 km at an effective altitude between 15 m and 22 km. Up to three missiles against a single target can be guided. The destruction zone is 17 m. In the electro-optical firing mode the missile uses radio-command guidance. The vehicle's weight is 32.34 t, its length 9.3 m, the height is 3.8 m in travel configuration and 7.72 m in operation.
The radar vehicle carries the 9S18M1 early warning and acquisition 3D radar (NATO: SNOW DRIFT) with a range of 100 km (ca. 35 km at target flight altitude of 100 m, 10 km against helicopters at an altitude of 30 m) at maximum height of 23 km operating at 3800 MHz. It has replaced TUBE ARM radars system initially used. The weight of the vehicle is 35 t, its travel height 3.25 m, in operation the vehicle is 8.02 m high.
The launcher-loader vehicle carries eight missiles. If necessary it can launch a

missile on command from the CP vehicle. It has the same dimensions as the launcher but is slightly heavier (35 t).
The battery command vehicle monitors the air situation and combat operation of the other systems. It is equipped with a computerized control and data display console, a secure communication and data transmission and a land navigation system. Up to 75 targets can be monitored and 15 of these can be followed automatically. Vehicle's weight is 29.87 t, travel height is 2.975 m, with radio mast erected at full height 8.785 m.
The GM-569 chassis is almost identical to that of the 2S6 system.
Under development is the BUK-2M URAL as an evolutionary successor model. It consists of a similar launcher vehicle, a command and control vehicle, a radar carrier with the "CHAIR BACK" system mounted on a telescopic arm and a loader/launcher vehicle. The 9M38M2 missile has a longer fuselage and shorter fins. It uses a more energetic solid-propellant rocket engine increasing acceleration speed to approx. 1200 m/s. The target acquisition radar on the launcher vehicle has a detection range of about 120 km. The surveillance radar has a maximum range of 160 km. The command centre is able to engage up to 36 targets simultaneously.

Role. In service with Finland (18), India (40), Poland (20), Russia (250), Syria (20) and Yugoslavia.

Assessment. More advanced low to medium surface-to-air missile system compensating SA-6' restriction not being servicable without direct link to remote radar guidance.

BUK-H1 AADML

BUK-M1 (9A39M1 PZU)

BUK-M1 (9S470 M1)

BUK-M1 (9S18M1 SOT)

ZRK-BD STRELA 10 armoured air defence missile launcher (NATO: SA-13 GOPHER)

Models.

9A34 STRELA 10
9A35 STRELA 10
PLRK STRELA 10M (Czech Republic)
9A34MZ STRELA 10MZ
9A35MZ STRELA 10MZ

Development and production.

1975 first appearance, initially also with SA-9 missiles.

Distinguishing features. Hull of MT-LB AGPC. Launch installation and missile containers similar to SA-9 GASKIN but larger. Radar between the two pairs of launcher ramps. When travelling they lay horizontally on top of hull.

Other details. The 9M37 missile has a weight of 55 kg with a warhead of 4 kg. It is 2.2 m long and has a diameter of 0.4 m. It is fitted with a cryogenically cooled passive infra-red seeker operating in two frequency bands. Range of the 1.5 mach missile is from 500 m to 8 km with an effective engagement altitude to about 3.5 km. The radar system is only for ranging purposes. Target is acquired via an optical-mechanical sighting system.
STRELA 10M is a Czech upgrade program provided with an integrated command and control system (notebook), the improved radar system 9S86 (NATO: SNAP SHOT), data processing electronics and an acoustic threat warning system.
STRELA 10MZ (sometimes called STRELA 10M3) operates the improved 9M37M missile provided with a three channel warhead (photo-contrast, infrared, interference channels) which can be chosen prior to launch. Its operational range is 5 km, the altitude level is 25 - 3000 m. (TABLE)
One vehicle (9A35 resp. 9A35MZ) out of four (forming one sub-unit) is equipped with four radio direction-finder antenna units (NATO: HAT BOX), one between the two front vision ports, one on either corner of the vehicle's rear deck and one facing backwards. This systems detects radar bomb sights, airborne radars and other emitters.
New modifications include missile resupply boxes on both sides of the hull carrying two reload missiles each.

Role. In service with Algeria (32), Bulgaria (20); Cuba (40), Czech Rep. (100), Hungary (12), Iraq (60+ delivered), Jordan (20), Libya (60+), Poland (60), Russia (350) and Syria (60).

Assessment. AADML of high mobility capable to follow mechanised formations on the battlefield. Low resistance against IR counter-detection measures.

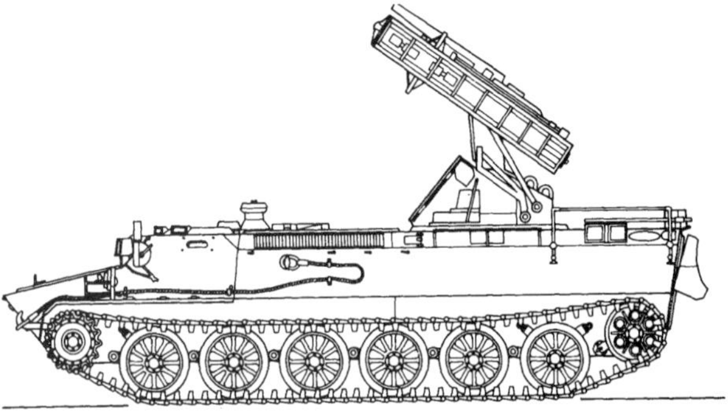

STRELA 10 AADML

STRELA10 AADML

STRELA 10MZ AADML

9A331 TOR-M1 armoured air defence missile launcher (NATO: SA-15 GAUNTLET)

Development and production.

1990 first public display at a defence exhibition in Manila.

Distinguishing features. GM-355 chassis almost identical with the 2S6 AADG with six roadwheels and three support rollers but with slightly raised center hull section. A long launching section with inclined top covers two third of vehicle's length with a square radar antenna in the front part and a rotating radar antenna mounted on the rear of the turret.

Other details. The TOR represents the first air defence system to use vertically launched missiles. The 9M331 missile has a single-stage solid propellant motor giving a maximum speed up to 850 m/sec. It carries a 15 kg HE fragmentation warhead with proximity fuze. Launch weight is 167 kg, the missiles length is about 2.9 m at a diameter of 0.35 m (calibre 235 mm). It is cold-launched to a height of 18 to 20 m. Then thruster jets turn the missile towards the target and the solid motor is ignited. Range is from 1 km to 12 km with altitude limits from 10 m to 6000 m. The missile can manoeuvre up to 30 G. The warhead has a weight of 15 kg and contains high density alloy fragments. Eight missiles in ready-launch containers are carried.
The 3D pulse Doppler surveillance radar operates in the C-band and provides range, azimuth, elevation and automatic threat information for up to 48 targets to a maximum range of 25 km. It also initiates the tracking of up to 10 targets. On the front of the turret a phased-array pulse Doppler tracking radar is mounted. It operates in the K-band and is capable of simultaneously tracking and initiating attacks on two targets travelling at a maximum speed of up to 700 km/h. A small dome-type antenna on the top left of the tracking radar may capture the missile in its launch phase. An autonomous TV system ensures operation in battlefield clutter and ECM environment. Its range is 20 km.
The initial TOR version used a different radar array than the digital systems of the recent TOR-M1.

Role. Introduced into Soviet service use (100+), delivered to China (14) Greece (21 plus 19 more on order) and offered for export.

Assessment. Very advanced launch concept and swift reaction time characterise this system. Homing-in capabilities are still unknown.

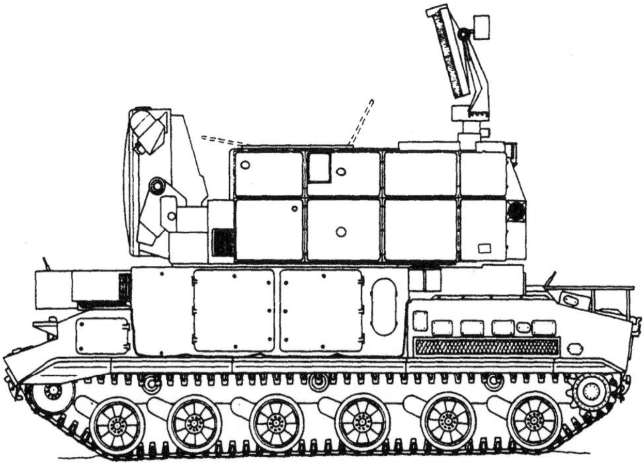

TOR-M1 AADML

IRM armoured engineer reconnaissance vehicle
Inzhenernaya Razvedyvatel'naya Mashina

Wt 17.2 t L 8.22 W 3.15 H 2.4 220 kW 52 km/h 500 km Crew 6

Development and production.

1984 first appearance of IRM.

Distinguishing features. Side. Vehicle uses the BMP suspension but with seven roadwheels, four return rollers, sprocket rear. Wedge-shaped front bow, flat hull roof with rear engine compartment sloping downwards. Two propulsive units on the hull rear with cowled propellers. Small box-like turret well forward. Front. Large splashboard covering whole width of upper front. Small turret central, retractable periscope on the front right. Sloped sides.

Other details. Two RShM-2 mine-detecting sensors on long beams are folded rearward attached on both sides of forward hull which can be extended forward. This system detects ferromagnetic materials at depths of up to 0.3 m and activates an autostop system which halts the tank. The tank also carries an echo sounder for measuring underwater ground profile, a clinometer using an artificial horizon, a land navigation system and a PR-1 penetrometer. Portable engineer instruments include hand-held mine detectors, a tool for measuring the thickness of ice, a DSP-30 rangefinder, and a PIR periscope. Crew protection is ensured by automatic NBC and fire retardant systems. It also carries a smoke generating device. Apart from its amphibious capability (water speed 10 km/h) the vehicle can also be employed submerged using a erectable snorkel. To extract the vehicle when it gets stuck in muddy terrain it can be equipped 9M39 solid-propellant rocket engines.

Role. In service with the CIS and Ukrainian Armies.

Assessment. Sophisticated system to speed up river crossing and sealanding operations.

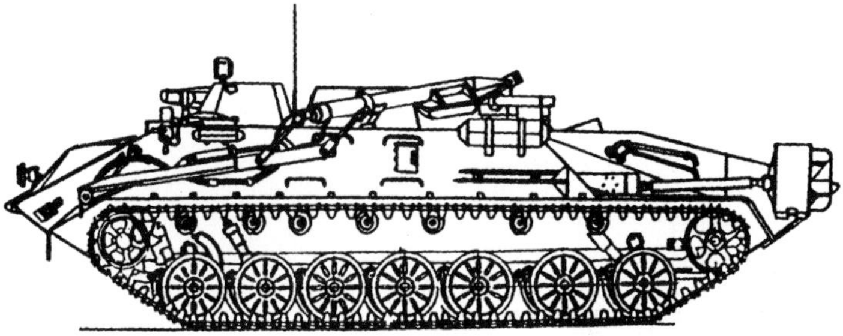

IRM AERV

IRM AERV

IMR armoured engineer vehicle
Inzhernaya Mashina Razgrazheniya

Models.

IMR
IMR-2
IMR-2M1
IMR-2M2

Development and production.

1969 service entry of the IMR in the Soviet Army.
1980 introduction of the IMR-2.
1987 IMR-2M1 enters service.

Distinguishing features. Side. T-55 resp. T-72 (IMR-2) hull with turret replaced by telescopic jib crane retracted when not in use and traversed to rear lock position. Dozer blade front mounted. Bucket for crane stowed on left side of chassis, cradle at hull rear. Front. Crane turret central. Crane normally fitted with pair of pincer-type crabs or buckets.

Other details. Crane turret can be traversed through 360 degrees. The front dozer can either be used in a V-configuration or as a straight dozer blade. Both the turret and the dozer blade are hydraulically operated. In travelling position the dozer is folded back on the front hull section. All versions are operated by a crew of two.
The IMR has a total weight of 37.5 t.
The initial IMR-2 was also equipped with mine clearing extended charges.
The IRM-2M1 did not retain these charges and had better armour protection for the hydraulic equipment. The boom mechanism is supplemented with a scraper-ripper. The dozer V.-blade width is 3.38 m.
The IRM-2M2 again has mine clearing extended charges in two boxes on top of the rear track guards. Using the dozer the vehicle develops a speed of 8 - 12 km/h. The boom load-carrying capacity with 8.4 m extension is 2 tons.
(TABLE)

Role. In service with the CIS armies, Yugoslavia and other former WARPAC countries.

Assessment. Versatile vehicle for obstacle clearance and for crane work.

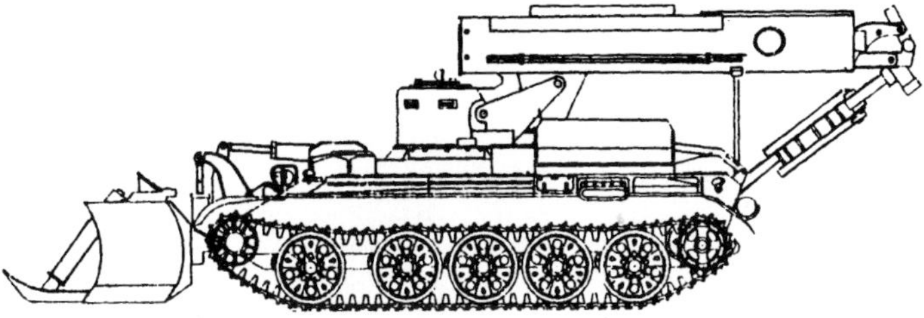

IMR AEV

IMR-2 AEV

IMR-2M1 AEV

IMR-2M2 AEV

MTU armoured vehicle launched bridge
Mostoukladchik Tankovyy Ustroystvo

Models.

MTU-20
MT-55 (Czech model)
MT-55A (MTU-55)
BLG-60M2 (formerly GDR)
BLG-67 (Bulgaria)

Development and production.

1957 introduction of MTU (later called MTU-12).
1960 joint development by GDR (bridge and bridge laying mechanism) and Poland (chassis) leading to BLG-60M2.
1967 issue of MTU-20.
1970 introduction of MT-55A (MTU-55) into Soviet service several years after its Czech forerunner MT-55.

Distinguishing features. All AVLB are based on turretless T-54/55 MBT chassis. MTU-12 had a short 12 m one-piece bridge with distinctive lattice work on the sides.
MTU-20 has bulky bridge with its ends folded back so they rest on top of the bridge ends. MT-55 and MTU-55 are fitted with scissors type bridges.

Other details. The MTU-20 bridge is 20 m long and 3.3 m wide. Before launch both folded ends are opened out. Then the bridge is also moved forward like the MTU-1 model. This bridge can carry vehicles up to 60 tons.
The MT-55 has a scissor type bridge launched hydraulically over front of tank while a launching girder foot is lowered to the ground. (TABLE)
The Soviet MT-55A is almost identical to the Czech version. It has a weight of 36.5 t, total length of the vehicle is 10.5 m.
BLG-60M2 has a smooth non-slip covering on bridge deck, no stiffeners on the girder webs, bridge end closed. It takes two or three minutes to lay and up to six minutes to recover from either end. A snorkel can be fitted for deep fording operations. This version has been derived from the BLG-60 and has widened bridge tracks and increased hydraulic power. The bridge has a span of 21.6 m. BLG-67 is almost identical.

Role. In service with Afghanistan, Bulgaria, Czech Rep. (MT-55), Egypt, Finland, India (MTU-20), Israel, Iraq (7 BLG-60 delivered), Poland, Russia (563 MTU-20, 568 MT-55A), Syria, Ukraine, Yugoslavia (MT-55) and some other countries.

Assessment. Scissor type bridge very conspicuous when raised to full length prior to laying.

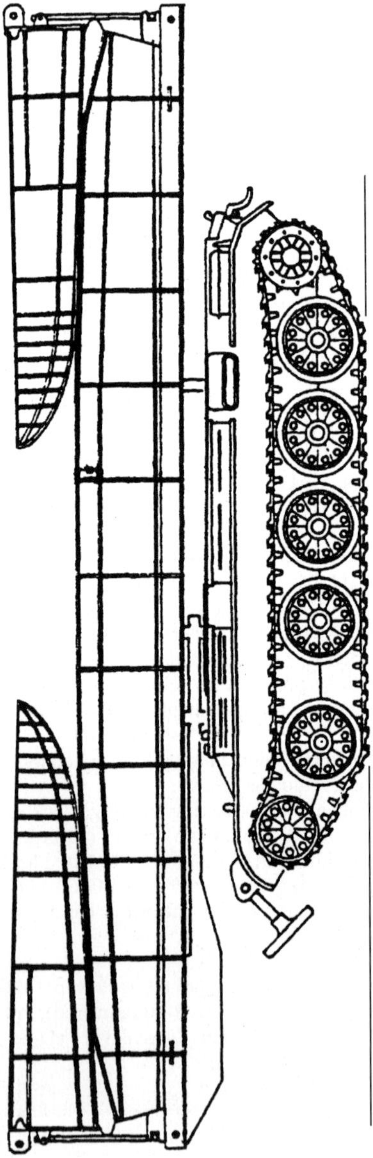

MTU-20 AVLB

MTU-20 AVLB

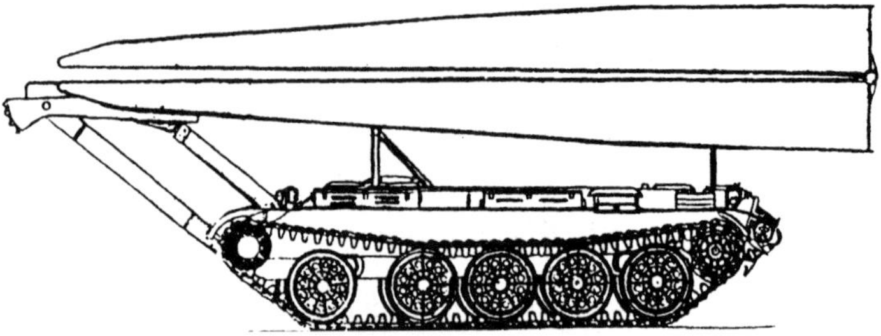

MT-55A AVLB

MT-55A AVLB

BLG-60M2 AVLB

MTU-72 armoured vehicle launched bridge

Development and production.

1974 service introduction and start of low-scale production.

Distinguishing features. Huge bridge on T-72 chassis with bridge ends folded up so they rest on top of the bridge ends.

Other details. The bridge is of aluminium alloy reducing the weight to 6.4 tons. A hydraulically operated stabilizer blade is mounted at the front under the hull nose. The folded bridge ends are unfolded and locked prior to launching. A hydro-mechanical drive mechanism then moves the extended bridge over the cantilever launching girder. The time for laying is 3 minutes. The bridge can be recovered from either side.
(TABLE)
The Polish version is called PMCz-94 and is now being converted to the PT-91 chassis configuration.

Role. In service with Poland and Russia.

Assessment. Up-to-date bridging system on a highly mobile chassis.

MTU-72 AVLB

GMZ armoured engineer minelaying vehicle
Gusenichnyy Minnyy Zagraditel'

Wt 28.5 t L 10.3 W 3.25 H 2.7 294 kW 60 km/h 500 km Crew 3

Development and production.

Development started in the early 1960s as a replacement for the PMR trailer systems.

Distinguishing features. Side. The tank uses hull and suspension of the SA-4 GANEF ADML system but with slightly elongated hull. Seven roadwheels and four return rollers, sprocket front. Distinctive gap between third and fourth roadwheel station. Higher superstructure with aft entrenching mechanism raised when in road movement. Front. Sloped glacis with driver left and engine right. Cdr's cupola behind driver. Square, high superstructure with rear compartment. Operator's cupola on right rear.

Other details. Work speed of the GMZ is 4 to 10 km/h resp. 3 km/h when mines are buried. Minelaying rate is eight surface or four burried mines. The tank carries 208 anti-tank mines. The mines are placed in spaces of 4 to 5.5 metres thus enabling the crew to lay a minefield of 1 km in 15 - 20 minutes. The tank has two large roof hatches for reloading (time 12 - 15 minutes). Armament of the tank consists of a 7.62 mm MG externally mounted on the cdr's cupola.
The GMZ-2 has a modified entrenching system.
The GMZ-3 is modified to lay TM-57, TM-62M and TM-62P3 anti-tank mines. This version also features a front entrenching blade.

Role. In service with Syria and the CIS countries.

Assessment. Minelaying vehicle of high service use, employment favoured by armour protection.

GMZ-2 AMLV

GMZ-2 AMLV

GMZ-3 AMLV

UR-77 METEORIT armoured engineer mine clearing vehicle
Ustanovka Razminirovainiya

Wt 15,5 t L 7.26 W 2.85 H 2.1 221 kW 60 km/h 500 km Crew 2

Development and production.

1970 introduction of UR-67.
1975 service entry of UR-77.

Distinguishing features. UR-67 (also called MTK) used chassis and hull of BTR-50PK APC. It had an elevatable rocket firing device on rear deck. This version was phased out.

UR-77 (also called MTK-2) is mounted on the 2S1 AAH chassis. In place of the removed gun turret there is a cylindrical flat cabin housing the commander and the operator. The erectable rocket launcher is behind this superstructure attached to the vehicle's rear. Square rear door with rounded ends.

Other details. Both MTK versions fire flexible tubes filled with explosives attached to rockets. After reaching the ground the line charge is then detonated, creating a cleared path.
With the UR-77 rocket system two shots can be fired simultaneously. The flexible tubes have a length of either 90 m (double tube) resp. 180 m. A path of 3 metres width can be cleared. The firing range is 200 - 500 m.

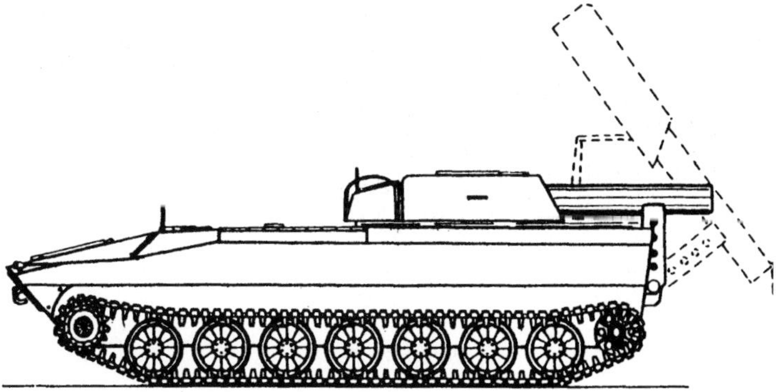

UR-77 AEMCV

UR-77 AEMCV

T-54/T-55 armoured recovery vehicle

Models.

BTS-1 (Bronetankovyy Tyagach Sredniy) (T-54-T)
BTS-2
WZT-1 (Czech Rep., Poland)
SPK-12G heavy lift crane vehicle
BTS-3 (WPT-2, Poland)
BTS-4
T-55-TK (WZT-2, Poland)
MT-55 (Czech Rep./Slovakia)
T-55 ARV (Iraq)

Non-serial versions.

KAM-1 Finnish conversion with a large central turntable mounted crane.
KAM-2 advanced Finnish model.

Development and production.

After the introduction of T-54 MBT construction work started in the late 1950s for the development of ARV based on the MBT chassis. Later the T-55 chassis was used, too. Modified ARV had been developed by the CSFR and GDR.

Distinguishing features. T-54/T-55 MBT chassis used with turret removed and recovery equipment installed.

Other details. Automotive details refer to the MBT chassis.
BTS-1 was the first version simply a MBT chassis with its turret removed.
The BTS-2 has a winch in lieu of the turret and above it is a small open container for tools and repair equipment. A small tripod jib crane is also carried. At the ARV's rear there is a large entrenching spade. The WZT-1 is a modified version of it.
The SPK-12G uses an unprotected operator's cabin with a 5 t crane.
The BTS-3 has a large 20-tons hydraulic crane on the left front of the hull. Other equipment include BTU-55 dozer blade, rear-mounted spade, winch and other specialised equipment. The WPT-2 was a similar Polish development, the Czech equivalent is the WZT-2.
(TABLE)
The BTS-4 is a light crane vehicle with the addition of a 1.5 t crane on the right rear. It also has a winch, rear-mounted spade and extensive external stowage.
MT-55, built in the CSFR, has a hydraulic 1.5 t crane on right side of hull, stowage platform, large spade at rear, main winch with a capacity of 44 t and an auxiliary winch of 8 t and special equipment. Mine-clearing rollers can be fitted at hull front and a deep wading snorkel.
Most of the ARV are armed with a 12.7 mm ADMG.

Iraqi conversion of T-55 hull with raised superstructure, dozer blade and turntable mounted crane on left front side.

Role. In service with almost any country using T-54/T-55 MBT, also in Turkey (150 T-54T ex-NVA stock).

Assessment. Early models are only useful for towing purposes. Only those fitted with high capacity cranes and main winches exceeding 20 t can be estimated as up-to-date ARV.

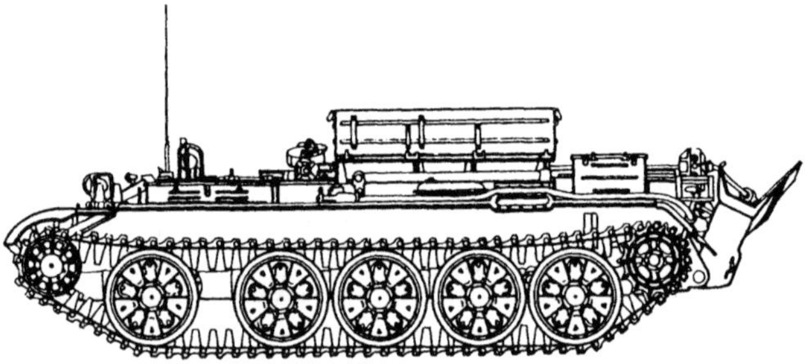

BTS-2 ARV

T-54 (B) ARV

T-55-TK ARV

MT-55 ARV

T-55 ARV (Iraq)

BREM-1 armoured maintenance and recovery vehicle
Bronirovannaya Remontno-Evakuatsionnaya Mashina

Models.

BREM-1
M-84 ARV (Yugoslavia)
WZT-3 (Poland)
VT-72 (CSFR; GDR: T-72 TK)
VT-72B (Czech Rep./Slovakia)

Non-serial version.

IWT-72 AEV (Poland)

Development and production.

1975 start of development.
1984 introduction into Soviet Army.

Distinguishing features. Side. Chassis and suspension of T-72 MBT. Front dozer blade. Crane pivoted at front left side of hull traversed to rear when travelling with jib in horizontal position. Large rectangular stowage box on roof towards rear. Front. Small dozer blade with front mounted winch behind it. Driver at front right with commander on his left.

Other details. The hydraulically operated crane with a telescopic jib has three lifting capacities: 3, 12 and 19 t. At 3 t the jib can be extended up to 4.4 m and traversed through 360 degrees. At 12 t the jib length is 2.97 m maximum with a slewing angle of 45 degrees. No extension is possible when the maximum capacity is used. The tank has two towing rods of 1.68 m length and two ropes of 5.5 m. The front dozer is 3.1 m wide and is mainly used to stabilise the vehicle's recovery operations. Inside the platform (1.7 x 1.4 m) a maximum load of 1.5 t can be carried. In addition to other tools carried on the BREM-1 there is a hydraulic jack with a capacity of 30 t and electric welding and cutting devices.
(TABLE)
The WZT-3 (Woz zabezpieczenia technicznego) offers a higher performance than the BREM-1 model. It has a forward operating winch with a 29.5 mm cable at 200 m length 300 kN). An auxiliary winch inside the tank between cdr's and driver's seats has a 400 m cable and a pull of 20 kN. Vehicle's height is 2.17 m, its weight is 41.5 tons. Upgrade program including installation of the S12-U 625 kW engine of the PT-91 in on the way.
The VT-72 (Vyprostovaci tank) resembles the WZT-3 but has a 19 t crane mounted on the right side instead of the left. Total weight is 46.5 t, the height is 2.64 m. The main winch can pull 300 kN, rope length is 200 m. The VT-72B is based on the T-72B chassis with V-84-1 engine (618 kW).

Role. In service with Belorussian, Russian and Ukrainian T-64/T-72/T-80 tank units, in the Czech Rep., India (35 VT-72B from Slovakia, more in license production), Poland and in Kuwait (15 M-84 ARV) and in Slovakia.

Assessment. First ARV of modern standard not only for towing purposes but also with crane capacity to lift power-packs.

BREM-1 ARV

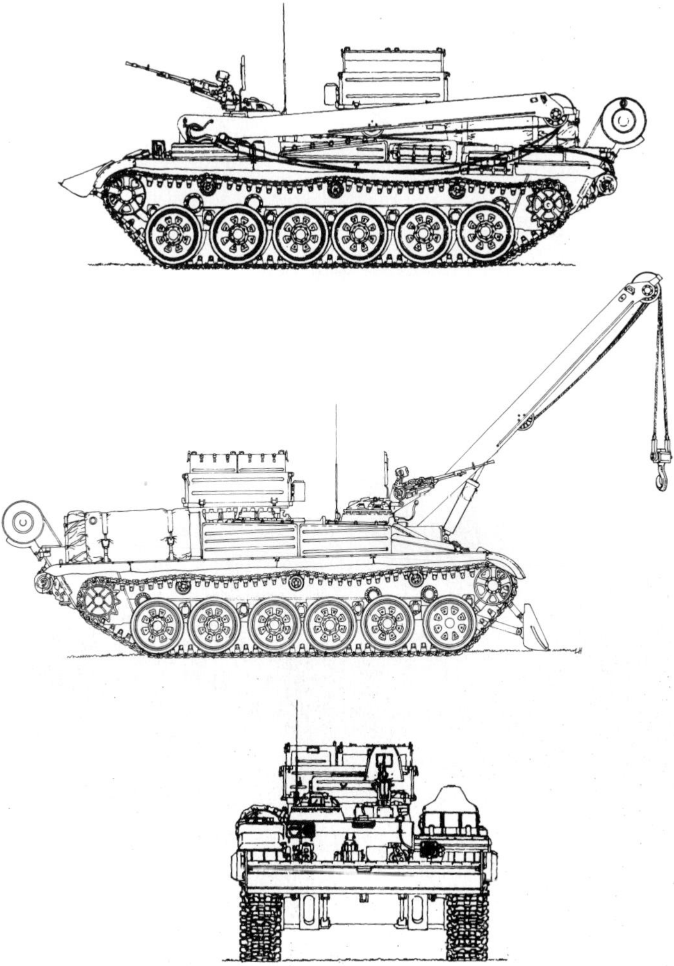

BREM-1 ARV

WZT-3 ARV

VT-72 ARV

AL FAHD AF-40 amoured general purpose carrier family

L 7.9 W 2.94 H 2.065 (roof) 90 km/h 800 resp. 600 km range

Models.

AF-40-1 APC and ACCC
AF-40-2 AVR

Development and production.

1977 technical studies started with different 6x6 vehicle layouts.
1981 first prototype issued.
1988 second prototype built with Cockerill LCTS 90 mm gun turret.
1997 first production AF-40-8 (8x8) issued by Abdallah Al Faris.

Distinguishing features. Side. 8x8 vehicle with flat superstructure, pointed hull bow and vertical rear wall. The wheelbase between second and third station is slightly longer. Front. Boat-shaped sponsoned hull with inclined upper and lower sections. Sloped front bow with folded-down trim vane and two crew stations protected with hatches. Small hatch on the right side underneath the hull bow protects winch.

Other details. The amphibious vehicle features a McPherson hydro pneumatic suspension adjustable to 450 mm on both sides (minimum roof height 1.81 m, maximum 2.26 m improving obstacle crossing. All wheels are steerable, when cruising on the road normally the four front wheels are steered. The steering radius is 7.5 metres. The wheelbase is 1.42 m, 1.8 m between the central stations. Producer data refer to a gradient of 80 % and a side slope of 55 % with auto levelling. A trench of 2 m can be traversed. There is no tyre inflation system. Fuel capacity is 550 litres.
The AF-40-1 vehicle has an empty weight of 16 t and a maximum payload of 3.5 t. It is powered by a 10 cylinder air cooled 294 kW Deutz engine positioned in the front. Up to 11 soldiers can be transported in the crew compartment.
The AF-40-2 has a maximum combat weight of 19.8 tons and will presumably carry a 105 mm gun turret. The Deutz engine is of a 12 cylinder type developing 404 kW. Both vehicles feature a ZF automatic transmission with 6 forward and 1 reverse gears. The engine is positioned in the rear, the vehicle lacks a rear ramp. Speed in water is 10 resp. 8 km/h.
Front protection level is 14 mm at 300 m and 7.62 mm at 25 m on the side. The bottom is protected against AP mines.

Role. First production order for Saudi Arabia amounts to 200 vehicles.

Assessment. Remarkable agile vehicle concept. Due to poor armour protection only suitable for police missions.

AL FAHD AF-40 APC

BIONIX Infantry Fighting Vehicle armoured personnel carrier

Provisional data: L 5.9 W 2.7 H 2.6 70 km/h Crew 10/11

Models

IFV 25 APC
IFV 40/50 APC

Development and production.

1991 studies on a M113 successor vehicle start.
1997 tested vehicles issued.
1997 start of series production.

Distinguishing features. Side. Tracked vehicle with six roadwheels and three support rollers, drive sprocket front. Bulky hull with sloped front section. Almost vertical rear. Front. Sponsoned box-shaped hull with vertical side walls. Turret installation moved to the left side.

Other details. Vehicle layout resembles German TH 495 with conventional IFV design, engine in the front. The prototypes are powered by a 354 kW Detroit Diesel 6V92TA two-stroke engine coupled to a GDDS MT-500-3EC hydromechanical transmission. The propulsion system for the series vehicles is not finally decided since a diesel-electric drive solution is under consideration, too. The vehicle features a hydropneumatic suspension.
The IFV25 is equipped with an electrically operated two-man turret armed with a stabilized 25 mm M242 Bushmaster cannon. It has a x8 main sight with thermal imager. Coaxial weapon is a 7.62 mm MG. Vehicle's weight is 22.9 tons.
The IFV40/50 has a weight of 21.5 tons and is armed with a one-man, open-topped cupola locally produced by CIS (Chartered Industries of Singapore) mounting a 40 mm grenade launcher on the right and a 12.7 mm MG on the left side. It is also electrically power operated but lacks a thermal sight.

Role. In production for Singapore forces.

Assessment. Remarkable indigenous vehicle project which could provide export chances in the Far East. Conventional infantry carrier with high mobility and reasonable armour protection against small calibre cannons.

IFV 25 APC

IFV 40/50 APC

ROOIKAT armoured vehicle reconnaissance

Non-serial versions.

ROOIKAT with 105 mm gun turret
ZA-35 AADG with twin 35 mm armament
ZA-HVM AADML with SAHV-3 AD missiles

Development and production.

1976 start of development of three different designs.
1987 five prototypes completed.
1990 series production stops.

Distinguishing features. Side. Eight large roadwheels with larger gap between second and third one and an escape hatch on both sides in this interval. Sharpened bow with slightly sloped upper side and steep lower part. Horizontal hull roof with raised engine deck. Long flat turret with sloped forward part and rear bustle. Long gun tube with cylindrical bore evacuator near centre point. Front. High front bow, slightly sloped glacis with driver's place right from centre. Flat pointed turret front with small mantlet. Wide periscope on top of right turret roof in front of cdr's hatch with external ADMG mounting.

Other details. Main armament is a fully stabilised LEW developed 76 mm L 62 gun firing either HE or APFSDS rounds. One 7.62 mm MG is mounted coaxially left of the gun, the ADMG has the same calibre. A computerized fire-control system is implemented. The optronic equipment includes a day and night image intensifier sight with laser rangefinder for the gunner and a panoramic sight for the commander, apparently identical to the one (German development) used in the Israeli MERKAVA MBT. The tank is powered by a V-10 cylinder diesel engine coupled to an automatic transmission with six forward and one reverse gears. Either four or eight wheels are propulsed, the front four ones are steered.
(TABLE)

Role. In service with the South African Army (200).

Assessment. Probably well suited for local terrain conditions with high mobility. Rather heavy design, in other countries used for tank destroyer missions. Armament not strong enough to fight up-to-date armour.

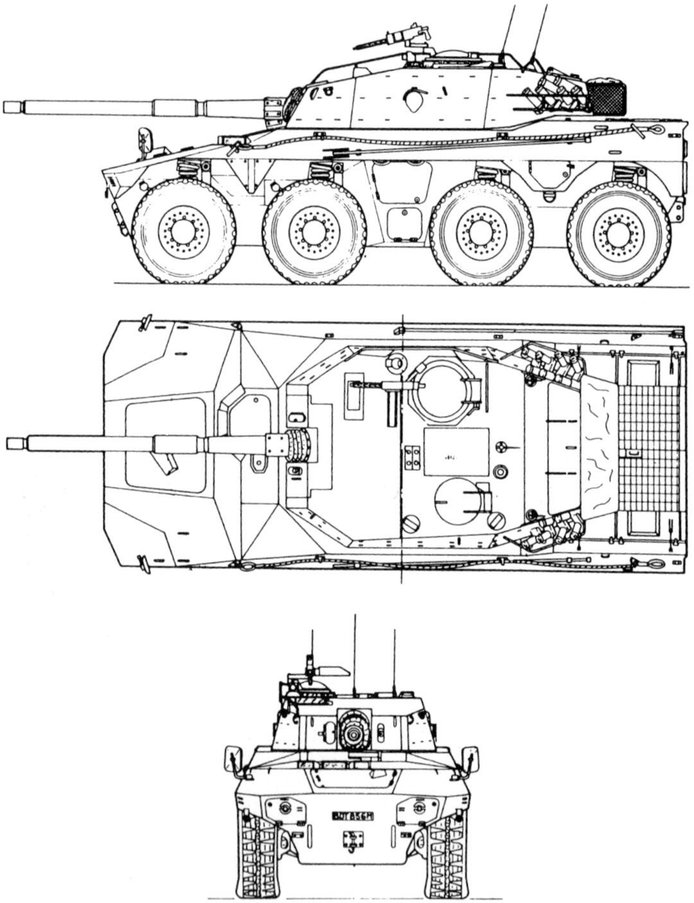

ROOIKAT AVR

ROOIKAT AVR

RATEL armoured general purpose carrier

Wt 18.5 t L 7.21 W 2.51 H 2.92 207 kW 105 km/h 1000 km

Models.

RATEL 20
RATEL 60
RATEL 90 FSV (Fire Support Vehicle)
RATEL TDM
RATEL CP AICC
RATEL AIMV
RATEL EAOS ACCC
RATEL Armoured Repair Vehicle

Non-serial version.

RATEL 120 mm AMC
RATEL with high-velocity missiles
RATEL LOG (8x8) Logistic Support Vehicle

Development and production.

1974 first prototype by Sandock-Austral to meet the requirements of the South African Army.
1976 start of series production of Mark 1, later Mk 2 and 3 with minor modifications, later all brought to latter standard.
1985 export order to Morocco.
1987 production stops with mortar carrier version.
1991 serial production of TDM with SWIFT missile starts.

Distinguishing features. Side. 6x6 vehicle with long distance between first and second axle. Forward-opening door to rear of first road wheel. Long box-shaped hull with pointed bow.
Horizontal roof and vertical rear. Engine on left rear side, access door on right rear side. Four firing ports and vision blocks in each hull side. Front. Sloped glacis. Driver central, large windows to his front and sides. Lower sides of hull vertical, upper sides sloped.

Other details. Four roof hatches over troop compartment. Passageway right rear above which two hatches and one circular mount for ADMG. Planetary gearing in wheel hubs.
RATEL 20 fitted with a two-man turret armed with a 20 mm F2 cannon (French M693) a coaxial 7.62 mm MG and an externally mounted 7.62 mm ADMG. Turret traverse and weapon elevation are manually. Elevation is -8 to +38 degrees. 1200 rounds 20 mm and 6000 7.62 mm are carried. Crew is 4 plus 7 infantrymen.

RATEL 60 is fitted with a two-man turret armed with a 60 mm mortar instead of the 20 mm cannon, firing HE, smoke or illuminating bombs up to a range of 1500 m. It carries 51 mortar bombs.
RATEL 90 is armed with an improved version of the ELAND 90 mm gun Elevation is from -8 to +15 degrees. 22 rounds are carried inside the turret, additional 40 inside the hull. The gunner sitting on the right has a M494 optical sight. It carries a crew of ten (five infantry-men). Its weight is 19 t.
RATEL TDM is armed with the ZT3 SWIFT 127 mm ATGMS in a three-round box launcher mounted above the turret. 12 rounds are carried inside, reloading is manual. Effective range is 4000 m, armour penetration capability is 650 RH steel.
Command version (Wt 18 t) has a crew of nine and is armed with a 12.7 mm MG in place of the 20 mm cannon. It is fitted with three radio sets.
The mortar carrier is armed with a 81 mm mortar firing through a two-part circular roof hatch which opens front and rear. It has a crew of five and carries 80 mortar bombs. This version has only one firing port on each side.
The EAOS vehicles (Enhanced Artillery Observation System) are rebuilt carriers provided with on-board and portable target acquisition, gun management, navigation systems and data transmission units. Pivoted on either side of the hull is a hydraulically operated mast with multi-sensor day/night observation camera with integrated LH-31 laser rangefinder.
The Repair Vehicle has a fixed lifting jib mounted on the rear hull.

Role In service with Morocco (80), Namibia and South Africa (1303 plus 53 TDM).

Assessment. Unconventional, relatively large vehicle for the special conditions in southern Africa. It offers much space for the crew and opportunity for fighting under-armour. Good mobility in open country.

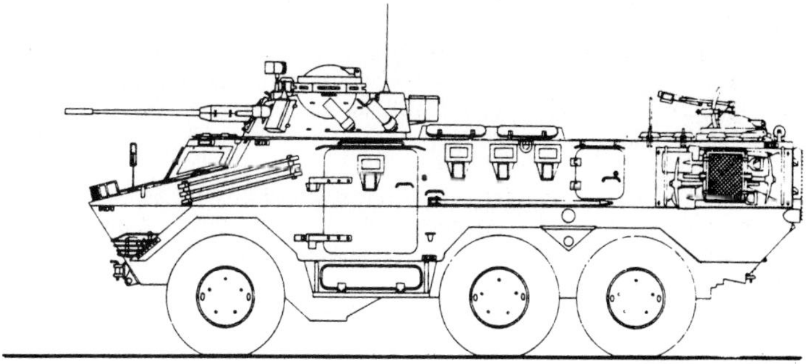

RATEL 20 APC

RATEL 20 APC

RATEL 60 APC

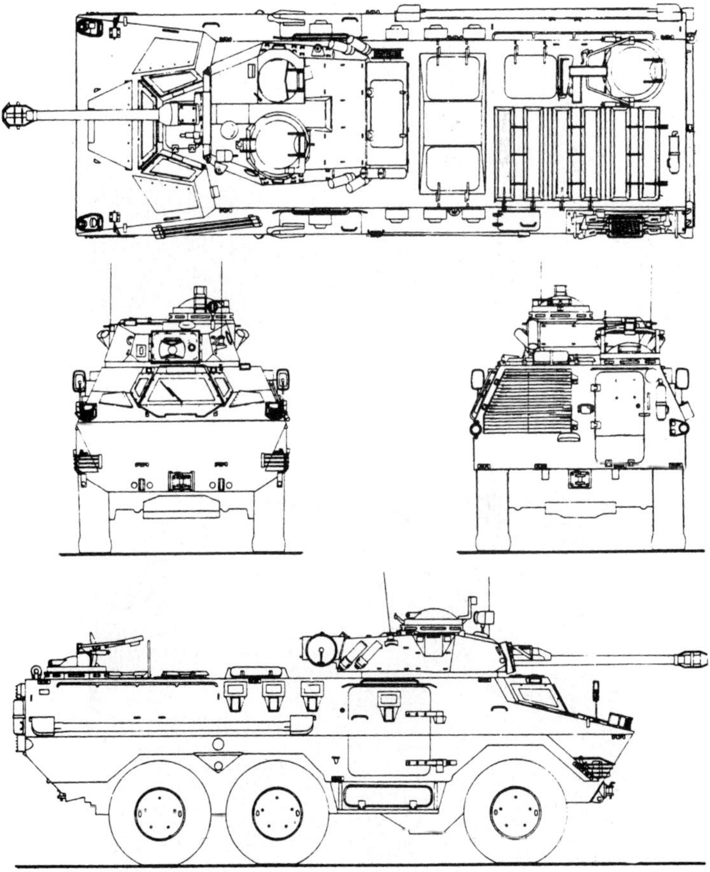

RATEL 90 FSV

RATEL 90 FSV

RATEL TDM

RATEL CP AICC

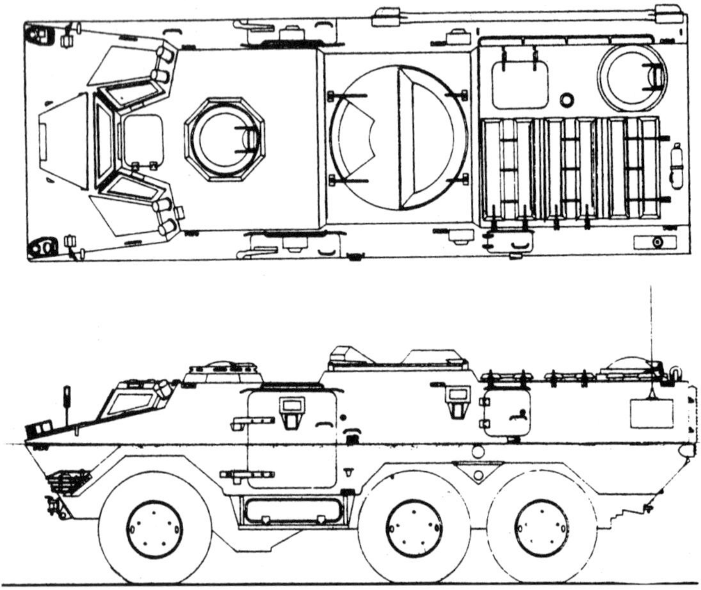

RATEL AIMV

RATEL AIMV

RHINO G6 armoured artillery howitzer

Development and production.

1975 considerations start in ZA to build a highly mobile SP howitzer.
1981 first prototype is issued by Lyttleton Engineering Works.
1984 first advanced development vehicle completed (6x6).
1988 series production starts.

Distinguishing features. Side. 6x6 vehicle with two roadwheels well forward just behind seperate driver's cab with three windscreens. Engine in centre. Crew compartment and turret in the rear over two axles. Two flat cupolas on turret. Gun in drum mantlet, single baffle muzzle break and fume extractor well forward, cradle and recuperator protruding. Front. Forward of first axle wedge shaped box containing 12 rounds and bush-clearing device.

Other details. Armament consists of the 155 mm howitzer which is a 45-calibre modified version of the G5 towed howitzer. On the roof top a 12.7 mm ADMG is mounted. Practical firing arc for the turret is 40 degrees to each side. 47 projectiles and 52 charges are stowed.
(TABLE)

Role. In service with the South African Army (43), Oman (24) and UAE (78).

Assessment. Unusual solution for the need to cover long distances and to cross bush country.

RHINO AAH

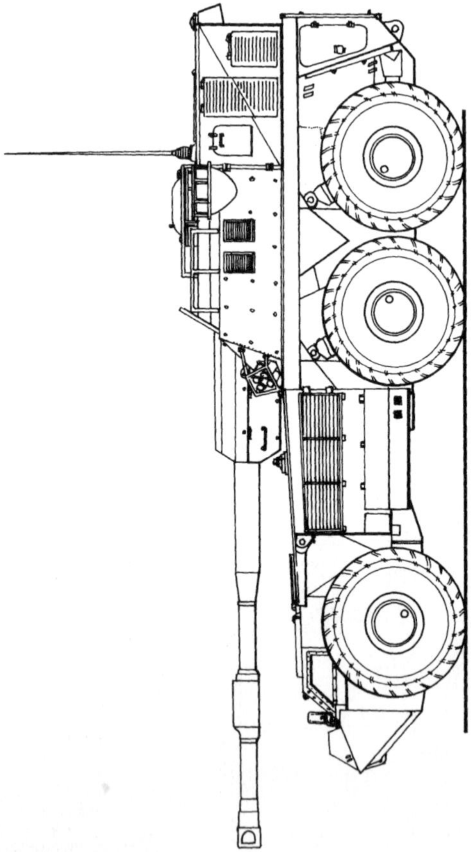

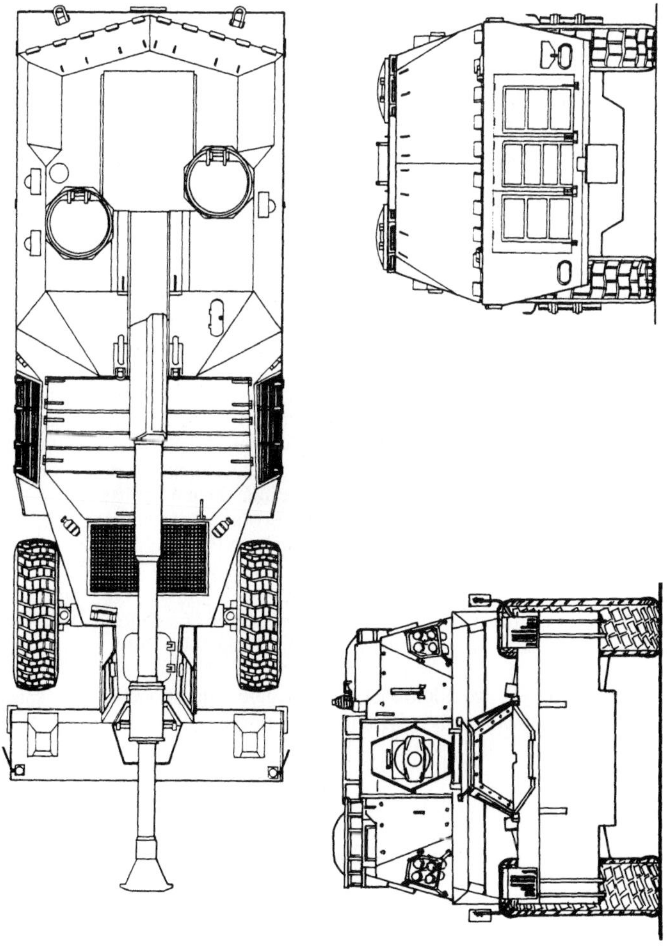

RHINO AAH

VEC armoured vehicle reconnaissance
Vehiculo de Exploración de Caballeria

Models.

VEC with TC-20 turret (Pegaso 3562.01)
VEC with TC-25 turret (Pegaso 3562.03)
VEC with French Hispano-Suiza Lynx 90 mm turret (Pegaso 3564)

Development and production.

1977 completion of first prototypes by Pegaso using many components of the BMR-600. Several armament installations tested.
1987 licence produced Italian OTO Melara T 25 turret accepted.

Distinguishing features. Side. 6x6 vehicle similar to APC, but glacis heavily sloped and rear hull top sloping back. Engine rear. Turret in forward half of the vehicle with sloping front, cannon in cylindrical vertical mantlet protruding in turret roof. Door in the left hull rear. Front. Driver in central position. Turret centre of hull with sloping sides and distinct chamfer between turret front and sides.

Other details. Main armament of the Pegaso 3562 versions consists either of the German Rh-202 20 mm Rheinmetall or the US M242 25 mm Bushmaster cannon, secondary armament is a coaxial 7.62 mm MG. 170 resp. 400 ready rounds are stowed inside the turret. Electric servomechanism gives a maximum angular traverse speed of 40o/s and 25o/s elevation. The gunner has an ENOSA P-204 periscope with x1 or x8 magnification, while the commander has an ENOSA PP-03 periscope and seven episcopes. NBC and land navigation systems are optional.
(TABLE)
The LYNX turret is described in the country section of France.Role. In service with the Spanish Army (32 3562.01, 208 3562.03 and 100 3564).

Assessment. Role of two additional crew members questionable. The space necessary for two men results in bulkiness of vehicle.

VEC AVR (TC-20)

VEC AVR (TC-25)

VEC AVR (Lynx)

BMR-600 armoured general purpose carrier
Blindado Medio de Ruedas

Models.

BMR-600 APC (company designation 3560.50) with MOWAG 7.62 mm remote-controlled MG mounting
BMR-600 APC with TC-3/A turret
BMR-600 ACCC (3560.51)
BMR-600 ACCC (3560.56)
BMR-600 AMC (3560.53)
BMR-600 AAV (3560.54)
BMR-600 recovery and maintenance vehicle (3560.55)

Non-serial versions.

BMR-600 TDM (3560.57) with Euromissile HCT turret (HOT)
BMR-600 AMC with 81 mm mortar (3560.53)
BMR-600 AMC with 120 mm mortar (3560.59)
BMR-600 APC with French GIAT Toucan I turret
BMR-600 TDG with CETME TC-7 turret with two 106 mm recoilless rifles
BMR-600 AADG with 20 mm Meroka turret
BMR-600 ANBCRV

Development and production.

1972 development started by ENASA in cooperation with the army.
1975 first prototype (Pegaso 3500) issued with engine behind the driver on the left side of the hull, armament on the right.
1979 vehicle is selected for the Spanish Army in a redesigned version (3560), series production starts.
1982 first export order by Egypt, one year later by Saudi Arabia.

Distinguishing features. Side. 6x6 vehicle with evenly spaced axles. Long box-shaped hull with pointed front. Exhaust pipe right side of hull. Two rectangular roof hatches above troop compartment and large rear ramp. Front. Sloped glacis with driver's compartment offset to left side, large engine access door right of it. Hull sides above wheel line slopes slightly inwards. MG weapon station behind the driver on left side.

Other details. Basic model is either armed with a 12.7 mm MG (2500 rounds) or a 7.62 mm MG. Many different weapon installations are optional. The vehicle is fully amphibious and has a water speed of 9 km/h propelled by two water jets or 4.5 km/h when driven only by its wheels. NBC system optional.
(TABLE)
ACCC versions with additional radio equipment.

AAV has a crew of four and can carry four stretcher or two stretcher/four seated patients.
Recovery and maintenance vehicle has a roof-mounted hydraulic crane with a lifting capacity of 18 t and a winch on the rear of 7 t traction power (100 m cable length). Four hydraulic stabilisers can be lowered to the ground.

Role. In service with Egypt (217 APC, 13 ACCC, 10 AAV and 10 Rec/Maint), Morocco (100), Peru (24), Saudi Arabia (140) and in Spain (682 APC/ACCC plus 30 AMC).

Assessment. Lightly armoured carrier with similar performance data as VAB

BMR-600 APC

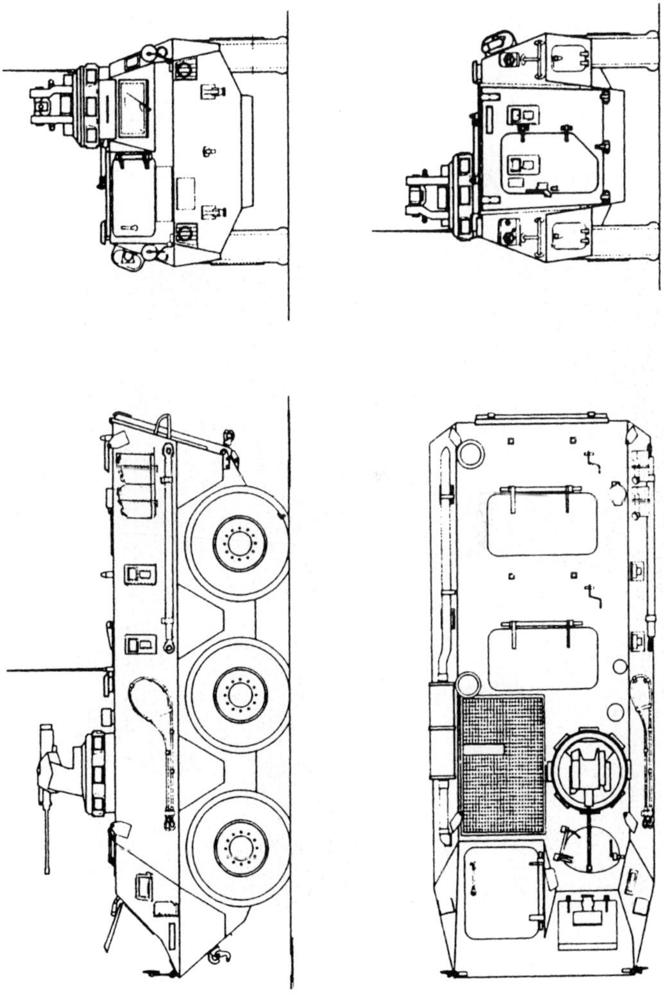

BMR-600 APC (TC-3)

BMR-600 APC (TC-3)

BMR-600 ACC

BMR-600 MERCURIO ACCC

BMR-600 AMC

BMR-600 AAV

BMR-600 ARRV

BLR armoured personnel carrier
Blindado Ligero de Ruedas

Wt 11.6 t L 5.65 W 2.5 H 1.99 154 kW 93 km/h 570 km Crew 15

Non-serial version.

BLR with TC-9/OP turret

Development and production.

1980 presentation of an ENASA developed light APC.
1985 production for Spanish Marines and Guardia Civil.

Distinguishing features. Side. 4x4 vehicle with box-like hull with boat-shaped nose, flat roof, vertical rear with two doors. Two large doors between road wheels on each side opening forward. Two firing ports beside side door with vision block above. Octogonal cupola on forward turret roof with external MG mount with shield. Front. Two large windows with upwards opening shutters. Cupola in central position. Vertical hull sides with top sloping slightly inwards.

Other details. Main armament comprises the CETME AC-10/556 M-1 external mounting with either a 7.62 mm or a 5.56 mm MG (1000 rounds stowage). The MG has an elevation of -15 to +68 degrees. Different anti-riot equipment can be mounted. The Pegaso 9220 6-cylinder diesel engine is situated in the middle of the crew compartment, which is gas tight and has a ventilation system. Fire suppression system for the engines and the wheels. Fuel capacity is 200 litres.

Role. In service with Spain (28 with the Marines and 22 with the Guardia Civil) and Ecuador (10).

Assessment. Lightly armoured vehicle only suitable for low-intensity warfare.

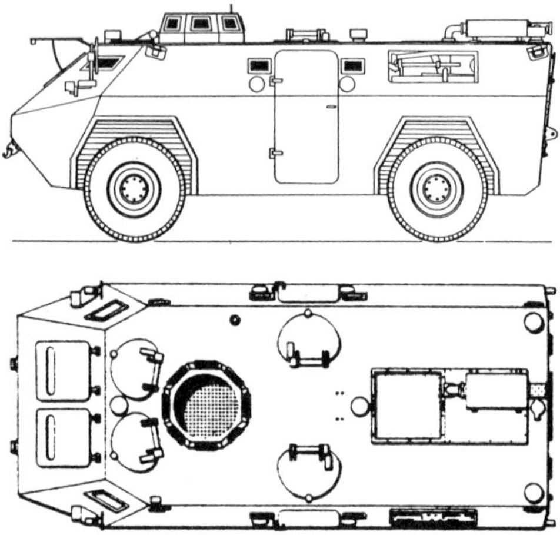

BLR APC

BLR APC

Ikv 91 tank destroyer
Infanterikanonvagn

Development and production.

1968 contract with Hägglund and Söner for the Ikv 91 development.
1969 three prototypes completed followed by extensive trials.
1974 two pre-production vehicles issued.
1975 series production starts running until 1978.

Distinguishing features. Side. Six large smooth-faced roadwheels, rear sprocket, no return rollers, boxed-in top run with wavy-edged skirt. Low front, long hull. Sloped hull rear. Turret with sloping front, small mantlet and undercut rear. Long thin gun barrel with fume-extractor half-way along barrel. Front. Rounded front, driver's hatch on the left side. Sponsoned hull with vertical sides. Turret sides almost vertical. Low-set tapering mantlet. Commander's cupola right, loader's cupola left with external ADMG.

Other details. Main armament consists of the KV 90 S 73 low-pressure gun. No APFSDS rounds fired. The gunner uses a Jungner Instrument AB monucular periscope TP-1050L (x10 magnification at a 6o field of view) including a Bofors laser-rangefinder linked to a fire-control computer. Two erectable LYRAN launchers are fitted on the rear turret roof. Natural swimmer, split splashboard at front and vertical extension on driver's hatch for swimming. Unusually low ground pressure.
(TABLE)

Role. In service in Sweden (280).

Assessment. Experiences with the Strv 103 tank with fixed gun seems to initiate the adoption of a turreted tank destroyer. The Ikv 91 has a very mobility especially in rough, muddy terrain and is useful against light armour. Thanks to standardization and the use of commercial parts it is easy to operate. Against modern MBT the 90 mm gun has become obsolescent.

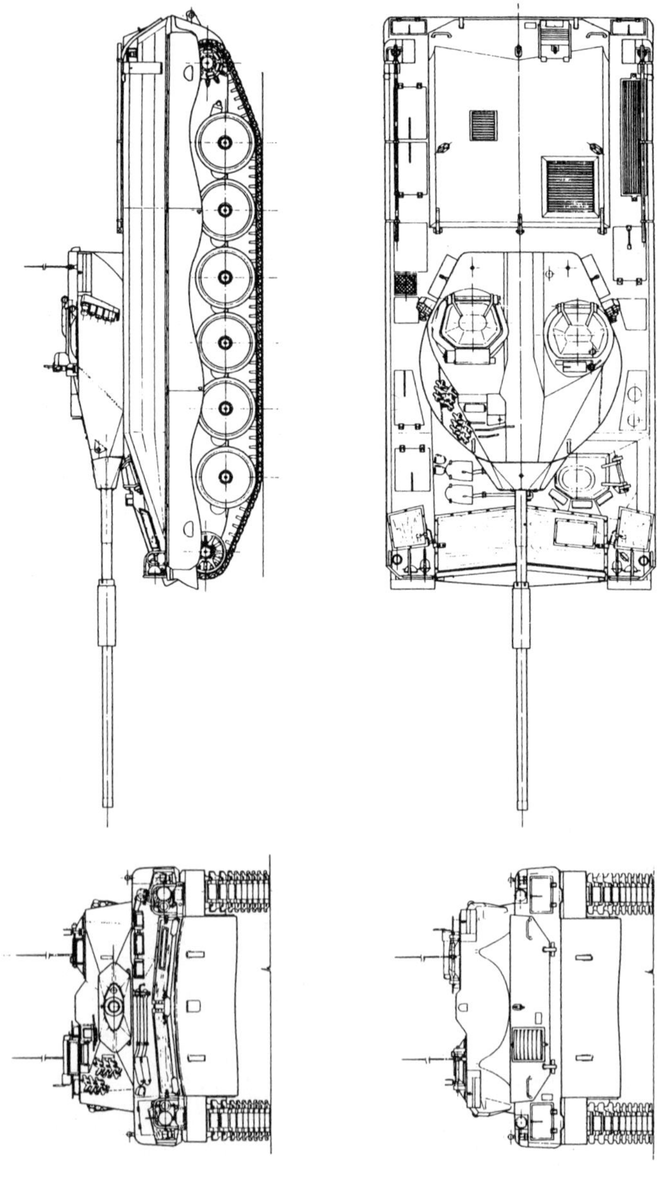

IKv 91 TDG

IKv 91 TDG

Combat Vehicle 90 armoured general purpose carriers
(Stridsfordon 90)

Models.

CV 9040
CV 9030
CV 9040 AV Lvkv (Luftvärnkanonvagn) AADG with 40 mm Bofors gun
CV 90 FCV Stripbv (Stridsledningspansarbandvagn) ACCC
CV 90 FOV Epbv (Eldningspansarbandvagn) AAOP
CV 90 ARV Bgbv (Bärgningsbandvagn)

Non-serial versions.

CV 9025 with 25 mm M242 chain gun
CV 90105 with GIAT TML 105 turret
AMOS (Advanced Mortar System) with 120 mm twin mortar

Development and production.

1984 development starts at HB Utveckling AB for automatic test rig and 40 mm two-man turret.
1985 contract for five prototypes with 40 mm gun. Test firing chassis Ikv 91 TDG with 40 mm L/70 gun in firing trials.
1988 further prototype versions programmed.
1991 production order of CV 9040.
1993 first deliveries to the Swedish Army.
1995 Norway selects the CV 9030.
1999 Switzerland's decision in favour of CV 9030CH.

Distinguishing features. A family of light carriers with the general lay-out: engine in front right, driver front left, radiators right rear, seven roadwheels, front sprocket. Sloped front engine deck, turret offset to the left.

Other details. Conventional APC layout with troop compartment in the back accessible by a large side opening door. The two-man turret has electric power traverse and elevation. The gunner is provided with a Celsius Tech Electronics UTAAS sight including a thermal sight manufactured by Kollsman. The commander has a cupola with six periscopes. The Swedish APC version is armed with the 40 mm L70 Bofors cannon. It can be fired either in the single shot mode or automatic at 300 rpm. A 7.62 mm MG is mounted coaxially on the left turret side. Two Bofors Lyran 71 mm twin mortar for illumination purposes.
(TABLE)
The Swiss version will have a slightly higher hull.
The Norwegian version has the same turret, is armed with the 30 mm Bushmaster II cannon and is fully stabilized.
The FCV and FOV versions are only armed with a 7.62 mm MG in the two-man turret and is provided with further communication resp. observation equipment.

The AA tank has a modified turret armed with the 40 mm Bofors cannon, too. The elevation angle is from -8 to +50°. This tank has a crew of seven, beside the commander and gunner in the turret and driver in the front there are four in the rear compartment, the radar operator, the combat controller, the external co-ordinator and the observer/loader.

Role. In production for the Swedish (600) and the Norwegian (104) Armies. Swiss requrement amounts to 186 vehicles.

Assessment. Agile and versatile carrier family with only light armour protection. CV 9040 with substandard firepower due to lack of weapon stabilization.

CV 9040 APC

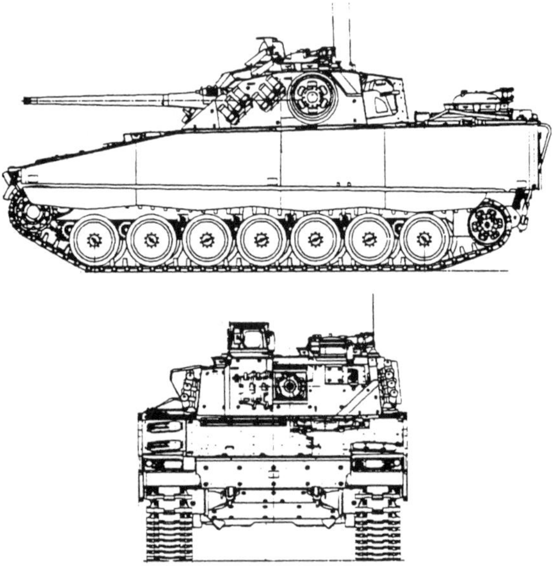

CV 9030 APC

CV 9040 AV Lvkv AADG

CV 90 FCV Stripbv ACCC

CV 90 FOV Epbv AAOP

CV 90 ARV Bgbv

Pbv 302 armoured general purpose carrier
Pansarbandvagn

Models.

Pbv 302 APC
Pbv 302 Mk 2 APC
Stripbv 3021 ACCC
Epbv 3022 AAOP
Bplpbv 3023 AAFDV

Development and production.

1961 requirement issued by the Swedish Army, development contract made with Hägglund and Söner.
1966 first production vehicles built, production ceased in 1971.
1974 Bplpbv 3022 issued followed by other models.
1978 presentation of Pbv 302 Mk 2.

Distinguishing features. Side. Five large roadwheels, sprocket front, no return rollers. Upper part of track boxed in. Sloped front with rounded edges. Box-shaped superstructure, flat roof with large roof hatches towards rear. Vertical rear curves to join sides. Cylindrical turret on forward edge of hull front. Front. Steep glacis with 2-part trim vane. Driver sits front central with roof hatch and three episcopes. Small turret with vertical sides on the left, its roof slopes down towards front. Cannon in sleeve and drum mantlet. Commander's cupola on the right. Twin doors rear.

Other details. 20 mm Hispano-Suiza main armament. The gunner has a monucular sight of x8 magnification. Against aerial targets an external open sight has to be used. Twin Lyran launcher at right side of hull rear.
Pbv 302 Mk 2 with additional cupola at rear hull roof for the rifle-squad leader. Left rear door with vision block.
(TABLE)
Vehicles employed in UN missions reveived add-on armour protection.
Stripbv 3021 with additional radio equipment.
Epbv 3022 with prominent cupola on forward right with combined binocular and optical x4 or x10 coincidence rangefinder. Azimuth sensor linked to traversing cupola.
Bplpbv 3023 has a crew of three and also carriers a ranging section of seven. Four radio sets and fire direction computer. Additional equipment on the roof top.

Role. In service with the Swedish Army (350).

Assessment. A rather high, lightly armoured carrier with good mobility. No fighting mounted under full armour protection possible.

Pbv 302 Mk 2 APC

Epbv 3022 AAOP

Pbv 302 Mk 2 APC with add-on armour

Pvrbv 551/Lvrbv 701 AT/AA missile carriers

Models.

Pvrbv 551 (Pansarvärnrobotvagn) TDM
Lvrbv 701 (Luftvärnsrobotbandvagn) AADML

Development and production.

1983 contract awarded to Hägglund for the conversion of obsolete Ikv-102/103 TDG into anti-tank and anti-aircraft missile carriers.
1984 first deliveries to the Swedish Army, production until 1986.

Distinguishing features. Side. Six small roadwheels, two return rollers, rear sprocket. Upper part of suspension covered by wavy skirt on forward half of the tank. High superstructure with steep front, commander's and driver's cupolas front, horizontal hull roof, vertical rear. Missile launcher retracted into the tank when not in use. Front. Box-like superstructure with sponsoned hull.

Other details. The Pvrbv 551 is armed with a Hughes TOW ATGM with a maximum range of 3750 m. In elevating position vehicle's height amounts to 2.46 m. The Lvrbv 701 is armed with the Bofors RBS 70 low altitude air-defence missile system with an operating range up to 5 km. With the launcher in firing position vehicle's height is 2.89 m including an IFF antenna.
Both versions are powered by a Ford 2658E V-6 petrol engine coupled to a Mercedes W 4A-018 transmission. The vehicle lacks a NBC system.

Role. In service with the Swedish Army (57 resp. approx.20).

Assessment. Combination of modern missile system with refurbished old carriage of low combat value. Launchers are unprotected in firing position. Low mobility.

Pvrbv 551 TDM

Lvrbv 701 AADML

155 mm Bandkanon 1A armoured artillery gun

Development and production.

1961 first prototype completed after several years of development (six roadwheels, three return rollers) followed by six pre-production vehicles.
1966 production starts running until 1968.
1988 contract awarded with Bofors for modernisation.

Distinguishing features. Side. Six large roadwheels, last one acting as idler, front sprocket, no return rollers. Extended low front. Fully enclosed turret on rear top of hull. Long gun with pepperpot muzzle brake extends over front of vehicle. Cantilever beam over gun breech carrying large enclosed magazine over vehicle rear. Front. Shallow glacis, driver left. Sponsoned hull. Cleft turret with gun mounted centrally.

Other details. 155 mm L/50 ordnance developed by Bofors. Magazine loaded at a rate of fire of 15 rounds/min, intense 3/sec. 7.62 mm ADMG on right half of turret. Rotating-head periscope on left half of turret with small cupola behind it. Loading system. A loading tray, swinging below the magazine, transfers the rounds from the magazine holding 14 rounds to the ramming position, from which they are rammed by a rammer mounted on the loading tray. Both loading tray and rammer are spring actuated. The springs are energised during the return of the piece to battery and by the movements of the loading tray.
Magazine. There are seven tubes holding two rounds each. Rounds fall from these tubes onto the 2 spring-loaded feed tray. Opening of the armoured hatches above the magazine is powered electrically. Magazine can be lowered on outrigged legs. Reloading by exchange of magazines of fixed ammunition takes two minutes using the vehicle-mounted crane of the limber.
Modernisation includes the installation of the Detroit Diesel DDAD 6V-S3T engine with 221 kW and a new automatic transmission.
(TABLE)

Role. In service with the Swedish Army (26).

Assessment. An AAG of still advanced concept, great range with conventional rounds and high rate of fire. High weight reduces mobility.

VK 155 1A AAG

Brobv 941 armoured vehicle launched bridge
Brobandvagn

Development and production.

1968 first prototype built by Hägglund and Söner.
1973 series vehicles produced.

Distinguishing features. Side. Suspension similar to Pbv 302 but with six road-wheels. Sloped glacis with dozer blade on lower part. Shallow box-shaped hull with fully enclosed crew compartment front and open topped cargo room rear. One-piece bridge carried on telescopic beam on top of hull. Front. Driver central flanked by commander and operator. Hull sides and rear vertical.

Other details. Class 50 girder bridge launched over front of vehicle. To launch the bridge the hydraulically operated telescopic arm is extended across the gap and the bridge slides forward on it. The telescopic arm is retracted once the bridge is laid. While laying the bridge the tank is supported by its front spade. The bridge can be laid and recovered from either end. For swimming (water speed is 8 km/h) the bridge, which like the vehicle has natural buoyancy, is towed behind. Weight of bridge is 7 t. The tank has a 7.62 mm ADMG.

Role. In service with the Swedish Army (17).

Assessment. A novel laying system similar to the rejected A prototype of German BIBER AVLB. First ever MLC 50 bridge on an APC hull.

Brobv 941 AVLB

Bgbv 82 armoured recovery vehicle
Bärgningsbandvagn

Development and production.

1968 first prototype completed by Högglunds.
1973 production models issued.

Distinguishing features. Side. Hull and suspension almost identical to Brobv 941 AVLB. Hull front with splashboard and front dozer blade. Two rear spades. Flat top with fully enclosed crew compartment at front and open-topped stowage space in rear half of the tank. Turret on front edge of hull same as Pbv 302 APC's. Folding crane fixed in rear compartment. Front. Commander's (right) and driver's cupola (centre) protruding in upper part of glacis. Turret on left, crane on right side. Fourth crew member behind cdr's station also with roof hatch. Main winch central.

Other details. HIAB-Foco 6000 C crane lifts maximum load of 8 t. Main winch has 145 m cable length and leads from the rear of cab through the hull rear with maximum load of 60 t with spades used. A complete supplement powerpack of the Strv 103 can be transported in the cargo bay.

Role. In service with the Swedish Army (24).

Assessment. A large measure of standardisation has been achieved by the use of production components of the Pbv 302 and the Ikv 91. The vehicle is capable of recovering all Swedish armoured vehicles, though the engine output should be too low for towing over bigger distances.

Bgbv 82 ARV

EAGLE light reconnaissance vehicle

Wt 3.8 t L 4.9 W 2.28 H 1.75 117 kW 125 km/h 450 km Crew 4

Models.

EAGLE I
EAGLE II

Development and production.

1995 production start.
1996 final deliveries to Denmark.
1997 contract signed for the supply of 175 EAGLE II.

Distinguishing features. Side. 4x4 vehicle with box-like truck-shaped superstructure. Two doors between wheel-stations with small windows in upper half. Small turret on top in the rear roof section. Front. Car-like front with prominent engine grill between front lights. Vertical sides with inverted roof section.

Other details. The vehicle is based on the US HMMWV (Heavy High Mobility Multi-purpose Wheeled Vehicle). The armoured superstructure provides protection from small arms. The commander sitting left of the driver has a circular roof hatch. The observation cupola MBK 2 is fitted with an externally mounted 7.62 mm MG on its right hand side and a vision block in the left cupola roof with a thermal imager in between. Traverse and elevation is manually. A bank of six smoke dischargers is arranged at the cupola's rear. The EAGLE I is powered by a GM 6.5 L NA diesel engine with 117 kW coupled to a Hydra-Matic 4 L 80 E automatic transmission. The EAGLE II differs in having a turbo-charged engine developing 139 kW and a 400 kg higher payload; total weight is 4.1 t. Both versions have an NBC protection system.

Role. In service with Denmark (36) for UN missions and in Switzerland (154, 175 EAGLE II on delivery until 2001).

Assessment. Lightly protected agile vehicle well suited for liaison and peacekeeping missions. Economic solution using a large scale production vehicle.

EAGLE I AVR

EAGLE II AVR

Panzer 68 main battle tank

Models.

Panzer 68/75 (Pz 68 AA2)
Panzer 68/88

Non-serial version.

Panzer 68/88 with 120 mm gun

Development and production.

1964 start of series production of Panzer 61.

1968 first PT of improved Pz 61 (Pz 68) with weapon stabilization, altered transmission and wider tracks.
1971 delivery of first batch of Pz 68 (170 vehicles).
1977 production of second batch (50) with gun thermal sleeve.
1978 production start of Pz 68 AA2 (110) with enlarged turret.
1983 delivery of fourth batch (60) identical with Pz 68 AA2.
1988 modernisation program for 195 Pz 68 initiated due to start in 1992 (fire-control system with laser rangefinder).

Distinguishing features. Side. Six roadwheels, sprocket rear, three return rollers. Cast hull with rounded front and well sloped glacis. Engine deck sloped downwards, undercut rear. Three stowage boxes above track guards. Rounded cast turret with small bustle in centre. Long gun with fume extractor in centre. Front. Driver's hatch in upper part of sloped glacis. Turret central with loader's cupola on the left side with external ADMG.

Other details. Main armament is the 105 mm L7 series gun (56 rounds stowage). Besides the 7.5 mm ADMG there is a 7.5 mm MG coaxially left of the gun. MG ammunition stowage is 5200 rounds. (TABLE)

Role. In service with the Swiss Army (195).

Assessment. From the beginning an ill-fated tank project with many deficiencies and malfunctions. Below average mobility, high silhouette.

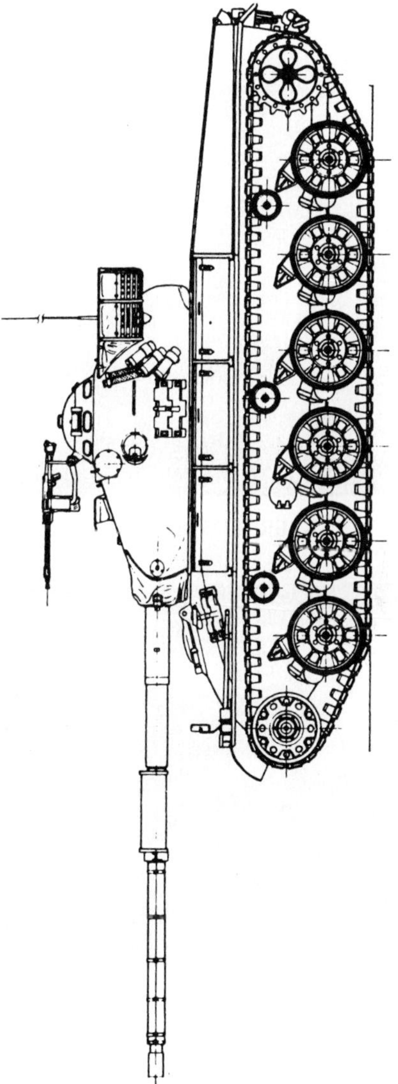

PANZER 68/75 MBT

PANZER 68/75 MBT

PANZER 68/88 MBT

PIRANHA armoured general purpose carrier

Models. (all 8x8 except when otherwise stated)

PIRANHA (4x4) with 7.62 mm MG armament
GRIZZLY (6x6) APC
PIRANHA (6x6) with GAD-AOA turret
PIRANHA (6x6) with V 041 turret
PIRAÑA (6x6) with 12.7 mm MG (Chile)
PANZERJÄGER 90 (6x6) TDM
COUGAR (6x6) TDG
PIRAÑA (6x6) with French F1 90 mm gun (Chile)
PIRANHA/PIRAÑA (6x6) ACCC with 7.62 mm MG armament
PIRANHA (6x6) AMC with 81 mm mortar
PIRAÑA (6x6) AMC with 120 mm mortar (Chile)
PIRANHA (6x6) ARV
HUSKY (6x6) AMV
PIRAÑA 8x8D APC (Chile)
LAV-25 APC (Light Armored Vehicle) (Australia: ASLAV-25)
LAV with Delco turret
LAV-AT (Anti-tank) TDM
LAV with LANCELOT HOT ATGMS turret
LAV with MECAR LCTS Mk8 90 mm gun turret
LAV with TS 90 F-4 turret
LAV with GIAT TS 90 F-4 90 mm turret
LAV-C (Command) ACCC
LAV-M (Mortar) AMC
LAV with 120 mm AMS (Armoured Mortar System)
LAV-AD (Air defense) AADG/ML
LAV-L (Logistics)
LAV-R (Recovery)
LAV-MEWSS (Mobile Electronic Warfare Support System)
PIRANHA II AVR
BISON APC (some used as ambulance vehicle)
PIRANHA II with KUKA 606 turret
ASLAV-PC
PIRANHA II with MECAR LCTS Mk8 90 mm gun turret
ASLAV-S (Surveillance)
PIRANHA II / ASLAV-C ACCC
BISON Command Post
PIRANHA II AAOP
PIRANHA II / ASLAV-A (Ambulance)
PIRANHA II AMC
BISON Mortar (Carrier)
BISON Mobile Repair Team
PIRANHA II / ASLAV-F (Fitter)
COYOTE (LAV-RECCE)

PIRANHA III (10x10) ASV (Armoured Sensor Vehicle)
PIRANHA III (10x10) ACV (Armoured Command Vehicle)
PIRANHA III (8x8) AEV (Armoured Escort Vehicle)
PIRANHA III APC
KODIAK APC

Non-serial versions.

All PIRANHA configurations have been equipped with numerous weapon installations ranging from light MG to 105 mm guns. It is also offered for special tasks like TDM, AAV, ARV etc.

Development and production.

1972 issue of first prototype after private venture development of MOWAG.
1976 start of series production for export.
1977 licence to General Motors of Canada.
1979 first COUGARs and GRIZZLYs issued.
1981 licence to Cardoen (6x6) and FAMAE (8x8) in Chile.
1982 licence production of 8x8 versions as LAV for the USMC in Canada. Two different prototypes for a Swiss requirement as TDM.
1988 prototype of BISON APC by GM Canada built.
1989 Australia purchased 14 LAV-25 and one ARV from USMC stock.
1990 start of series production of Panzerjäger 90.
1991 licensee GKN receives first Saudi order (due to Swiss arms ban).
1992 deliveries for second Saudi order starts at GM Canada.
1993 203 COYOTE ordered by Canada.
1995 first deliveries to Oman by GKN licensee.
1996 two pre-production PIRANHA III (6x6 and 8x8) built.
1997 first PIRANHA III (10x10) issued to Swedish Coast Artillery.
1998 license-production of KODIAK for Canada starts.

Distinguishing features. Side. Either two, three, four or five large wheels on each side. 6x6 configuration with wide gap between first and second wheel station. Sloped front, boat-like bow with lower part sloping back to front wheel. Horizontal hull roof, vertical rear. Screws for water propulsion behind rear wheels. According to configuration firing ports in side walls. Different weapon/turret installations. Front. Pointed bow with sloped upper and lower part. Driver's position front left with weapon applications resp. cdr's hatch behind. Different arrangement of roof and rear hatches.
The first PIRANHA generation is now unofficially called PIRANHA I. The follow-on category II and III models differ in payload and slightly in size resulting in larger tyres and higher weight. Additionally they feature technical improvements like ABS, differential locking, central tyre inflation system etc.

The 8x8 variants have the following outline specifications:

	PIRANHA I	PIRANHA II	PIRANHA III
Weight (t)	12.5	14	16.5
Payload (t)	3	4	6
Length (m)	6.3	6.8	6.93
Width (m)	2.5	2.6	2.66
Height of hull (m)	1.88	1.92	1.98
Tyres	325R16	325/85R16 or 335/80R20	365/80R20 or 12.00R20

The PIRANHA 4x4 has a slightly shorter hull (4.52 m) and a weight of 7.5 t. Its height is 1.66 m (without armament). Armament is either a 7.62 or 12.7 mm MG. This version is non-amphibious.
The other exported PIRANHA I 6x6 vehicles are armed either with 7.62 mm MG, the Oerlikon-Bührle GAD-AOA turret with 20 mm cannon, or the MOWAG Type V 041 twin MG turret (12,7 and 7.62 mm MG).
The Australian ASLAV differ from PIRANHA I / LAV having a swivel driver's hatch, revised exhaust, smoke projector system and engine louvres. The 25 mm APV variant has additional stowage bins around the rear turret half.
Canada operates six models, three 6x6 and three 8x8 ones. The 6x6 were produced between 1979 and 1982. The COUGAR features the SCORPION 76 mm gun turret while the GRIZZLY APC is equipped with a Cadillac Gage turret with a 12.7 mm and a 7.62 mm MG in a twin mount. It has also two firing ports on each side of the troop compartment. The HUSKY AMV has a HIAB 650A 4.5 t crane and an 8 t winch. The 8x8 line consists of the BISON with a raised troop compartment (height of hull 2.21 m) as APC with a section of 8 soldiers, in a command version, as 81 mm mortar carrier and as a maintenance carrier with a 5 t boom crane and specialised equipment, the reconnaissance variant COYOTE in three different sub-versions, one as a battle group kit consists of two man-portable dismounted surveillance sensors, one Racal MSTAR radar the other one with TV camera, thermal imager and laser-rangefinder; the second version is the fixed brigade kit with an electrically powered extendable 10 m mast comprising both sensor suites. Radar range is about 24 km detecting moving targets, the camera has a maximum of 18 resp. 12 km day/night mode. Additionally the vehicles have GPS navigation and laser warning. The third sub-version is only prepared to receive the sensor equipment. Fully equipped the vehicle has a weight of 13.4 tons. The third model is the KODIAK APC in the PIRANHA III configuration equipped with the LAV-25 turret. It is powered with a 257 kW Caterpillar 3126 diesel engine coupled to an Allison MD 3066 P fully automatic transmission.
Chilean PIRAÑAs differ from the regular PIRANHA I in fuel capacity (400 litres) increasing the range to 1000 - 1200 km, a turbo-compressor for operations above 5000 m altitude, and the 8x8 version lack the rear propellors. There are four 6x6 variants (APC, command, 120 mm mortar carrier and fire-support with

low pressure 90 mm F1 gun (Vo 750 m/sec) for HE and HEAT ammunition) and the 8x8 APC.
Two orders were placed by Saudi Arabia. The Ministry of Defence and Aviation ordered 300 vehicles splitted between the British GKN Defence comprising six versions, an APC for VIPs, surveillance radar carrier, internal 81 mm mortar, external 120 mm mortar (transported on the rear hull roof and lowered to the ground behind the vehicle by means of hydraulic arms), ambulance and recovery versions and GM Canada comprising three versions, LAV-25, anti-tank with LANCELOT HOT launcher and an assault gun version with the GIAT TS-90 turret with a 90/52 mm CS-90 F4 gun. The second order of 1117 LAVs concerned the Saudi National Guard. The two APC versions either have an open MG mount or the Delco LAV-25 turret with TOW launchers on either side. The anti-tank, command, ammunition carrier and recovery versions are almost identical to the US LAVs. The mortar carrier has a turret-mounted Royal Ordnance 120 mm breech-loaded mortar. The assault gun features the Cockerill LCTS Mk8 turret with the 90 mm MECAR gun and Kollsman day/night range sight. Additionally an engineer and an ambulance version were delivered.
The Panzerjäger 90 is fitted with a Norwegian Thune-Eureka ALT turret with twin TOW-2 launcher (eight missiles on stowage). The vehicle's weight is 11 t, its height is 2.97 m.
The USMC operates eight versions. The LAV-25 is armed with a 25 mm M242 chain gun and a coaxial 7.62 mm M240 MG in a Delco two-man turret.
(TABLE)
The AT version is provided with the same TOW launcher in use with the M901 (two missiles ready-for-launch, 14 inside the LAV). It is operated by a crew of four. The mortar version is equipped with an internal 81 mm mortar with a stowage of 94 rounds and a crew of five. The AD version is armed with a five-barelled 25 mm GAU-12/U cannon and two pods of four STINGER missiles on each side. The command versions have additional radio equipment and the logistic carriers have a higher superstructure with twin hatches and a loading crane at the left hull rear. The recovery version mounts a 1.8 t crane and a rear mounted winch with a capacity of 13.6 t. The MEWSS has a jammer system, detector radio sets and a land navigation system. Ten lines of antennas arranged irregularily on both sides are mounted on top of a telescopic mast.

The Swiss PIRANHA II APC is armed with the manually operated KUKA 606 one-man turret with 12.7 mm M2 MG and 210 ready use rounds and PERI Z16 day sight.
The PIRANHA III generation has a new basic light-weight hull design with add-on armour, increased internal volume, a hydropneumatic suspension at all wheel stations, a high-torque driveline common between all versions and new engines. The front section of the 6x6, 8x8 and 10x10 versions is identical.
Sweden is being delivered by three variants of the PIRANHA III, a 10x10 command vehicle, a 10x10 surveillance and artillery targeting radar carrier and an 8x8 APC version for escort missions. The 10x10 vehicles have a weight of 20 t and are 7.45 m long.

Role. In service with Australia (61 ASLAV-25, 33 PC, 9 C, 10 F, 10 S, 2 A, 1 LAV-R, 150 more on order: 82 -25, 18 S, 16 C, 13 PC, 11 F, 5 R, 5 A), Canada (269 GRIZZLY, 195 COUGAR, 27 HUSKY, 149 BISON APC, 18 C, 16 M, 16 R; 203 COYOTE, 651 KODIAK on delivery as APC, 71 to be fitted with TUA turrets removed from M113, 39 engineer, 47 artillery, mortar observer), Chile (150 APC, 20 AIFSV, 5 ACCC, 50 AMC), Danmark (18 APC, 2 CP, 2 ARV on order, all III), Ghana (20 4x4, 30 6x6/Oerlikon, 2 ARV), Ireland (34 PIRANHA III APC and 4 C3 with Helio FVT800 turrets plus one ambulance and one recovery on delivery), Ireland (34 PIRANHA III APC and 4 C3 with Helio FVT800 turrets plus one ambulance and one recovery on delivery), Liberia (24), Nigeria (82 6x6/V 041, 12 AMC, 10 ACCC), Oman (80 PIRANHA II APC, C, M, R, AOP, Amb), Qatar (PIRANHA II: 36 Cockerill, 2 C, 2 R on delivery by GKN), Saudi Arabia (300 for the MODA incl. 144 vehicles with TS-90 F-4 gun and 98 LANCELOT and 384 LAV-25, 47 APC, 111 TOW, 182 C, 130 Cockerill, 73 AMS, 67 R, 71 ambul., 34 engineer, 18 L for the National Guard), Sierra Leone (10), Sweden (PIRANHA III: 27 ASV, 17 ACV, 10 AEV), Switzerland (310 PzJg, 515 PIRANHA II/KUKA on delivery), USA (422 LAV-25, 96 AT, 50 C, 50 M, 125 AD, 94 L, 46 R, 12 EW and 12 BISON APCs for the National Guard).

Assessment. Versatile and proven design for cost effective mass production of a great variety of models. Light enough for helicopter transport, beach-landing capability. Successful for special requirements in low-intensity

PIRANHA (4x4) APC

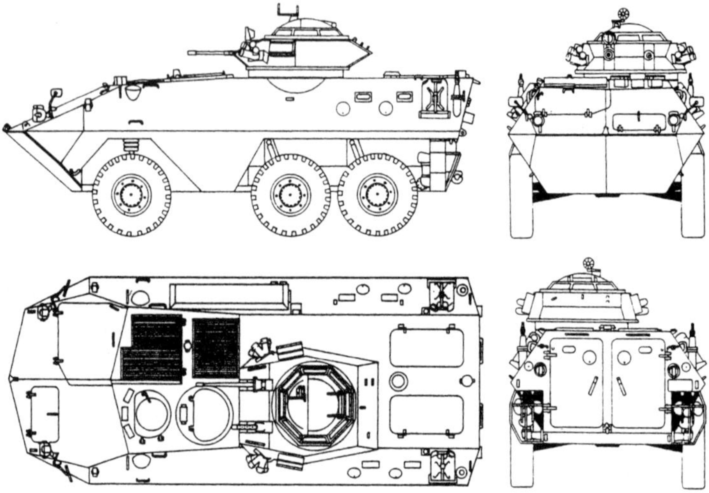

GRIZZLY APC

PANZERJÄGER 90 TDG

COUGAR TDG

PIRANHA (6x6) ACCC

PIRANHA (6x6) ARV

HUSKY AMV

PIRAÑA 8x8D APC

ASLAV-25 APC

LAV with Delco turret

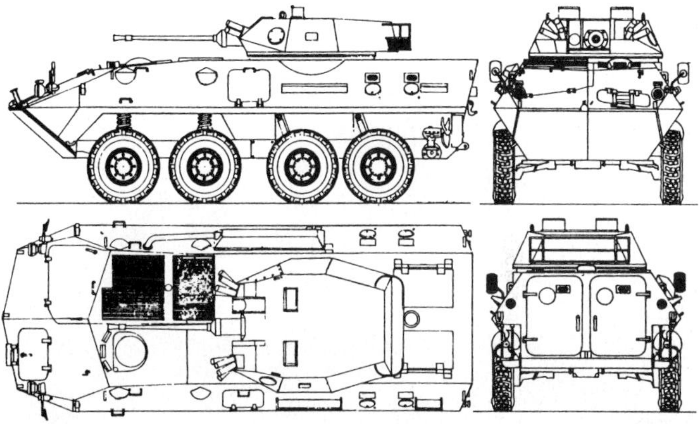

LAV-25 APC

LAV-AT TDM

LAV TDM (HOT)

PIRANHA TDG (LCTS Mk 8)

LAV TDG (TS-90)

LAV-C ACCC

LAV-M AMC

LAV with 120 mm AMS

LAV-AD AADG/ML

LAV-L ACC

LAV-R ARV

LAV-MEWSS

PIRANHA II AVR

BISON APC

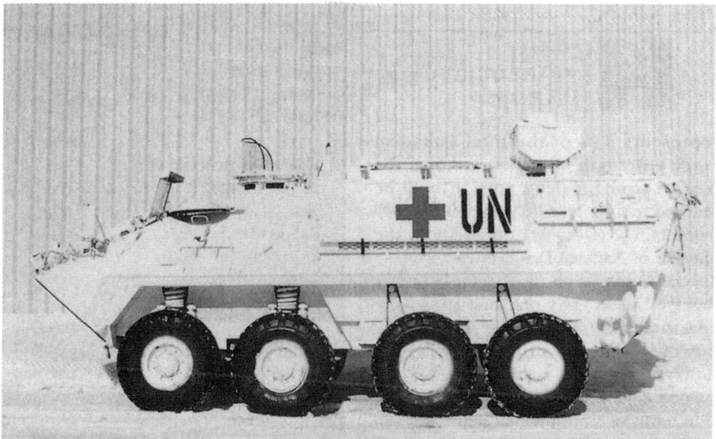

BISON AAV

PIRANHA II APC (KUKA 606)

BISON AMC

BISON ARV

COYOTE AVR

PIRANHA III (10x10) ACV

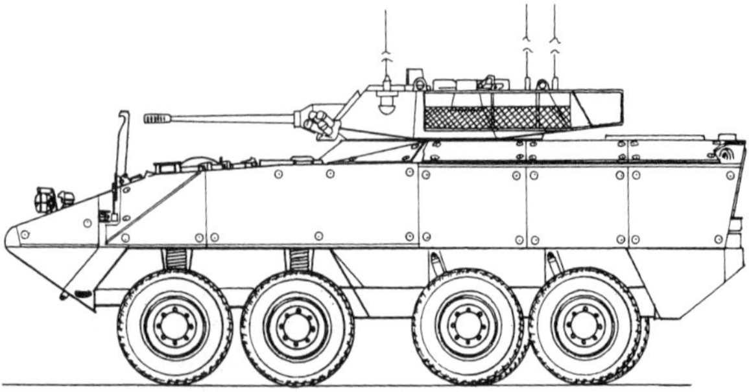

PIRANHA III APC

KODIAK APC

BrüPz 68 armoured vehicle launched bridge
Brückenlegepanzer

Development and production.

1965 development contract for Konstruktionswerkstätten Thun.
1972 version on Pz 61 chassis with aluminium bridge accepted.
1974 start of production.

Distinguishing features. Unmodified Pz 68 chassis. Bridge is carried on top of hull in open position extending well to front and rear.

Other details. MLC 50 bridge is supported by a long cantilever which is hydraulically operated to lay the bridge across the obstacle. When the bridge is laid, the boom is withdrawn and returned to its travelling position, still extending front and rear of the tank. Time to lay the bridge is about two minutes. The bridge can be recovered (within about three minutes) from either side. The tank has no armament.
(TABLE)

Role. In service with the Swiss Army (30).

Assessment. A novel laying system similar to that of the rejected A prototype of the German BIBER. Here, too, the silhouette is kept very low by avoiding the use of a system based on a scissors-bridge.

BrüPz 68 AVLB

Entp Pz 65 armoured recovery vehicle
Entpannungspanzer

Development and production.

1965 start of development by Konstruktionswerkstätten Thun.
1967 troop trials completed.
1972 first series vehicles issued to the troop.

Distinguishing features. Based on Pz 61 MBT chassis with long, almost rectangular armoured superstructure with slightly sloped front and vertical sides. Front dozer blade. Crew and winch compartment at front, engine compartment rear. Large A-frame jib folds back onto hull roof for travelling. Air intake louvres at front of right side. Commander's cupola front right with smoke dischargers, cupola for winch operator with ADMG left. Driver's hatch in front of cdr's position. Entrance door front left. Exhaust rear right.

Other details. (TABLE)

Role. In service with the Swiss Army (69).

Assessment. A sound conversion of an available MBT chassis. Use of old-fashioned A-frame limits versatility considerably.

Entp Pz 65 ARV

COBRA armoured general purpose carrier

Wt 6.2 t L 5.0 W 2.22 H 2.1 140 kW 115 km/h 725 km

Development and production.

1995 first prototype completed by OTOKAR as a private venture.
1997 eight prototypes built, evaluation by the Turkish Army.
1998 production order placed for APC and AVR versions.

Distinguishing features. Side. 2x2 coffin-like monocoque superstructure with inclined rear section and long flat front. Front. Inclined side walls. Two-piece glass windscreen, one engine grill on front bow flanked by two air-intake grills aside. Side doors behind the driver's compartment.

Other details. The armour superstructure is mounted as a one-piece steel hull on the helical coil suspension of the US HMMMV in its ECV version (Expanded Capacity Vehicle), 3.3 m wheelsbase, using all its automotive components including the GM V-8 turbo-charged engine and the 4 speed gearbox. The vehicle can climb a slope of 70 %, has a ground clearance of 325 mm and fords up to 1 m. Side slope is 40 %. It can be provided with a winch on the front (increasing the length by 229 mm). The turning radius is 7.67 m, fuel capacity 145 litres. The vehicle can be equipped with different weapon stations and has mission specific internal layouts.

Role. In production for the Philippines and Turkey.

Assessment. Vehicle concept based on proven US automotive components minimizing development risk. Due to light armour proctection only useful for reconnaissance and police missions.

COBRA APC

COMMANDO SCOUT armoured reconnaissance scout car

Wt 7.24 t L 5 W 2.06 H 2.16 110 kW 96 km/h 1287 km Crew 2-3

Models.

COMMANDO SCOUT with 1 m turret
COMMANDO SCOUT with 12.7 mm open MG mount
COMMANDO SCOUT commando pod

Development and production.

1977 development as a private venture of Cadillac Gage Textron.
1983 start of series production until 1987.

Distinguishing features. Side. Two large roadwheels each side. Very low, sloped front. Low superstructure with slightly sloped rear. Front. Sloped glacis extending half-way along vehicle. Driver's hatch upper left side with air intakes to right, exhaust pipe right side of hull. Lower front plate arched out over wheels. Hull sides sloping inwards top and bottom.

Other details. Different types of armament. 1 m turret either with twin 7.62 mm MG or 12.7 mm MG with one 12.7 mm and one 7.62 mm MG which can be elevated from -10 to +55 degrees. Ammunition stowage ranges from 1200 to 2400 rounds depending on calibre. Egyptian vehicle has a 12.7 mm MG in an open mounting. The command version has either a 12.7 mm or a 7.62 mm MG without a turret (weight is 6.8 t). Fuel capacity is 378 litres. The vehicle is not amphibious and lacks a NBC system.

Role. In service with Egypt (112) and Indonesia (28).

Assessment. Commercial attempt to build a light 4x4 ARSC. Low silhouette, limited cross-country performance, low level of protection.

COMMANDO SCOUT ARSC

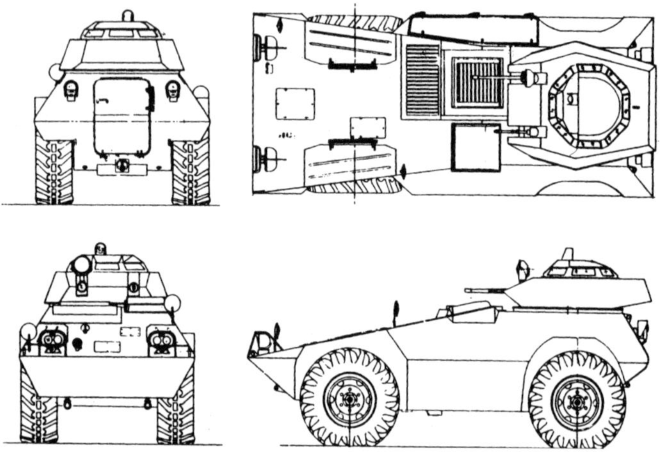

COMMANDO SCOUT (1 m turret)

COMMANDO SCOUT (1 m turret)

M48 PATTON main battle tank

Models.

M48A3
M48A5
M48T5 (Turkey)
M48A5T1 (Turkey)
M48A5T2 (Turkey)
M48H
MAGACH 5 (Israel)
NAGMACHON (Israel)

Non-serial versions.

Numerous modernisation kits offered by several producers.

Development and production.

1951 T 48 90 mm gun tank issued to overcome M47 MBT deficiencies.
1952 start of series production of the M48.
1963 2000 M48 and M48 A1 brought to A3 standard (diesel engine).
1975 conversions to A5 standard (105 mm gun) start, also with new M2 cdr's cupola.

Distinguishing features. Side. Six roadwheels, three/five return rollers, sprocket rear. Bulky hull with rounded protruding front nose. Raised engine compartment. Front. Well sloped glacis plate with driver centre, turret with curved side centre of hull. Gun barrel (90 mm) with T-shaped smoke suppressor at muzzle or (105 mm) with cylindrical fume extractor centre of longer gun tube. Either raised or flat cdr's cupola on turret right with ADMG.

Other details. 48 A3 has the AVDS-1790-2A diesel engine with 551 kW, a fuel capacity of 1420 litres and a range of 463 km. Wt 47,17 t, L 8.69 m, H 3.12 m. Main armament is the 90 mm gun with 62 rounds stowage. One coax 7.62 mm MG is fitted (6000 rounds), for the 12.7 mm ADMG 630 rounds are carried. (TABLE)
M48 A5E2 is the Spanish conversion with Wegmann smoke dischargers, a new Hughes fire-control system with laser rangefinder and a 559 kW AVDS-1790-2D diesel engine coupled to CD-850 6A transmission.
M48 A5T1 is local refurbishment in Turkey with Cadillac Gage add-on stabilization. M48 A5T2 are further improved tanks with Texas Instruments thermal sight and new ballistic computer.
M48H is a Taiwanese modification using new M60A3 hulls and M48 A5 turrets.
MAGACH are totally refurbished Israeli M48/A1/A2/A2C brought to A5 standard with low Urdan cdr's cupola. Many tanks fitted with reactive add-on armour and third MG externally behind the gun mantlet.

NAGMACHON is a turretless M48 with add-on special armour similar to the NAGMACHON/PUMA based on CENTURION.
M48T5 is a German conversion of Turkish M48 to the M48 A2GA2 standard including a MTU MB 837 Ea-500 diesel engine with 551 kW.
All M48 can be fitted with front dozer blade, anti-mine rollers/ploughs.

Role. In service with Greece (799 A5 of which 493 are retrofitted with German MOLF fire-control system), Israel (500), Iran (about 200 A5), Jordan (200), Lebanon (104 A5), Morocco (184 A5), Pakistan (200 A5), Portugal (86 A5), South Korea (416 A3, 597 A5), Taiwan (450 H, 60 A5), Thailand (75 A5), Turkey (658 M48A3, 183 M48T5, 1369 M48A5T1, 751 M48A5T2), Vietnam (several hundred).

Assessment. First post-war generation tank with considerable growth potential allowed many improvements and upgradings. No adequate armour protection against latest high-velocity ammunition. Below average mobility.

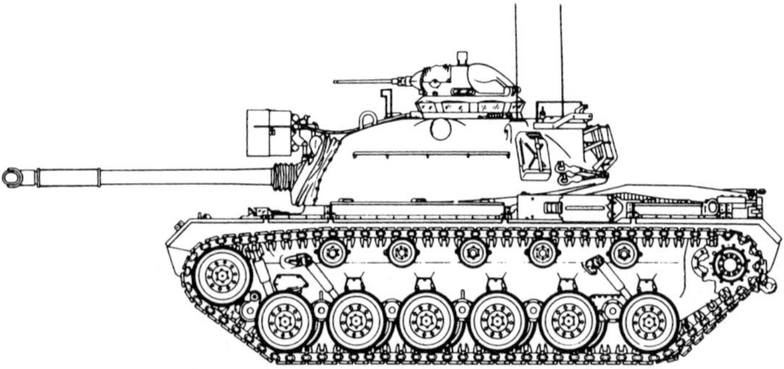

M48A3 MBT

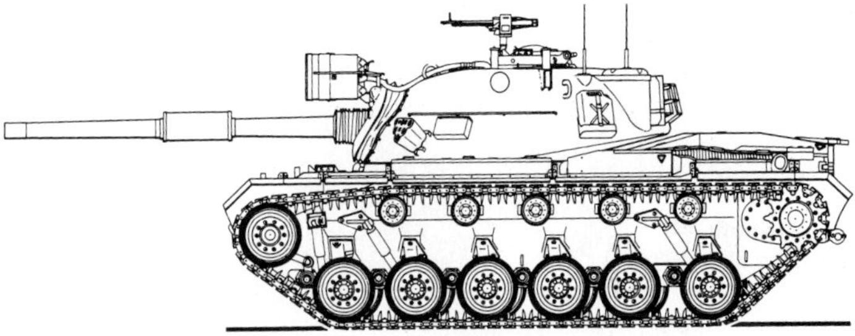

M48A5 MBT

M48A5 MBT (South Korea)

M48T5 MBT

M48A5 T2

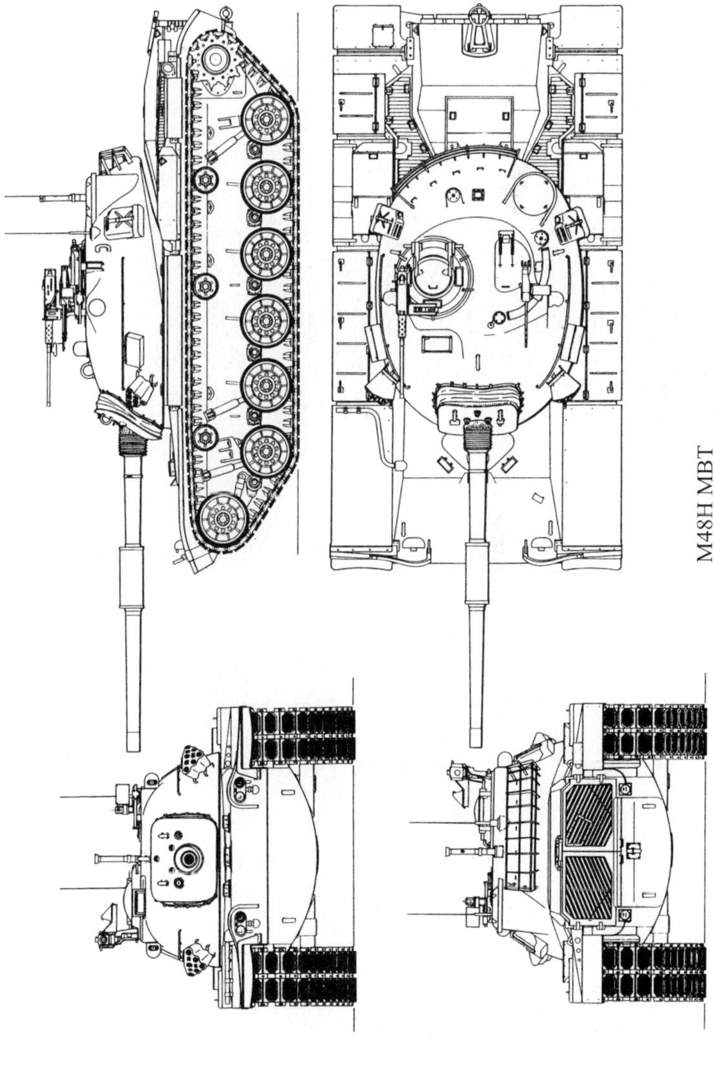

M48H MBT

M48H MBT

M60 PATTON main battle tank

Models.

M60 A1
M60 A3
M60 A3Ö
MAGACH-6M (Israel)
MAGACH-7 (Israel)

Non-serial versions.

Several up-grading kits offered.

Development and production.

1956 further development of the M48 A2 leading to XM60.
1959 standardisation of M60 with diesel engine, 105 mm gun and welded steel hull. Start of M60 production (still with cast turret resembling M48, 1535 built).
1962 M60 is superseded by M60 A1 with reconfigurated better armoured turret, later the M60 A1 AOS was built (add-on stabilisation).
1965 licence production of the M60 A1 in Italy.
1973 production start of M60 A2 (armed with 152 mm gun/ Shillelagh missile launcher in redesigned turret. 526 produced until 1975, all phased out).
1975 production start of M60 A1 RISE (Reliability Improvement Selected equipment Engine)
1977 production start of M60 A1 RISE/PASSIVE with passive night sight. Old M60 A1 brought to this standard.
1978 introduction of the M60 A3 with a Hughes fire-control system with laser rangefinder and a tank thermal sight (TTS). 1686 produced, later 3714 M60 A1 were brought to this standard. Production ceased 1982.

Distinguishing features. Side. Similar in appearance to M48 MBT with six roadwheels, three return rollers, sprocket rear. Turret with new tapered design. Front. Sloped glacis plate, driver central. Large external gun mantlet with rounded corners. Cdr's cupola on right side with ADMG.

Other details. M60/A1 with AVDS-1790-2A diesel engine, other data corresponding with M60 A3 (data table).
M60 A3 with cross-wind sensor on turret roof, optical rangefinder exit closed by flap, new M239 smoke dischargers.
(TABLE)
Israeli M60 with several modifications including Urdan cdr's cupola. Most tanks have Blazer explosive reactive add-on armour and third external 12.7 mm MG behind the gun mantlet and fourth 7.62 mm MG in front of loader's hatch.

Latest upgrade (MAGACH-7) include a passive armour package fitted to the front parts, turret front and sides and forward roof section and comprises a new type of skirts. To compensate the 5 tons add in weight an upgraded AVDS-1790-5A engine with 668 kW is installed. The Elbit/ELOP Matador computerized fire control system is also being installed.
M60 A3Ö, Austrian conversion of M60 A1 roughly identical with A3 standard with further hull/suspension modifications.
The M60 can be fitted with front mounted dozer blade or mine-clearing rollers/ploughs.

Role. In service with Austria (168 A3ô), Bahrain (180 A3), Brazil (91 A3 TTS), Egypt (835 A3 plus 680 ex-US A1), Greece (357 A1, 312 A3), Iran (about 150 A1), Israel (950 A1, 450 A3), Jordan (118 A1, 150 A3 TTS), Morocco (60 A1, 240 A3), Oman (6 A1 + 47 A3), Portugal (93 ex-US A3), Saudi Arabia (460 A3), Singapore (6), Spain (50 A1E, 154 A3E, 106 A3E TTS), Sudan, (20 A3), Taiwan (460 A3), Thailand (178 A3), Tunisia (30 A1, 54 A3), Turkey (274 A1, 658 A3 TTS, all ex-US), Yemen (140 A1), USA (approx. 2700 A3, no longer in service).

Assessment. Improvement in many respects over the M48 series. Not up-to-date against latest generation MBT. Still poor mobility, high silhouette.

M60A1 MBT

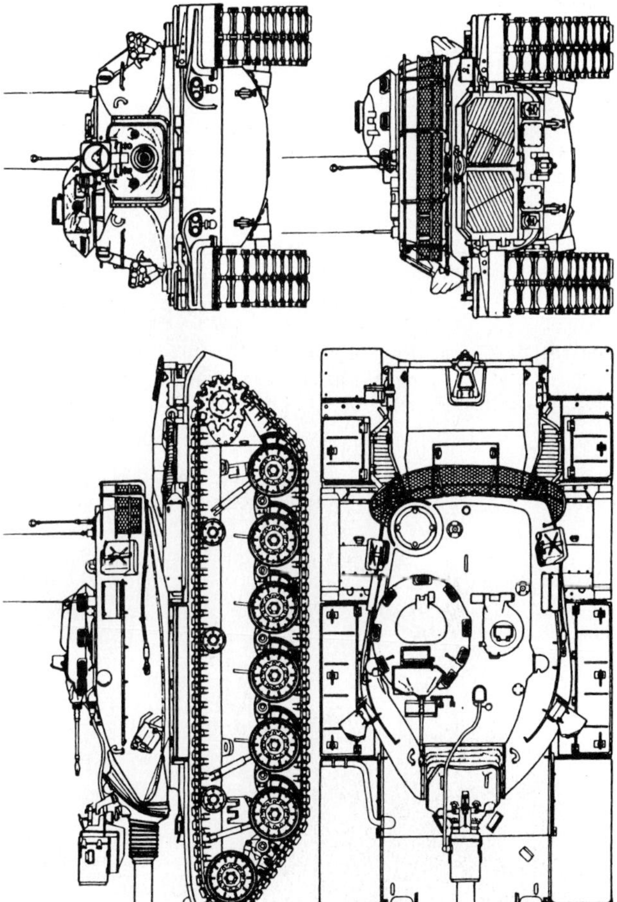

M60 A3 MBT

M60A3 MBT

M60A3Ö MBT

M60A3 (IL) with Blazer

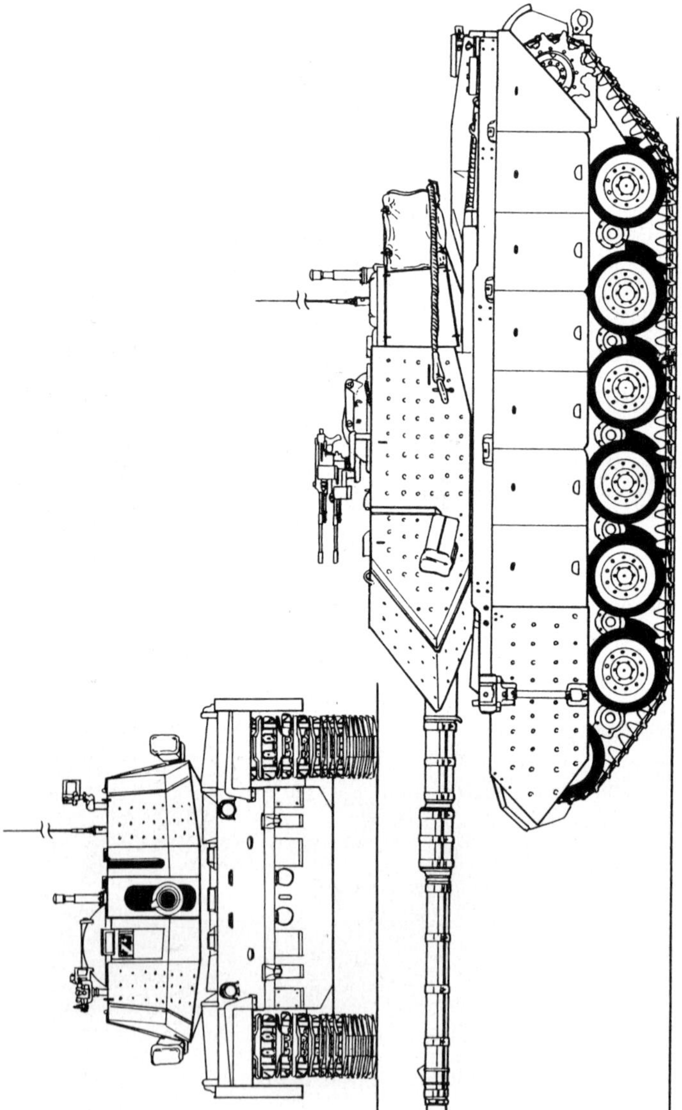

MAGACH-7 MBT

MAGACH-7 MBT

M1 ABRAMS main battle tank

Models.

M1
IMP M1 (Improved M1)
M1A1
M1A1 HA (Heavy Armor with depleted Uranium)
M1A1D
M1A2
M1 AVLB WOLVERINE

Non-serial version.

M1E2 with longer 105 mm gun tube by Watervliet
M1A1 with Multi-sensor Target Acquisition System (MTAS)
Combat Mobility Vehicle (CMV) BREACHER
ABRAMS Recovery Vehicle

Development and production.

1971 termination of MBT 70 follow-on project XM803.
1973 development contract awarded to Chrysler and Detroit Diesel Allison Div.
1976 one prototype from each firm delivered (XM1). Chrysler type accepted.
1980 production start at Lima Army tank plant, two years later also at Detroit Arsenal Tank Plant (2374 built until 1985).
1984 production of IMP M1 until 1986 (894).
1985 M1A1 issued.
1992 first M1A2 manufactured.
1993 the last of 4796 M1A1 tanks for the US Army produced.
1995 development of M1A2 SEP starts.
1996 last new M1A2 produced.
1998 M1A2 SEP upgrade program starts.

Distinguishing features. Side. Seven roadwheels, two return rollers, sprocket rear. Top run of track boxed in by skirting plates. Nose slopes back under hull, slightly sloped glacis, very low hull, markedly raised engine compartment with vertical rear. Extended turret with almost vertical front that slopes to rear, bustle extending well to rear. Gun with fume extractor and muzzle reference system. Low-profile ring-mounting for ADMG. Front. Driver's hatch central in upper hull front under gun mantlet. Narrow rectangular gun mantlet. Turret overhangs hull, sides sloped inwards. Rectangular sight on right side of turret roof in front of cdr's cupola, which has externally mounted ADMG. Hull rear vertical with air louvres in upper part.

Other details. Main armament of the M1 and the IMP M1 is the US L7 105 mm gun derivative M68E1 (55 rounds stowage). A 7.62 mm MG is mounted coaxially (11400 rounds). The 12.7 mm ADMG has 1000 rounds stowage. Stabilised integrated day/night (thermal) sight with laser rangefinder. Digital fire control computer. Turret stabilised in azimuth, sights and mounting in ele

vation. IMP M1 with increased armour front protection, internal resolver system, suspension improvements, redesigned driver's panel. Weight increased to 55.3 t compared to the 54.4 t of M1.
M1A1 is armed with the Rheinmetall 120 mm smooth-bore gun. Many internal improvements. Collective NBC and ventilation system initiated automatically by a sensor. New improved two-stage heating system. Modified final drives, transmission upgrading, stronger shock absorbers. M1A1 for the USMC (182) have the deep water fording kit with two snorkels. (TABLE)
M1A2 features important modifications. A cdr's stabilised independent day/thermal sight on the front left turret roof is integrated. A carbon dioxide laser rangefinder, an improved cdr's weapon station and a battlefield management system is in use. Armour protection is further enhanced (depleted uranium) comparable with M1A1 HA.
Current M1A2 are produced in the "SEP" (System Enhanced Package) configuration. Previously converted A2 will be modified, too. The SEP include upgraded digital electronics, color digital map for the tank command and control system, 2nd generation FLIR (forward looking infra-red), UAAPU (Under Armor Auxiliary Power Unit) and environmental cooling.
1535 M1A1 will be converted to M1A1 D standard with appliqué C3 digitization, EPLRS (Enhanced Position Location Reporting System), improved navigation, eye safe laser, external APU.
WOLVERINE is a M1 chassis with the LEGUAN bridge.

Role. In service with Egypt (555 M1A1), Saudi Arabia (315 M1A2) and the USA (2344 M1 of which 796 are earmarked for AVLB and BREACHER and 998 for upgrade to A2, 892 IPM1, 4509 M1A1 of which nearly 50 % have heavy (DU) armor, 65 new M1A2 plus 1014 M1A1 to be converted to A2). Final decisions on BREACHER and WOLVERINE is pending, 22 AVLB are being delivered).

Assessment. Together with the LEOPARD 2A5 and the LECLERC the M1A2 marks the highest current standard in MBT technology. Good firepower, excellent armour protection and high mobility. High fuel consumption, no submerged crossing capability.

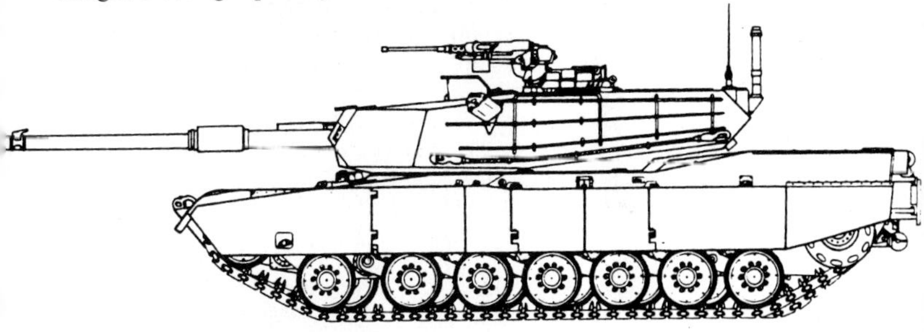

M1 ABRAMS MBT

M1 ABRAMS MBT

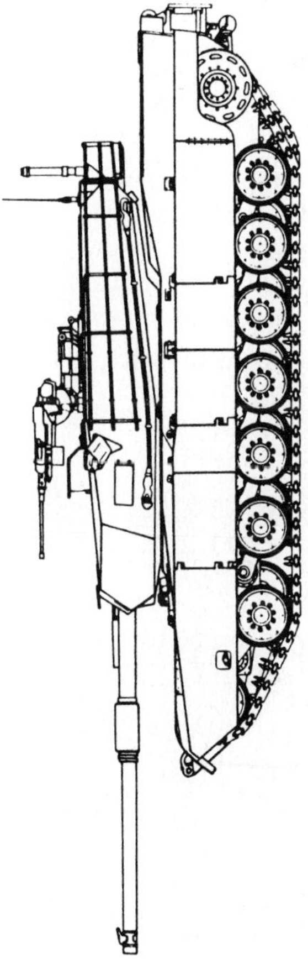

IMP M1 MBT

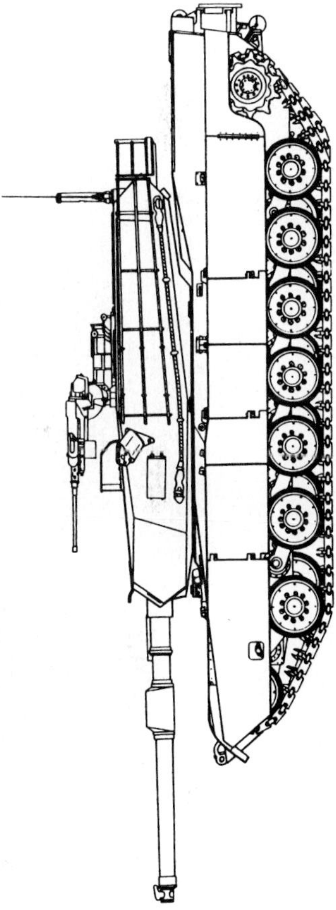

M1 A1 MBT

M1A1 MBT

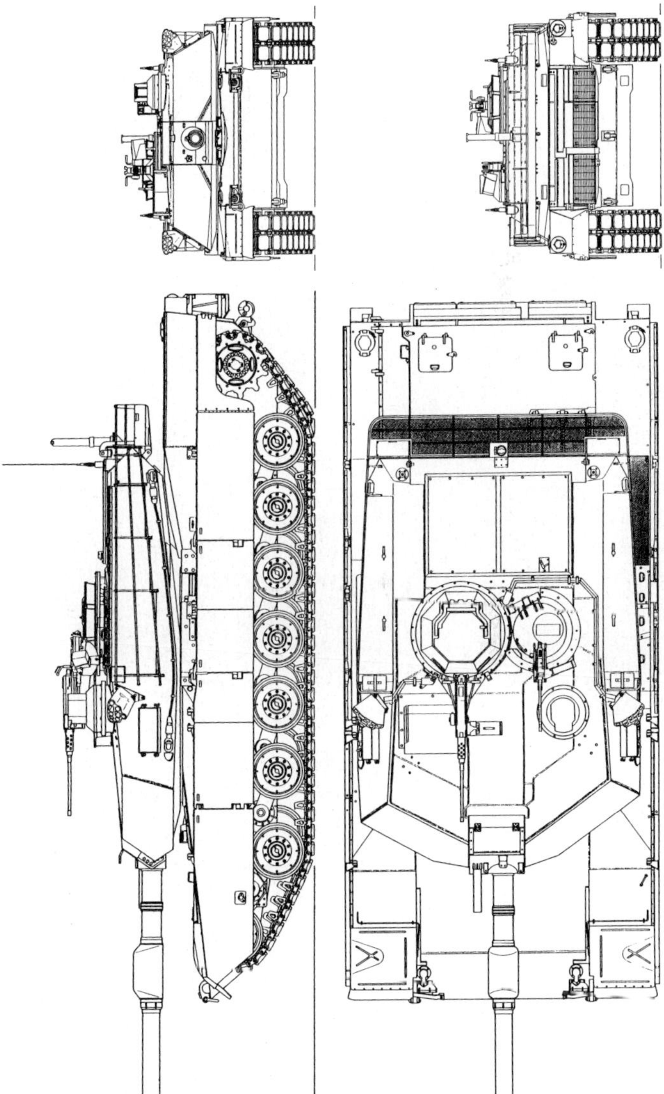

M1A2 ABRAMS MBT

M1A2 MBT

M1 AVLB WOLVERINE

CMV BREACHER

STINGRAY light tank

Wt 21.2 t L 9.35 W 2.71 H 2.55 410 kW 70 km/h 450 km Crew 4

Non-serial versions.

STINGRAY II
STINGRAY II with LAV-105 turret

Development and production.

1983 concept by Cadillac Gage for a light tank in 20t-class primarily for export. Turret prototype for firing trials on M551 SHERIDAN chassis.
1984 integration of turret to new chassis.
1988 start of series production.

Distinguishing features. Side. Six roadwheels, three return rollers, sprocket rear. Heavily sloped front glacis. Hull roof with three steps ending into raised engine deck. Flat turret with wedge-shaped gun mantlet. Cdr's cupola with prominent periscope head in front of it. External ADMG mount, cross wind sensor on rear turret roof. Front. Driver on upper centre part of front. Vertical sides, sloped turret sides. Cdr's cupola on right side of turret.

Other details. Main armament is a modified 105 mm L7A3 gun with longer recoil and muzzle brake. Gun elevation is from +20 to -7.5 degrees. Any type of NATO standard 105 mm ammunition can be fired. 36 rounds are carried. A 7.62 mm M240 MG is mounted coaxially with 2400 rounds stowage, the 12.7 mm ADMG has 1100 rounds stowage. The M36E1 roof-mounted day/night sight is linked with a Marconi DFCS fire-control computer with laser rangefinder. The commander is provided with a NV-52 day/night sight. The tank is powered by a Detroit diesel engine 8V-92 TA and has a fuel capacity of 757 litres.

Role. In service with Thailand (106).

Assessment. Agile light tank with limited protection and below-average armour-piercing capability.

STINGRAY light tank

BRADLEY Fighting Vehicles

Models.

M2 Infantry Fighting Vehicle
M2A2
M2A3
M3 Cavalry Fighting Vehicle
M3A2
M3A3
M6 LINEBACKER
M7 FIST (Fire Support Team)

Non-serial versions.

High Survivability BRADLEY
M2 with 30 mm Bushmaster II cannon
M2 with 35 mm Ares Talon cannon
M2 with LOSAT anti-tank missile launcher
M2 with Blazer turret

Development and production.

1964 requirement for a replacement of M113 by a real AIFV. Some abortive projects issued (XM701).
1970 new order for a lighter AIFV.
1973 contract awarded to Food Machinery Corp. for the construction of prototypes after adoption of the XM723 concept. (Chrysler and Pacific Car proposals rejected).
1978 two PT handed over (XM2).
1981 start of series production of both versions M2 for infantry, M3 for recconnaisance missions.
1985 M2A1/M3A1 in production.
1988 start of production of M2A2/M3A2.
1990 M2A1/M3A1 conversion to to A2 standard starts.
1996 M7 production starts.
1997 remanufacture of 1602 A2 vehicles to A3 status is launched.

Distinguishing features. Side. Six roadwheels with larger gap between third and fourth wheel, two return rollers, sprocket front. Upper part of track and hull sides with appliqué armour. High hull line with sloped front and horizontal roof and bulged hatch above rear troop compartment. Two circular firing ports in each side (sealed on M3 version). Vertical hull rear with large ramp, two firing ports. Large angular turret with overhang. Front. Glacis with trim vane, driver's hatch on left side. Sponsoned hull, thick skirting plates. Turret central with wedge-shaped mantlet, periscope head on the left side of the turret roof, twin ATGM launcher on left side swung into horizontal position prior to launching.

Other details. Main armament consists of the M242 25 mm Chain Gun and a 7.62 mm M240C MG coaxially on the right side. There are electric turret drives and a weapon stabilisation system. The 25 mm cannon has dual feed and fires either single shots or bursts at 100 or 200 rpm. The gunner uses the combined day/thermal sight ISU (Integrated Sight Unit) x4 or x12 magnifications with light pipe to the commander. 5.56 mm M231 submachine guns are used with the firing ports.

(TABLE)

The M3 is nearly identical to the M2 having a crew of five, more ammunition stowage (1500 rounds 25 mm, 4540 7.62 mm and 3200 5.56 mm, M2 has 900/2340/4200), 12 TOW missiles are stowed (M2 has 7). The BRADLEY is amphibious after preparation. A flotation screen is contained along the hull sides and erected prior to entering the water. Speed in water is 7.2 km/h.

The block I improvements (A1 vehicles) standard tanks fire the TOW 2 missile and a long rod 25 mm penetrator. Driver, commander and gunner are provided with a collective NBC system. The rear roof hatch has vision blocks. The hull bottom is armour strengthened by an anti-mine blast plate. A weapons interlock is integrated activated when hatches are open. All A1 BRADLEYs were converted to A2 standard.

The block II improvements (A2 standard vehicles) are fitted with improved armour protection, has an enhanced survivability due to restowing fuel and ammunition and using spall liners and additional armour protection of internal key components. Weight has increased to 29.94 t. The drive train was improved and the suspension,too. It has no side firing ports and carries an infantry squad of six soldiers (M2 seven). 662 A2 vehicles are equipped with the Cummins VTA-903 engine with 441 kW.

1433 of the A2 tanks are now in an upgrade program to the ODS standard as a result of Operation Desert Storm. Improvements include driver's thermal viewer AN/VAS-3, an eye-safe LRF, a missile countermeasure device mounted in a box on top of the roof on the main sight exit, GPS, battlefield combat identification and a complete interior restowage. For 175 vehicles an ERA set from RAFAEL was procured to improve protection against HEAT charges. The set consists of 105 ERA tiles to be fitted on the hull front and flanks and on the turret.

1602 A2 BRADLEYS will be converted to A3 status. This upgrade program comprises new ballistic fire-control with auto-tracking and automatic boresighting, digital vehicle electronics and information displays, combat ID, commander's independent thermal viewer, a FLIR monitor for the infantry squad, BITE and additional roof protection. Conversion program will last until 2004.

85 M3A2 vehicles were converted to the LINEBACKER model. The TOW launcher is replaced by a launch pod with four STINGER surface-to-air missiles.

266 A2s are converted to mount two STINGER gunners seated in the rear facing the rear. Six missiles are carried. The vehicles are designated Bradley Stinger Fighting Vehicle.

The M7 is also derived from the M3A2 BRADLEY. In place of the TOW launcher on the left side of the turret the box now houses the targeting equipment.

It consists of the AN/TVQ-2 Ground Vehicle Laser Locator Designator and the AN/TAS-4B thermal sight. The vehicle is also equipped with an inertial navigation system, an eyesafe LRF, the AFADTS (Advanced Field Artillery Tactical Data System) and additional radio equipment.

Role. In service in the US Army (4050 M2A2, 421 M3A2, 49 FIST, the remaining approx. 1400 BRADLEYs, subsequently designated A0, are not any more in front-line service) and in Saudi Arabia (400).

Assessment. The BRADLEY has a good overall performance but lacks sufficient fire-power and armour protection, even when upgraded to A3 standard.

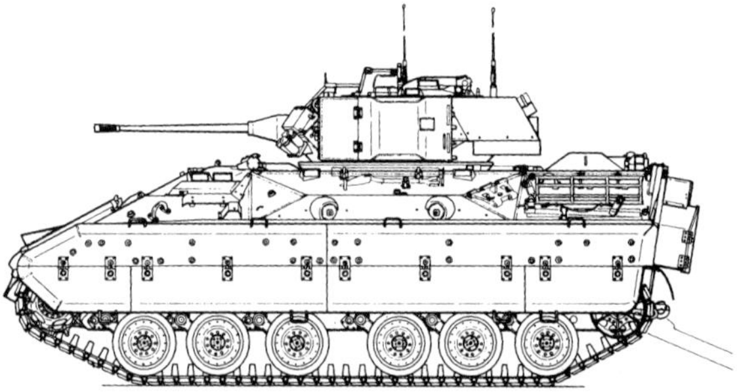

M2 IFV

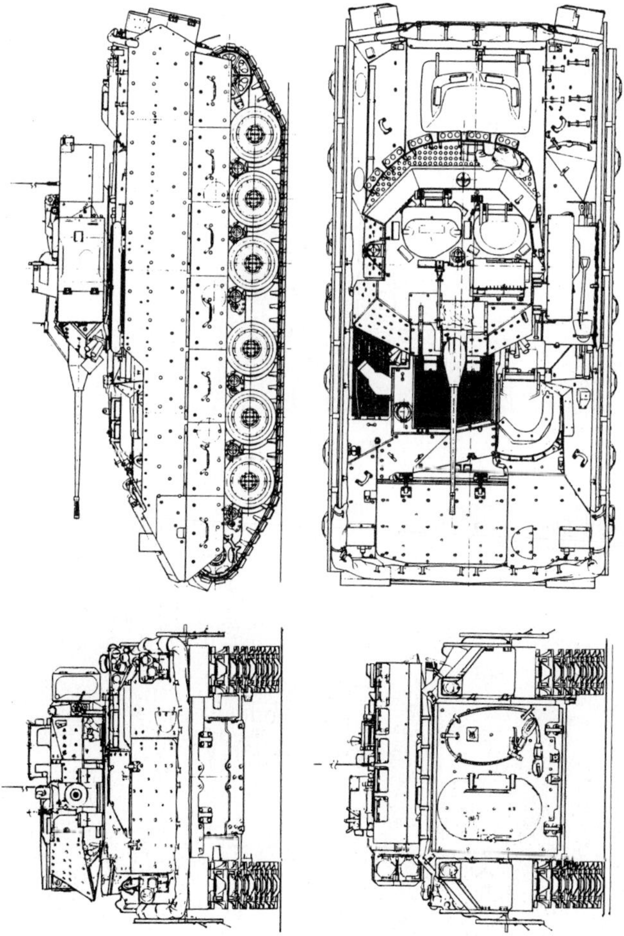

M2A2 BRADLEY IFV

M2A2 IFV

M3 CFV

M3A2 CFV

M6 LINEBAKER

M7 FIST

AIFV
Armoured Infantry Fighting Vehicle

Models.

AIFV (Dutch version: YPR 765 PRI) (Belgium AIFV-B.50)
K200 / K200A1 (South Korean basic version)
ZPT with 12.7 / 7.62 mm MG turret (Turkey)
YPR 765 PRI/I (AIFV-B-25)
ZMA with DAF SP turret with 25 mm Contraves cannon (Turkey)
ZMA with 25 mm DRAGAR turret (Turkey)
YPR 765 PRAT TDM
ZTA (Turkish Armoured TOW Vehicle)
AIFV-B-MIL TDM
K216 ANBCRV (South Korea)
YPR 765 PRCO-B AICV (AIFV-B-CP)
YPR 765 PRCO-C1, -C2,- C3, -C4 and -C5 AICV
K277 ACCC (South Korea)
YPR 765 PRRDR AIRV
YPR 765 PRRDR-C AIRV
YPR 765 PRGWT AIAV (AIFV-B-AMB)
K242 AMC (South Korea)
K281 AMC (South Korea)
YPR 765 PRMR mortar tractor
YPR 765 PRVR-A and PRVR-B ACC
K263 AADG (South Korea)
BI-HO AADG (South Korea)
YPR 806 ARV
AIFV-B-ENG engineer APC
AIFV-B-REC ARV
K288 ARV (South Korea)
AIFV-B-MTC AMV (Turkey: ZHT)
AIFV with 12.7 mm MG instead of 25 mm cannon (Philippines)

Numerous non-serial versions (also in South Korea and Turkey)

Development and production.

1967 two MICV under the designation XM765 built by Ordnance Division (FMC) based on M113 APC.
1970 further improved vehicles (M113A1 PI) as a private venture.
1978 first deliveries to the Netherlands (ordered 1975).
1981 Dutch follow-on order including TOW-version.
1982 licence production starts in Belgium.
1985 South Korean licence production begins.
1990 licence production in Turkey starts.

Distinguishing features. M113 hull with numerous improvements. About 85 % spares commonality with M113A1. Side. Glacis plate with trim vane. Hull roof horizontal to hull at rear angled downwards. Two firing ports with vision blocks above at each side of fighting compartment. Large ramp in hull rear. Front. Sides vertical half way up, then sloped inwards in area of fighting compartment. Engine front right, driver front left. Commander behind him with his own cupola.

Other details. Hull made of welded aluminium with an additional layer of spaced laminate steel armour bolted onto the hull.
YPR 765 PRI-I and AIFV-B-25 with centrally located 1-man turret with Oerlikon 25 mm KBA-B02 cannon with dual feed (180 rounds and 144 in reserve). Elevation is from -10 to +50 degrees. Mounted coaxially to the left is a 7.62 MG with 230 rounds (1610 in reserve).
(TABLE)
YPR 765 PRAT (Pantser Rups Anti-Tank) has the Emerson ITV kit as adapted on M901 TDM with twin TOW ATGM launcher (10 missiles). Weight is 11.5 t. It has a crew of 4.
Command versions with additional radio equipment in different configurations. YPT 765 PRCO-B is the company cdr's version, armed with 25 mm turret, too. The South Korean K277 has a raised rear superstructure.
YPR 765 PRRDR with British ZB 298 battlefield surveillance radar.
Ambulance versions (11.6 t) with four stretchers.
YPR 765 PRMR tows a 120 mm Thomson Brandt mortar. Crew 3 + 4 with 51 mortar bombs in rear compartment. Combat weight is 13 t.
Cargo versions have a crew of two and a weight of 13.4 t maximum.
YPR 806 PRBRG has a HIAB 3 t crane mounted on the roof, an internal 9 t winch and land anchors on both sides of rear ramp. It is also fitted with additional buoyancy pods. It has a regular M113 box-shaped superstructure.
Belgium version almost identical. Suspension is similar to M113A2, it has an individual NBC system and an external air intake/exhaust personnel heater.
AIFV-B-MIL carries a quick-disconnect mount for the MILAN ATGM on the left vehicle side.
The AIFV-B-ENG carries an engineer team and has a winch.
The cargo carriers have a capacity of 2 t payload.
The AIFV-B-REC has a hydraulic winch and a fairlead assembly for vehicle's recovery. A manually operated crane is mounted on the left hull top.
AIFV-B-MTC has a HIAB hydraulic crane on the mid-left top, a winch and special maintenance tools.
The basic South Korean version is designated K-200; it has a different type of armour (British firm Alcan) and is powered by a German MAN D-2848M V-8 diesel engine with 206 kW coupled to a British semi-automatic transmission T-300 with seven forward and reverse gears. Upgraded versions (A1 suffix) are powered by the MAN D-2848T engine developing 257 kW; the Allison X200-5K transmission is now fully automatic. The NBC version has NBC detection equipment including a MM-1 mass spectrometer and a marking set. The mor-

tar veriants carry either a 107 mm mortar with 88 rounds (K242) or a 81 mm mortar with 114 rounds (K281). The anti-aircraft version mounts a Vulcan 20 mm gatling cannon with a stowage of 2000 rounds; weight of the vehicle is 13.2 t. The K288 has a hydraulic winch with a capacity of 10 t and a 6 t crane mounted on the left hull roof.
Using a modified chassis (six roadwheels, sprocket front) Daewoo Heavy Industries developed an AADG now called BI-HO. It carries a two-man turret armed with two 30 mm Oerlikon KCB cannons on both sides. This weapon has a rate of fire of 600 rpm and an effective range of 3 km. Surveillance radar is a Siemens x-band system. Targeting is by using a SFIM stabilized panoramic VISAA sight combined with a Hughes Aircraft TV/thermal imaging camera. The vehicle's weight is 25 tons; it is powered by a Daewoo 10 cylinder D2840L engine with 382 kW coupled to a ZF LSF 1500 automatic drive.
The Turkish versions are powered by the Detroit Diesel Silver 53T with 221 kW coupled to an Allison X-200-4 transmission. The basic version is called ZPT (Zirhli Personel Tasiyici), the versions armed with a 25 mm gun turret are designated ZMA (Zirhli Muharebe Araálari) and are additionally equipped with Texas Instruments thermal sights. Following Swiss armaments ban Turkey switched from the Oerlikon cannon to the GIAT M811 25 mm weapon. The ZTA (Zirhli TOW Araci) version is fitted with the Norwegian ALT turret with two ready to launch missiles; vehicle's weight is 13.69 t. The ZHT (Zirhli Havan Araci) carries a 81 mm mortar.

Role. In service with Bahrain (25 YPR-765), Belgium (236 25 mm, 222 MILAN, 56 CP), Egypt (304 YPR 765 PRI, 6 PRCO-B, 210 PRAT, 79 PRCO-C), Malaysia (111 K200 incl. 4 AMC, 5 ARV, ACCC and AAV each), Netherlands (1426 incl. 42 PRAT, 135 PRMR, 45 YPR 806), Phillipines (45 incl. 6 recovery), South Korea (2000+, still in production, 150 BI-HO on order), Turkey (830 AAPC, 515 AIFV DRAGAR, 135 AIFV DAF, 48 ATV, 170 AMV) and the United Arab Emirates (53 APC, 75 ACCC and 8 ARV on order from Turkey).

Assessment. Cost-effective AIFV based on proven components with average fighting power. Section can fight under armour. Suitable for many variants.

YPR 765 PRI

K200 AIFV

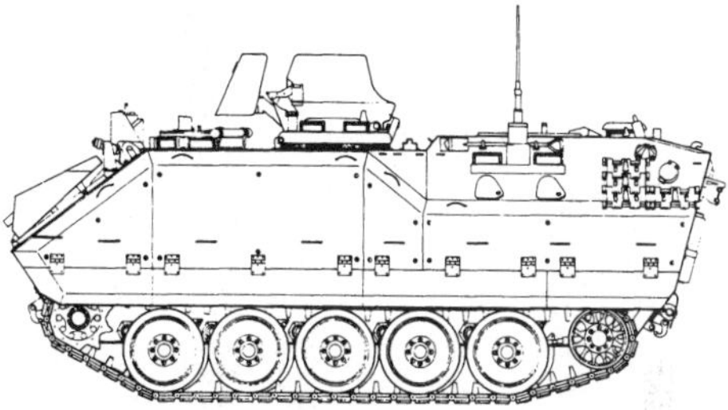

K200 AIFV

AIFV-B-25

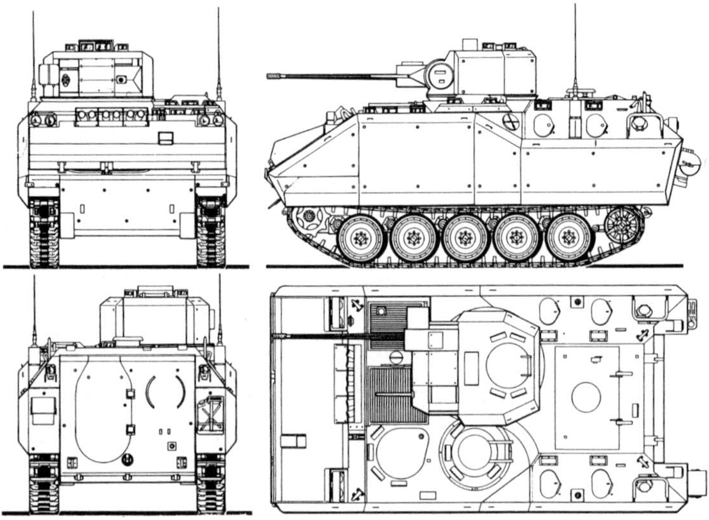

YPR 765 PRI/I

ZPT AIFV

ZMA AIFV (DAF SP)

ZMA AIFV (DRAGAR)

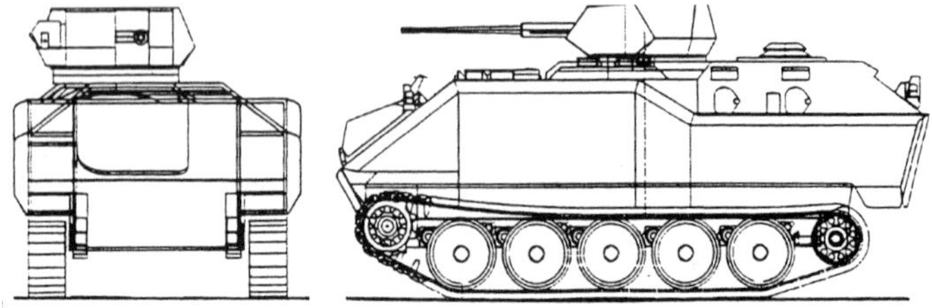

ZMA AIFV (DRAGAR)

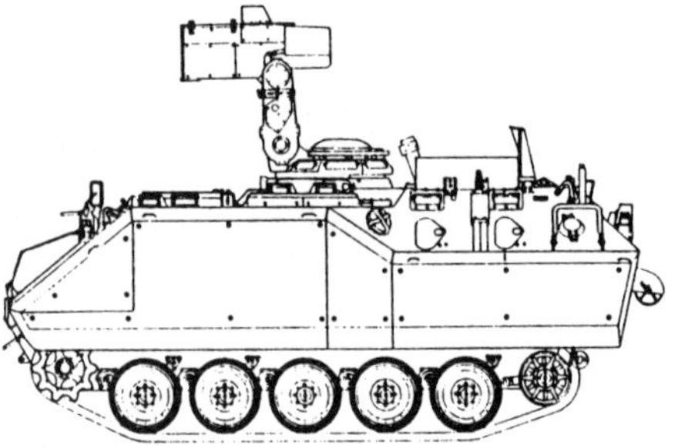

YPR 765 PRAT TDM

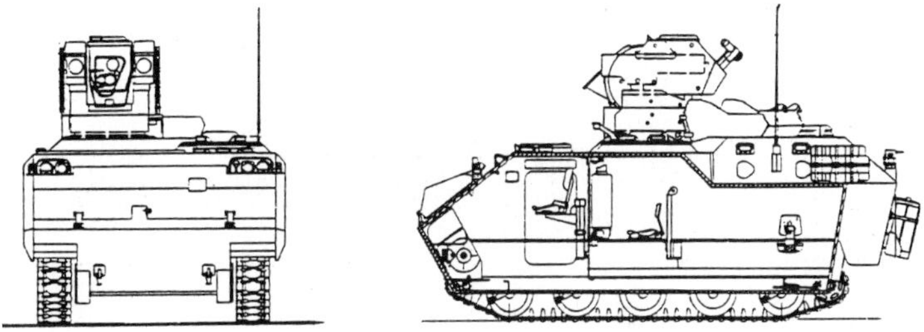

ZTA TDM

ZTA TDM

YPR 765 PRCO-C1 AICV

YPR 765 PRGWT AIAV

K 281 AMC

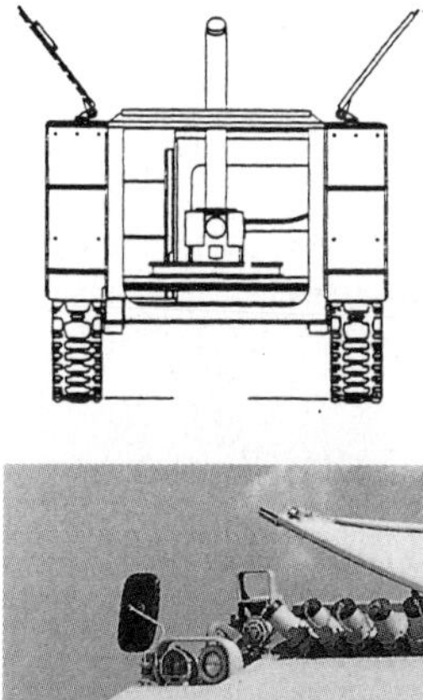

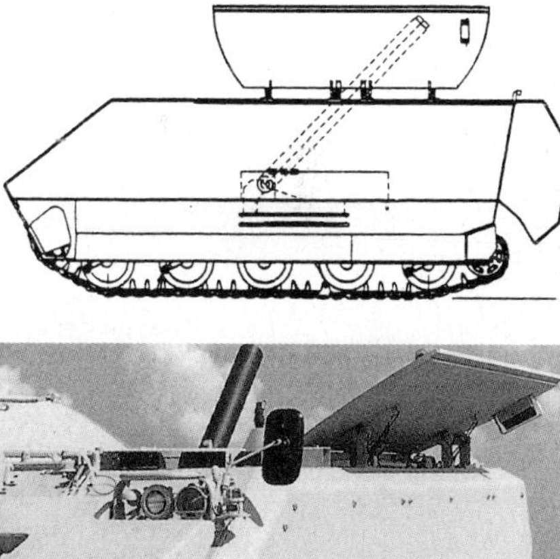

ZHT AMC

YPR 765 PRMR AMTC

K263 AADG

BI-HO AADG

M113 armoured general purpose carrier

Models.

M113 APC
M113 ACCC (all versions)
M113 AAC (all versions)
M113A1 APC
M113/M113A1G with GREEN ARCHER artillery radar system
M113A2 APC
M113A3 APC
M106 AMC (all versions)
M125 AMC (all versions)
M113A1/A2 with different 120 mm mortar systems
M577 ACCC (all versions)
M579 Fitter's Vehicle
M741 AADG (also known as M163)
M806 ARV (formerly M579)
M901 ITV (Improved TOW vehicle)
M981 FIST (Fire Support Team Vehicle)
M1064 (type classification for 120 mm AMC)
M1059 Smoke Generating System
M1068A3 ACCC
M1070 EFWS
M1101A3 Smoke Obscurant Carrier
M113A1 FSV with SKORPION turret (Australia)
M113A1 LRV (Light Recce Veh.) with MG turret (Australia)
M113A1B (Belgium, different variants)
M113A2 with TUA (TOW Under Armour) (Canada)
M113A2 EVSEV (Engineering Variant Specially Equ.Veh.) (Canada)
M113A1 with OTO Breda T25 turret (Denmark)
M113A2 with Serge Dassault TA-23S turret (Sinai 23, Egypt)
M113A2 with Serge Dassault RA-20S radar system (Egypt)
M577A2 with TRACKSTAR air-defence radar (Egypt)
M113A1G (Germany, different variants)
M113A1IL ZELDA (Israel, different variants)
VCC-1 APC (Italy)
VCC-1 with Improved TOW System (Saudi Arabia)
VCC-2 APC (Italy)
SIDAM 25 AADG (Italy)
M113A1 with Contraves SHORAD AADRV (Italy)
NM 135 (Norway, M113A1 with Swedish HÑgglunds 20 mm turret)
NM 142 with Kvaerner-Eureka ALT (TOW ATGM launcher) (Norway)
M113A2 with Swedish RBS 70 SAM (Pakistan)
M113A2 with ANZA SAM and twin Type 75-1 14.5 mm MG (Pakistan)
SPz 63/73 (Switzerland, different variants)
SPz 63/89 Swiss modernisation of SPz 63/73

ADATS AADG/TDM (Switzerland)
M113 with KUNG FENG multiple rocket launcher (Taiwan)
M113 AIFV, Taiwanese version (different variants)
M548 is a non-armoured carrier

Non-serial versions.

Vast number of modifications and different applications.

Development and production.

1956 start of development. FMC (aluminium) prototype T113 in favour of T117 steel PT adopted.
1960 start of series production.
1962 production of M577 starts.
1963 diesel engined version M113A1 adopted.
1978 production of M901 starts.
1979 Product Improved M113 (A2 standard) in production/retrofit.
1986 ADATS selected by Canada.
1987 production/conversion of M113A3 ends.
1988 introduction of M1059.
1990 last production vehicles for the US Army.
1994 reopening of production line at San José for foreign sales.

Distinguishing features. Side. Five roadwheels, no return rollers, sprocket front, upper part of suspension covered by skirt. Box-shaped hull with sloped front, horizontal roof and vertical rear with large ramp. Front. Sloped glacis with trim vane. High superstructure with vertical sides.

Other details. The M113 is used in numerous roles and has been modified many times. Generally a single 12.7 mm or a 7.62 mm ADMG is used in an external mounting in front of the central cdr's hatch. Total production exceeds 80.000 vehicles.
The basic M113 (14.813 produced) was powered by a Chrysler 75M V-8 petrol engine of 154 kW. Most of these vehicles have been brought to the A1 standard with GMC Detroit Diesel engine of 158 kW (25.323 newly produced vehicles). Of the M113A2 version 6141 had been issued with improved engine cooling and suspension. The single air heater is replaced by a dual air one. Many other family vehicles had been brought to the A1 resp. A2 standard.
(TABLE)
The M113A3 is equipped with a stronger GMC engine (202 kW) coupled to a new X-200-4 transmission, has internal spall suppressant armour, a steering yoke instead of the former steering laterals, and two external rear fuel tanks (also offered as a retrofit option).
Vehicle's weight of the baseline versions (including fuel and crew with equipment) rised from 9.1 t, 11.27 t (A1), 11.33 t (A2) to 12.27 t (A3). Using external fuel tanks weight grows by 414 kg with spall liners by 336 kg. Different

external armour kits are offered resp. used increasing the weight by 2 - 3 tons. Different turret options result in weight increases up to 3 tons.
Numerous countries use different dozer or mine clearing devices and weapon stations in open mount.
The M106 and the M125 series are US-built mortar carriers with the 107 mm resp. 81 mm mortar. Their weight is 12,16 resp. 11.46 t as A2s. Different mortar installations are built in Egypt, Germany, Singapore, Spain and Switzerland with 120 mm weapons.
The M741 is armed with a six-barreled M61A1 20 mm gatling gun (VULCAN) with a rate of fire of 3000 rpm. Most of them had been upgraded with PIVADS kit with a director-type sight, the AN/VPS-2 ranging radar system and a fire-control computer. Its weight is 11.7 t.
Several armies use the M113 as a stop gap solution with a pedestal-mounted TOW resp. pintle mounted MILAN ATGM system.
The 11.7 t M577 series is a special command and control version of the M113 with a higher roof to the rear of the driver's position. An improved depot conversion using the SICPS (Standard Integrated Command Post System) is designated M1068A3.
The M806 has a 3 t crane, a rear mounted winch and special maintenance equipment. Its weight is 11.57 t.
The M901 has an armoured retractable launcher for two TOW missile containers. The elevation arc is -31 to + 38 degrees, 360¯ traverse. 12 missiles are carried, to be loaded by hand via a rear roof hatch.
The M981 is externally identical with the M901 and is used to locate and designate artillery targets. It has a GVLLD (Ground Vehicle Laser Locator/Designator), a north-seeking gyrocompass, day/night sights and a digital data transmission device. The vehicle has a weight of 12.7 t.
The M1059 has two M54 smoke generators on top of the hull rear. Most are converted to the A3 Plus standard with new M56 diesel powered smoke generators.
The M1101A3 is based on a modified M901A1 with its launcher modfied to accomodate two M264 70 mm Smoke Projecting Rocket launchers. Each container houses 19 HYDRA-70 red phosphorus rockets with a range up to 6 km.

The M113A1B were license-built in Belgium in several variants. The M113A1-B-ATK has the 12.7 mm MG on the right and an unarmed cupola on the left. The M113A1-B-MIL has the 12.7 mm MG on the left cdr's cupola and an external mounting for the MILAN ATGM system. Two Bofors LYRAN launchers are on top of the rear left side. The M113A1-B-REC has a hydraulic winch and a manual crane on the top left. The M113A1-B-MTC has a HIAB crane on the left top (similar to the M579). 21 M113A1-B are fitted with the Israeli Elta EL/M-2130 artillery fire control radar system. Other vehicles are used in the engineer, command, ambulance and driver's training role.
Canada has procured 64 TUA turrets produced in licence of the Norwegian Thune-Eureka for the integration into M113 brought to A2 standard. Canada has also modified 70 of their existing engineer vehicles to the EVSEV configuration with a hydraulic earth auger and other special equipment.

Egypt introduced a low-level air-defence system using the Soviet-designed 23 mm ZU-23 cannon. The firing unit consist of a modified TA-20 turret (TA-23E) with two 23 mm cannons and three Egyptian SAKR EYE (copy of the Soviet SA-7 GRAIL) anti-aircraft IR-homing missiles on either side of the turret. The system has an Officine Galileo fire-control system. For every three or four units an acquisition/surveillance vehicle with the Dassault Electronique RA-20S radar is assigned. This system has a detecting range of 12 km operating in the E-band.
Germany is also an intensive user of the M113. Beside its employment as an APC, an ACCC, a 120 mm Tampella mortar carrier (500) and an AAC (407) the vehicle is used in several C3 roles, e.g. in the artillery fire control or air control mission, as an armoured aiming circle vehicle. The M113A1 ABRA is the artillery observation vehicle with the radar surveillance system RATAC mounted on top of an elevatable mast. 320 artillery observation vehicles are in use with a PERI D11 periscope with laser rangefinder, a land navigation system and a data transfer system. 346 vehicles are being upgraded with a hydraulic two-cycle brake system, 143 carriers additionally receive a new Euro-II engine and an automatic steering system with steering wheel.
220 M577A1G are in use as command post facility, information cell or air support cell.
The Italian VCC-1 has the upper side walls of the rear compartment sloped with two firing ports on each side with observation windows above. The MG mount consists of collapsible shields. The TDM version was delivered to Saudi Arabia and has the elevatable TOW launcher of the M901. The VCC-2 has vertical side walls.
The Israeli Army employes the M113 mostly as APC but also in many specialized versions. The anti-tank versions is equipped with the TOW ATGM system on an retractable telescopic mast. The artillery ACCC uses the PADS (Position and Azimuth Determining System). An ECM version carries the RAJ 101 acquisition and jamming system. Another version carries the EL/M-2106 point defence alert radar system. Most of the Israelian APC were retrofitted with add-on armour.
SIDAM 25 is a M113A2 with a high turret with externally mounted four 25 mm Oerlikon-Italiana KBA-B cannons, two of the guns having dual feed. Each barrel has a rate of fire of 570 rpm, single shot, 15 or 25 shot bursts. There are 600 HEI-T ready-use rounds plus 30 APDS-T for ground defence. Maximum range is 2500 m. The turret is power-operated and can be traversed 360 degrees (120/sec). Maximum elevation is +87, max. depression -5 degrees. The turret has day clear weather and LLLTV camera sighting system which tracks targets automatically. The digital fire control computer is linked to a laser rangefinder and an IFF system. 280 systems are ordered.
Singapore upgraded most of its fleet with, add-on armour, an enlarged trim vane to improve amphibious stability and a CIS 40/50 cupola with a 40 mm automatic grenade launcher and a 12.7 mm MG in the same weapons cradle.
Switzerland was one of the first foreign users. In 1973-75 625 APC were equipped with the Swedish Hagglund turret mounting the 20 mm HS804 cannon from

obsolete fighter aircraft. Other versions include a mortar carrier with the 120 mm PzMw 64, the command version (with/or without 20 mm turret), artillery control, engineer and fitter versions. From 1991 onwards 316 APC and 66 ACCC were modernized with additional armour and a new Detroit Diesel 6V53T engine with 195 kW (M-63/89).
ADATS was an Oerlikon-Bührle private venture project of a modular missile launching system for the engagement of both aerial and ground targets. This system engages aerial targets up to an altitude of 6 km from ranges of 1 out to 10 km and ground targets from 500 m to 8 km. The Contraves Italiana I/J band radar has a detection range of 24 km. The system also has a thermal sight and TV trackers. The missiles are laser tracked. The missile is 2.057 m long, has a diameter of 0.152 m and weighs 12.5 kg. Its warhead has an electro-optical proximity and an impact fuze. Speed is up to Mach 3, armour penetration exceeds 900 mm RHA.
Taiwan has developed a slightly modified version with five firing ports around the fighting compartment. It is powered by a British Perkins TV8.640 diesel engine with 158 kW coupled to a Taiwanese transmission. Taiwan employs also versions with Kuen Wu ATGM and the Kung Feng IV 126 mm multiple rocket launcher (range 9 km).
Several M113 derivatives have been dismantled or phased out (i.e. M113 ACAV, M132 flame thrower, M113 AVLB).
The unarmoured M548 carrier is a M113 derivative used in many roles (i.e. M474 PERSHING, M667 LANCE, M727 HAWK, M730 CHAPPARAL and different radar, mine scattering resp. transportation carriers).
The US Army converted numerous M113 for OPFOR (Opposing Forces) missions to simulate former Warzaw Pact vehicles e.g. BMP or ZSU-23-4.

Role. In service with Argentina (266, 18 M577), Australia (724 incl. 48 with SCORPION turret, T50 turrets are being replaced by a Transfield one-man turret, 22 M125), Bahrain (110), Belgium (525 incl. 72 MILAN), Bolivia (50), Brazil (614 upgraded with OM 352A diesel engine), Cambodia (175 delivered, status unknown), Canada (985 incl. 108 TOW, 61 M577, 30 M579, 70 EVSEV, 36 ADATS), Chile (115 + 128 ex-Italian), Colombia (60), Costa Rica, Denmark (640 incl. 56 TOW, 50 with T-25 turret, 9 GREEN ARCHER, 68 M106, 10 M125, 40 M579), Ecuador (64 incl. 44 M741), Egypt (2000 incl. 100 M106A2/M125A2, 26 FIST, 70 M577A2 plus 4 M577A2 with cranes, 95 M806, 40 M901), Ethiopia (70), Germany (3138, including 453 AAV, 500 AMC, 124 AMFDV, 86 AACPV, 329 AAOV, 68 RATAC, 152 AAFDV, 220 M577), Greece (1669 incl. 82 TOW, 40 M901A1), Guatemala (15), Haiti (12), Iran (200), Israel (5900 incl. 450 ex-US M901A1), Italy (3283 incl. 1002 M113, 1792 VCC-1/1, 270 TOW, 416 M106, 275 with SIDAM, 220 M577), Jordan (1200), Kuwait (231 incl. 50 M577), Laos (50), Lebanon (867 + 22 M125A2), Libya (100 VCC-1), Morocco (548 incl. 36 M125A1, 75 M577A1, 40 M741), Netherlands (514 incl. 53 RADAR, 12 M106), New Zealand (78), Norway (347, 97 NM-142, 53 NM-135, 28 M106, 12 M125), Pakistan (800), Peru (300), Philippines (100, 24 M741), Portugal (534 APC incl. 20 M106, 17 M125, 25

M901A1, 9 M577, 34 M741), Saudi Arabia (1700 incl. 224 VCC-1 TUA), Singapore (727 incl. 50 AMC and 10 M125A1, 175 each are retrofitted to IFV 25 and IFV40/50), Somalia (24 TOW), South Korea (560 incl. 60 M741, 140 M577), Spain (1313 plus 192 mortar carriers, 18 M577), Sudan (116 incl. 15 M125A1), Switzerland (192 M-63/73, 315 M-63/89, 836 M113, 132 M-64 120 mm AMC), Taiwan (1100, 31 M577A2 on order), Thailand (508 incl. 12 M106, 21 M125, 18 M901A3, 11 M577A2, 24 M741), Tunisia (120, 47 TOW, 12 M106A2, 25 M741, 18 M577), Turkey (2813 plus 365 with TOW), USA (28300 incl. 21775 APC, 350 TOW carriers, 1884 M106, 799 FIST, 3511 M577, 380 M741A1, 277 M1059, 250 M1070), Uruguay (15), Vietnam (approx. 100), Yemen (76 + 6 M577A1, 12 M106, 20 M741) and Zaire (12).

Assessment. Simple, cost-effective, versatile worldwide mostly built AGPC. In many armies in use as an infantry fighting vehicle though being unsuitable for that purpose. Too lightly armoured, section cannot fight mounted. Heavy weapon installations and many modification measures interfere with amphibious capability.

M113 APC (Greece)

M113A1 with MEXAS add-on armour (Canada)

M113A2 APC (Egypt)

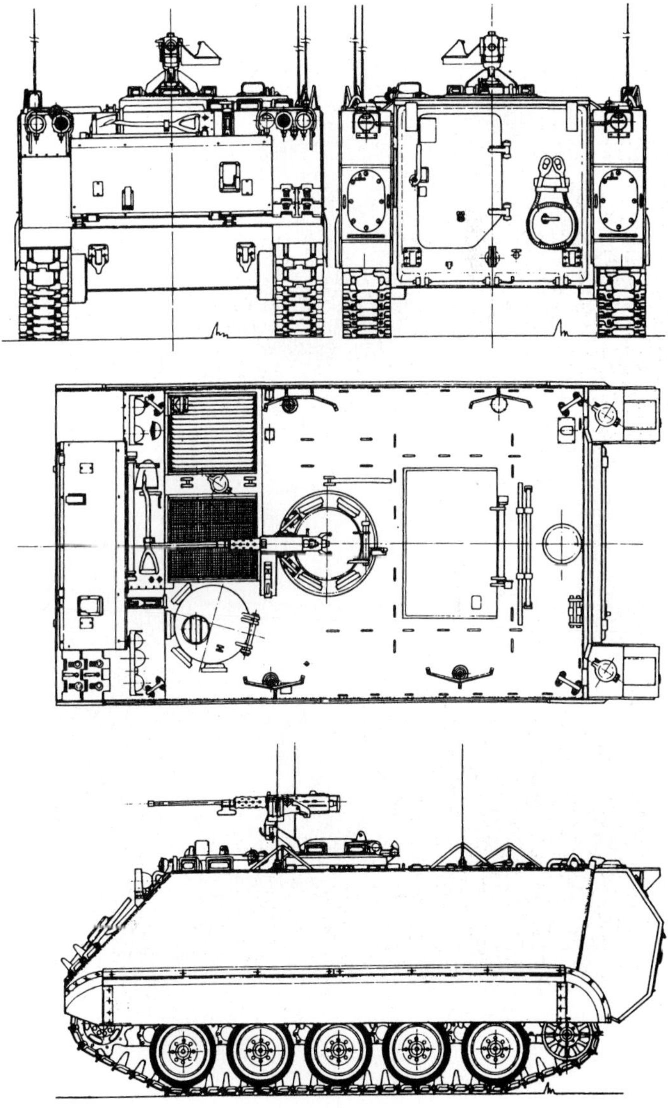

M113A2 APC

M113A3 with Armour Kit

M113 with GREEN ARCHER (Denmark)

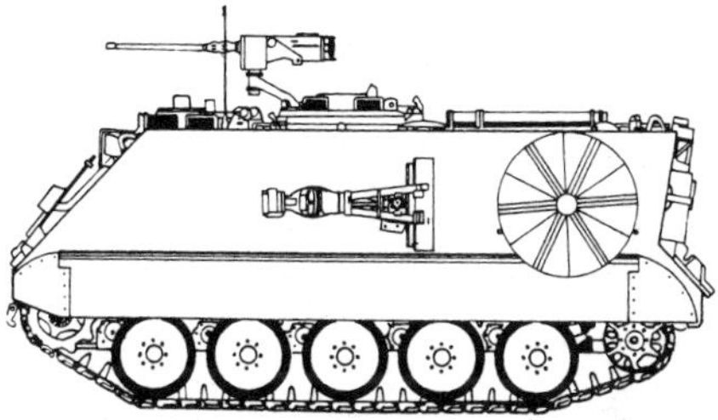

M106A1 AMC

M106A1 AMC

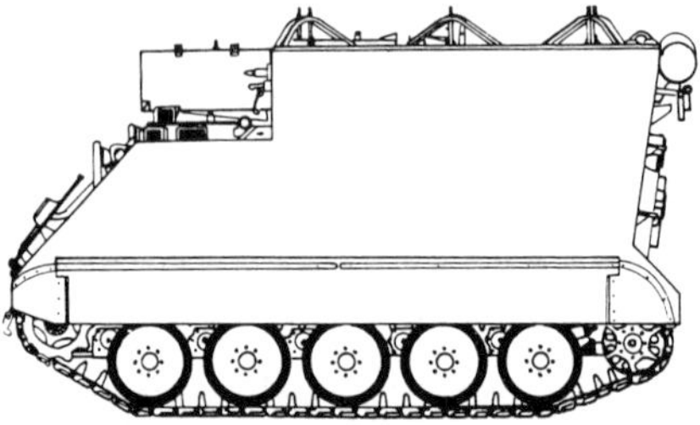

M577A1 ACCC

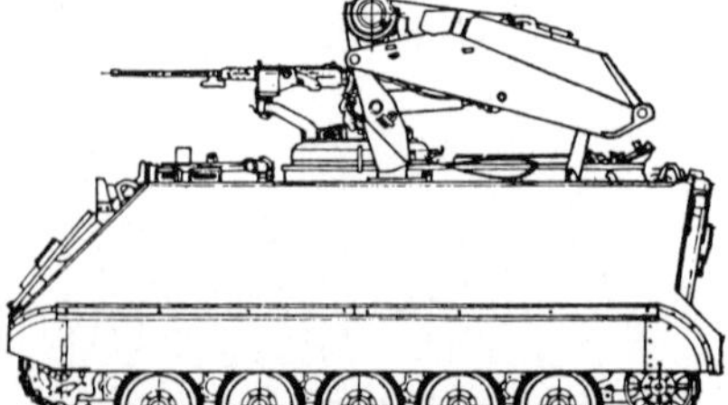

M579 Fitter (Australia)

M579 Fitter (Australia)

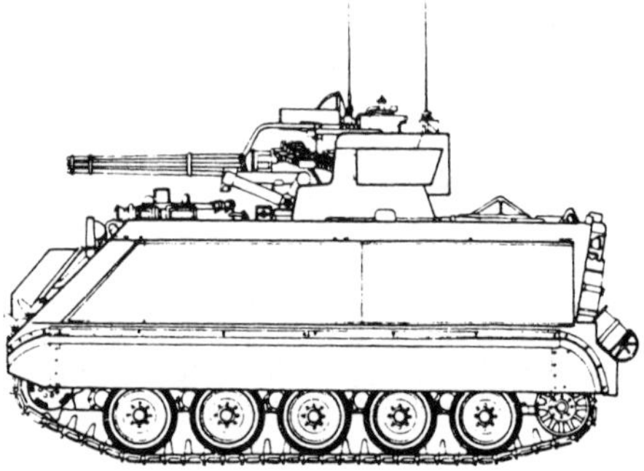

M741 AADG

M741 AADG

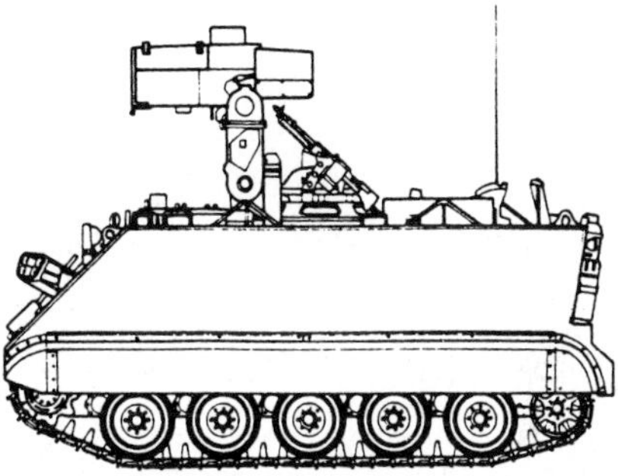

M901A1 TDM

M901A1 TDM

M981 FIST

M1064A3 AMC

M1059

M113A1 FSV

M113A1 LRV

M113A1B AMV

M113A2 TUA

M113A2 EVSEV

M113A1 (T 25 turret)

SINAI 23 AADG

M113A1G ACCC

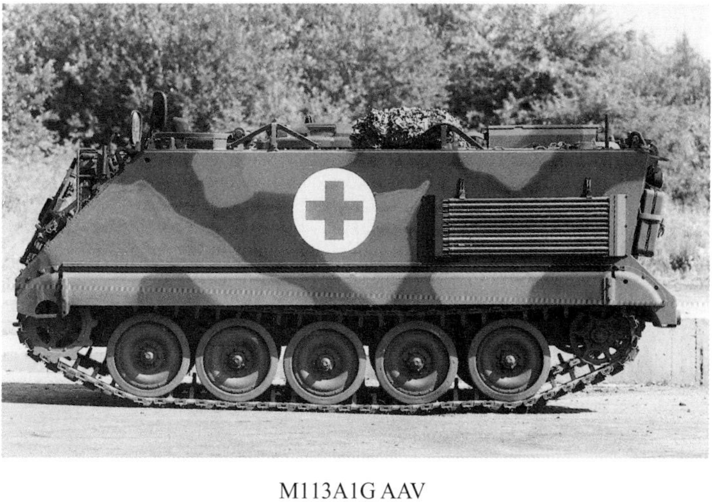

M113A1G AAV

M113A1G AMC

M113A1G AAOP

M113A1G AAFDC

M113A1G ABRA

M113A1IL ACCC ►

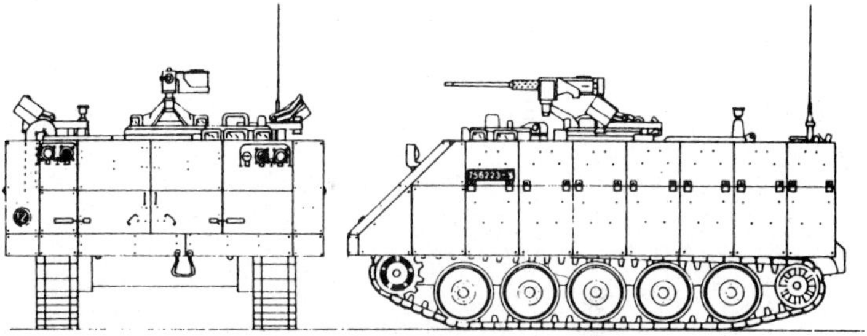

M113A1IL APC

VCC-1 APC

VCC-1 with ITOW

VCC-2 APC

SIDAM 25 AADG

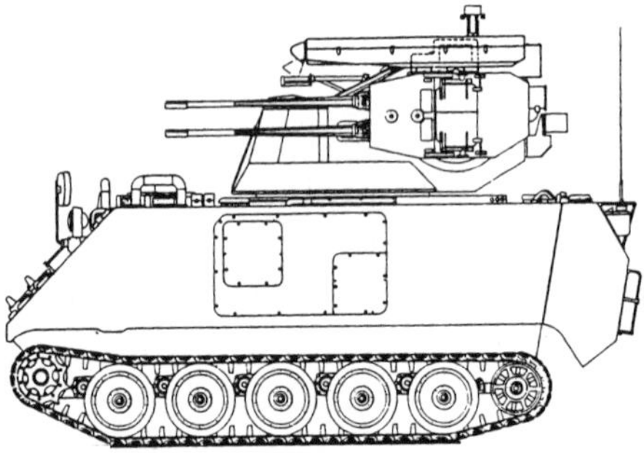

SIDAM 25 AADG

M113A1 with SHORAD (Italy)

NM 135 APC

M113A2 with RBS 70

SPz 63/73 AMC

SPz 63/89 APC

ADATS AADML/TDM

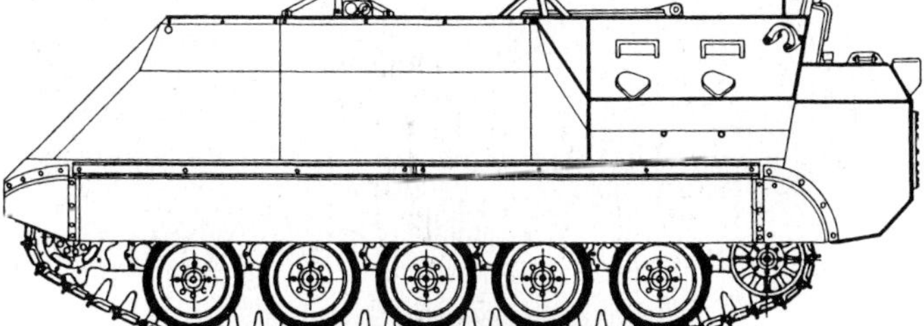

M113 AIFV (Taiwan)

M113 AIFV (Taiwan)

M113A1 with ZPU-4 MG turret (Lebanon)

M113A1 with ZU-23-2 turret (Lebanon)

M113 with TOW ATGMS (Italy)

COMMANDO V-100/-150/-200/-300 armoured general purpose carrier

Models.

V-100 APC
V-100 with 1 m MG turret
V-200 with 1 m 20 mm turret
V-200 with 20 mm two-man turret
V-200 with 90 mm Mecar gun
V-200 with Swedish RBS-70 SAM
V-150 APC
V-150 ACCC
V-150 Base Security Vehicle
V-150 with 90 mm Cockerill Mk III gun
V-150 with TOW ATGM system
V-150 with 107 mm mortar
V-150 AADG
V-150 Police Emergency Rescue Vehicle (ERV)
V-150 Recovery Vehicle
V-150 AAV
V-150S (variants)
V-300 (6x6) with 1m turret
V-300 (6x6) with 90 mm turret

Related versions.

ARMADILLO APC (Guatemala)
CHAIMITE APC (Portugal)
DN-IV/DN-V/DN-VI APC (Mexico)

Development and production.

1962 private venture development by Cadillac Gage starts.
1963 first prototypes of V-100 and CHAIMITE issued.
1964 start of series production of V-100 (M706), followed by interim vehicle V-200.
1971 introduction of the V-150 for export, produced until 1985.
1979 first two prototypes as private venture.
1981 prototype of ARMADILLO by Guatemala completed.
1983 production start of V-300 (Panama and Kuwait).
1985 V-150S in production.
1994 delivery of V-300 to the Philippines.

Distinguishing features. Side. Two large wheels each side with coffin-like superstructure. Pointed front with horizontal roof and pointed rear as well. Two-piece door in front of rear wheel arch. Different turret arrangements. Front. Slightly sloped glacis. Cdr and driver side by side in front hull step. Hull sides above curved wheel arches slope inwards.

Other details. Basic version armed with Cadillac Gage one-man turret with either twin 7.62 mm MG or one 7.62 mm and one 12.7 mm MG. Manual traverse and elevation from -10, resp. -13 to +55 degrees. Either 3800 rounds 7.62 mm rounds are carried or 1400 rounds 12.7 mm reducing the 7.62 mm stowage.
(TABLE)
The V-100 has a weight of 8.9 t and a crew of three with seven infantry soldiers.
The V-200 is larger and weighs 12.73 t. It is powered by a Chrysler 440 CID engine of 202 kW.
V-150 variants are powered by diesel engines compared to the former petrol ones.
The Command version has a fixed armoured pod in place of the turret and a pintle-mounted 7.62 mm MG with 2000 rounds stowage. Its height is 2.31 m.
The Base Security Vehicle is of similar appearance but has hatch covers which open to all sides. 3000 rounds of 7.62 mm are carried. Its weight is 9.07 t.
The V-150 with 90 mm gun has an elevation of -8 to +28 degrees and carries 39 90 mm rounds and 3200 7.62 mm. It has a weight of 9.89 t and carries a crew of three plus up to four infantry men.
The TOW version carries seven missiles. Its weight is 8.96 t and it is operated by a crew of four plus two infantry soldiers.
The 81 mm mortar carrier has a weight of 8.85 t. 62 mortar bombs are carried. The range is 4.4 km. It is operated by a crew of five.
The air defence version is armed with the 20 mm six-barrelled General Electric Gatling cannon. It has two outriggers at the rear sides and one on the front.
The ERV has a fixed pod with six vision blocks and eight gun ports. Its weight is 9.16 t, the crew consists of three plus nine additional men.
The ARV has a winch with a maximum capacity of 11.34 t with a cable of 60 m and 19 mm diameter. A 4.5 t boom is pivoted on the front of the vehicle with two stabilisers lowered. 2000 rounds of MG armament are carried.
The V-200 is powered by a Chrysler CID engine with 275 HP and has a weight of 12.7 t. The 20 mm version is armed with an Oerlikon 204 GK cannon (coaxial 7.62 mm MG) has an ammunition stowage of 400 20 mm resp. 3200 7.62 mm rounds.
The V-150S vehicles have an improved cooling system and other modifications.
The V-300 is a longer 6x6 version powered by a Cummins 6 CTA 8.3 with 275 HP. The vehicle is 6.4 m long and has a weight of 15 t.
ARMADILLO is very similar in layout to V-100 but slightly larger. It is powered by a Detroit Diesel 6V-53 212 HP (first 10 vehicles), later by model 6V-53T with 300 HP. Later vehicles are also provided with firing ports.
CHAIMITE is a Portugese V-150 derivative differing in having a vertical hull rear with negative inclination. It is either unarmed or equipped with a one-man turret with 12.7 mm and 7.62 mm MG armament in a twin mount.
DN-IV is a local copy of the V-150 powered by a Cummins V8-504 diesel engine and has a fixed circular open-topped turret with loped sides, armed with a 12.7 mm MG. The DN-V (TORO) has a 20 mm turret, the latest DN-VI has two externally mounted MG, is slightly longer and used as APC for up to 10 infantry soldiers.

Role. In service with Bolivia (10 V-100), Botswana (12 V-150), Cameroon (43 V-150 incl. 8 with 20 mm), Chad (8 V-150S 90 mm), Dominican Rep. (8 V-150), Gabon (8 V-150), Guatemala (7 V-150, 18 ARMADILLO), Haiti (6 V-150), Indonesia (28 V-150), Jamaica (14 V-150), Lebanon (V-150 and CHAIMITE), Malaysia (190 V-100/150), Mexico (26 LAV-150ST, 2 LAV-150ST ARV, DN-IV), Peru (15 CHAIMITE), Philippines (80 V-150, 25 V-150S, 20 CHAIMITE, 24 V-300), Portugal (15 V-150, 79 CHAIMITE), Qatar (8), Singapore (250 V-150, 200 V-200), Somalia (V-100/150), Sudan (30 V-150, 24 V-150S), Taiwan (300 V-150), Thailand (150 V-150), Turkey (124 V-150S), USA (140 V-100) and Venezuela (76 V-100/150 with Deutz BF GL 913 265 kW diesel engine).

Assessment. Attempt to develop a cheap wheeled carrier based on commercial assemblies. Limited capabilities for internal security and police roles. Versatile armament solutions, good mobility.

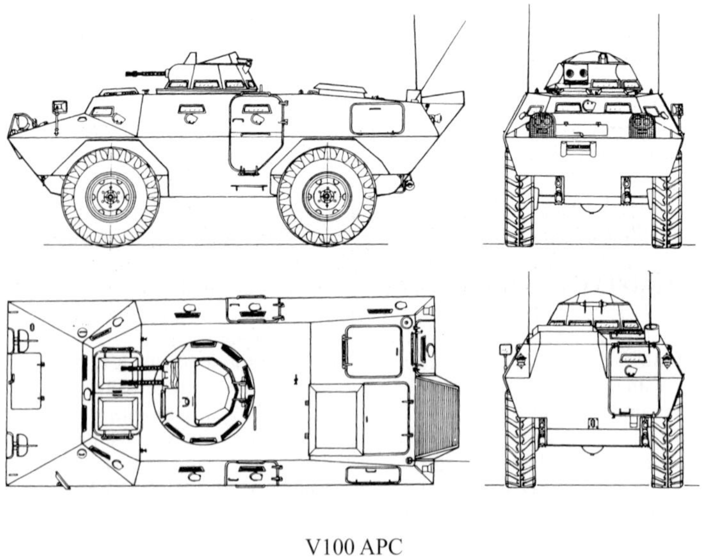

V100 APC

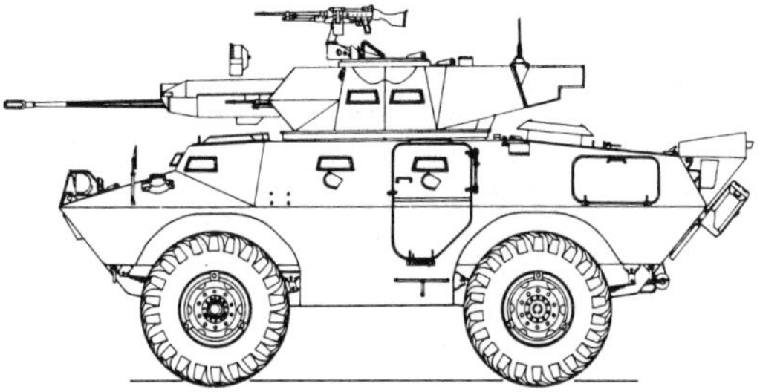

V-200 with Cadillac Gage 20 mm turret

V-200 with Cadillac Gage 20 mm turret

V-200 with 20 mm MECAR gun

V-150 APC

V-150 Base Security Vehicle

V-150 with 90 mm Cockerill Mk III gun

V-150 with 107 mm mortar

V-150 AADG

V-150 ARV

V-300 (90 mm turret)

ARMADILLO APC

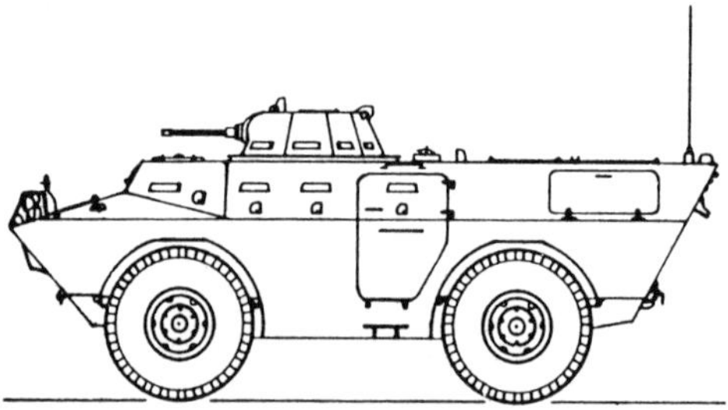

BRAVIA APC

DRAGOON/ASV 150 armoured personnel carriers

ASV 150: Wt 12.4 t L 6.25 W 2.57 H 2.59 191 kW 100 km/h

Models.

DRAGOON with Multi Gun Turret System
DRAGOON Patroller (ASV-XC)
DRAGOON AMV
M1117 (ASV 150)

Development and production.

1980 first production vehicle (DRAGOON 300) completed.
1988 introduction of law enforcement versions.
1995 follow-on version ASV 150 selected by the US Army.

Distinguishing features. The overall layout is similar to that of the COMMANDO (4x4) vehicle series. Different is the shape of the rear section which is straight with a negative angle. The DRAGOON features a rear door on the left side, the ASV 150 the engine grill.

Other details. The DRAGOON family (several version on offer) is based on the M809/M939 truck series. It has a light armour superstructure. Up to 12 persons can be seated according to mission layout. Two firing ports on each side and one to the rear are available. The vehicle is powered by a Detroit Diesel 6V-53T turbo-charged engine with 221 kW coupled to an Allison MT-653 automatic transmission. Maximum speed is 116 km/h, the cruising range is 900 km. The vehicle's weight ranges from 12 to 13 tons. The initial version was equipped with a one-man MGTS armed with a 12.7 and a 7.62 mm MG in twin arrangement.
The armoured security versions have an extended cab with slightly higher interior (2 metres headroom) and lack the MG turret. They are provided with obstacle removing ram and additional police equipment.
The armoured maintenance vehicle has a 2.3 t boom crane and a special maintenance tool set.
The ASV 150 has commonality with the LAV-150ST vehicle (S stands for stretched, T for turbocharged). But the solid axle construction is replaced by an independant coil spring suspension. The vehicle will be either equipped with the UGWS 40/50 (UpGunned Weapon Station) already in use with the AAV7A1 armed with a 12.7 mm MG and a 40 mm grenade launcher (Military Police) or a high-energy laser for destruction of explosive charges (US Air Force). The ASV 150 has a weight of 12.4 tons and is powered by a Cummins 6 CTA 8.3 diesel with 191 kW coupled to an Allison MD 3560 automatic transmission.

Role. DRAGOON was delivered to Canada, Thailand, Turkey, the US forces and Venezuela (100) and several federal police agencies. The ASV 150 will be delivered to the US Army (95 vehicles) and the US Air Force.

Assessment. More advanced LAV-100/150 successor, suitable for military police role.

DRAGOON APC

DRAGOON Patroler

M1117 APC

ASV 150 with HE-LASER

AAV7 Amphibious Assault Vehicle

Wt 22.8 t L 7.94 W 3.27 H 3.26 294 kW 64 km/h 482km Crew 28

Models.

AAV7
AAV7A1
AAV7A1 with 40 mm/12.7 mm gun turret
AAVC7A1 Command version
AAVR7A1 Recovery version

Development and production.

1964 requirement for a new LVTP (Landing Vehicle Tracked, Personnel) issued.
1967 15 prototypes (LVTPX12) delivered by FMC Corporation.
1970 start of series production, then called LVTP7, ending 1974.
1977 conversion of 14 LVTP7 to LVTP7A1 (later AAV7A1).
1983 restart of production of AAV7A1 until 1986.
1986 240 retrofit packages ordered with new Cadilac Gage turret.
1988 first appliquÇ armour plates ordered.
1995 license production contract with South Korea.

Distinguishing features. Side. Six roadwheels without return rollers, front sprocket. Boat-shaped hull with large front overhang. Vertical front with glacis almost horizontal. Mushroom-like driver's and cdr's cupolas in forward portion of turret roof. Front. High hull with vertical sides to the half-way point and sloped above it. Weapon station on the right, driver's cupola on the left hull roof.

Other details. Armament of the basic version is beside the engine on the right side of the hull top. One-man turret has a single piece hatch cover, eight vision blocks and a sight of x1 or x8 magnification. It is armed with a 12.7 mm M85 MG with 1000 rounds ammunition stowage. It has an elevation of -15 to +60 degrees.
AAV7A1 has a new powerpack, is slightly heavier (23.99 t) and later received a new Cadillac Gage one-man turret with 40 mm grenade launcher and 12.7 mm MG. The newly produced A1 can be distinguished from the earlier models by their square headlamps recessed into the nose.
The command version lacks the MG turret and is provided with additional radio equipment. It has a weight of 21.5 t and is armed with a pintle-mounted 7.62 mm MG (1000 rounds). It also has a navigation system.
The recovery version has a hydraulic crane which can lift 2.7 t and a recovery winch with a maximum capacity of 13.6 t. It weighs 23.6 t and has a crew of five.
All versions are fully amphibious propulsed in water by water jets at a maxi-

mum speed of 13.5 km/h. Fuel capacity is 681 resp 647 litres at the A1 version. Most of the US vehicles were retrofitted with add-on armour increasing the vehicle's weight by 2 t. The AAV7 can be fitted with a mine plough.

Role. In service with Argentina (19 AAV7, 1 C, 1 R), Brazil (20 A1, 2 C, 2 R), Italy (24, 1 C), Philippines (55), South Korea (53 plus 41 A1, 6 C, 3 R plus 57 currently in co-production), Spain (16, 2 C, 1 R), Thailand (12, 18 A1, 2 C, 1 R), USA (913 A1, 240 with new turret, 106 C, 64 R), Venezuela (9, 1 C, 1 R).

Assessment. Very large vehicle, particularly suitable for the special roles of the USMC.

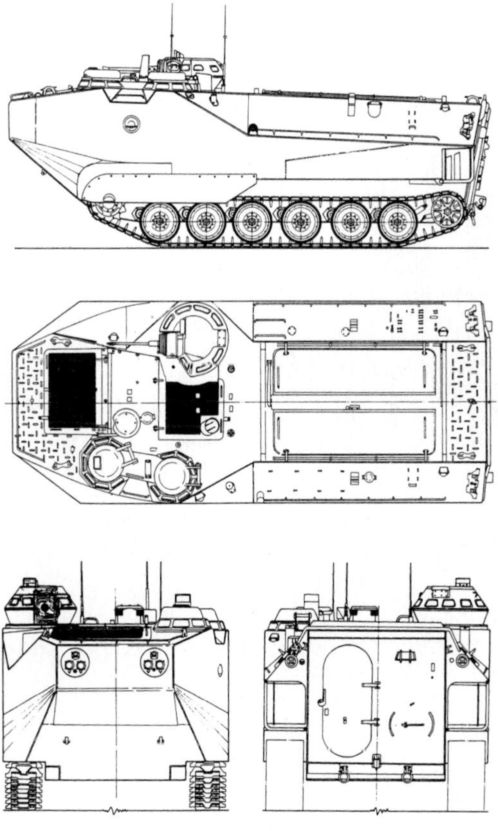

AAV 7A1 AAV

AAV 7A1 AAV

AAV7A1 (40 mm turret)

AAVC7A1 AACV (add-on armour)

AAVR7A1 AARV

M109 armoured artillery howitzer

Models.

M108
M109
M109A1B (converted M109 are called M109A1)
M109A2 (converted M109/M109A1 are called M109A3)
M109A6 (converted M109 variants are called M109A5)
M109A1C (Israel, A6 standard retaining former barrel)
M109A3G (Germany, M109A3GN same version in Norway)
M109AL (Israel, M109A1 with additional stowage racks)
M109L (Italy)
XT-69 AAH (Taiwan)
XT-69 Self-propelled Extended Range Gun (Taiwan)
ARE 122 PT with Soviet 122 howitzer for Egypt
Fire direction centre vehicle (FDCV)
M992 FAASV (Field Artillery Ammunition Support Vehicle)
VBCL AAFDV (Belgium)

Development and production.

1954 start of development of a new howitzer.
1959 completion of a first prototype T196 (155 mm) followed by the T195 (105 mm).
1962 start of series production of M108 (105 mm) and M109 (155 mm).
1974 M109A1B issued.
1979 production start of M109A2.
1983 issue of M992.
1984 approval to M109(A5) HIP (Howitzer Improvement Program) following M109(A4) HELP (Howitzer Extended Life Program).
1988 first roll-out of US and Israeli HIP conversions.
1991 conversion of 11 VBCL (a further 47 in 1992).
1998 last 122 mm vehicles for Egypt delivered.

Distinguishing features. Side. Seven large roadwheels, no return rollers, sprocket front. Bow front, sloped engine front glacis, horizontal hull top, vertical rear with spade (except M108). Large turret mounted rear with curved front and vertical rear. External ADMG mount on cdr's hatch. Front. Sloped glacis, driver on front left side. Hull with vertical sides. Wide turret with vertical sides over sponsons. Drum mantlet with external recuperators.

Other details. Main armament of the basic version was the 155 mm M126 barrel firing conventional rounds up to 14.6 km. The following versions are fitted with the longer M185 barrel with a maximum range of 18.1 km.
The M109A2/A3 have an improved recoil system, a redesigned rammer, improved hydraulic system, redesigned hatch and a bustle carrying additional 22 rounds. (TABLE)

The M109A5/A6 is fitted with a M284 barrel, automatic fire control system, navigation system providing automatic gun pointing, new gun/turret drives, increased ammunition stowage etc. It differs externally from older versions having a full width turret bustle.
The M109A3G has a new breech-block designed by Rheinmetall and a new barrel based on the FH-70 ordnance with a maximum range of 24.7 km. Ammunition stowage is 34 rounds.
The Israeli M109 have additional stowage racks on the hull and turret sides in order to increase ammunition stowage.
The M109L has a L39 barrel by OTO Melara with revised muzzle brake firing conventional rounds up to 24 km. Combat weight is 24.8 t, its length is 8.13 m.
The two Taiwanese XT-69 versions consist of the M108/M109 chassis with open topped rear compartment. A modified M114 155 mm howitzer resp. a 45-calibre 155 mm ordnance derived from the South African G5 are installed.
The Egyptian version mounts a licence produced 122 mm D-30 howitzer carrying 85 rounds. Vehicle's weight is 23.18 t, the length with armament is 6.96 m, height is 2.82 m.
The M992 has a large cargo compartment for 93 rounds 155 mm or 48 rounds 203 mm. Six rounds per minute can be supplied by hydraulically assisted horizontal/vertical stacker from storage racks to conveyor. Egyptian version has 0.7 t crane over front engine compartment for handling pallets of ammunition. Net ammunition payload is 5.4 t. The tank has a weight of 26.15 t, a length of 6.78 m, and a height of 3.2 m.
VBCL are converted M108 AAH with turrets replaced by an armoured superstructure.

Role. In service with Argentina (6), Austria (38, 18 A2 + 27 ex-British A2, 56 A5Ö), Belgium (108 A2, 41 A3), Canada (50 A1, 26 A2), Denmark (76 A3), Egypt (196 A2, 72 FDCV, 51 M992, 320 122 mm), Ethopia (12), Germany (555 A3G plus 22 with turrets removed serving as driver school tanks), Greece (51 A1B, 133 A2, 12 A5, 41 FDCV), Iran (50 A1, 390 A1B delivered, status unknown), Israel (530), Italy (280 M109L), Jordan (126 A1, 108 A2), Kenya (12), Kuwait (6 A1B, 31 A2), Libya (40), Morocco (36 A1B, 36 A2), Netherlands (126 A3), Norway (127 A3GN), Oman (15), Pakistan (152 A2), Peru (12 A1B), Portugal (6 A2, 14 A5), Saudi Arabia (280 A5 60 M992), South Korea (1040 A2 license produced), Spain (144 A5, 6 M992), Switzerland (581 A1B, 291 to be upgraded with a 47 calibre barrel starting in 1998), Taiwan (197 A2, 6 FDCV, 45 XT-69), Thailand (20 A5, 20 M992), Tunisia (19 A2), Turkey (42 A1), UAE (110 A3 with Swiss L47 barrels), USA (approx. 1500 A5, 950 A6, 789 M992), Venezuela (5), Vietnam (?).
The M108 is still in service with Brazil (124), Taiwan (120), Tunisia (40), Turkey (26).

Assessment. One of the most widely used SP artillery system which proved to offer considerable growth and upgrading potential in order to cope with current requirements.

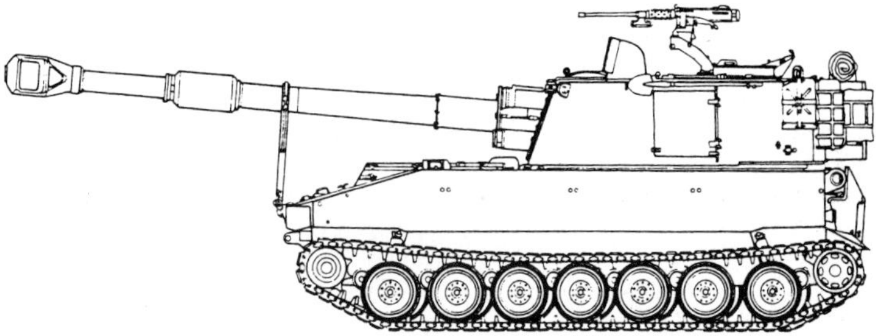

M109A2 AAH Steven Zaloga

M109A2 AAH

M109A6 AAH

M109AL AAH

M109L AAH

ARE 122 PT AAH

FDCV

FAASV (Egypt)

M992 FAASV

M107/M110 A2 armoured artillery howitzer

Models.

M107
M110A2

Development and production.

1956 concept study for a new self-propelled artillery family.
Contract with Pacific Car and Foundry Co for the development and production of six prototype vehicles, two 175 mm SP guns (T235), three 203 mm SP howitzers (T236) and one 155 mm (T245).
1957 expansion of the program to include ARV (T120) and T119 and T121 (unarmoured).
1961 standardisation of three systems, M107 175 mm SP gun, M110 203 mm SP howitzer and M578 ARV.
1962 first production vehicles completed.
1976 M110 A1 produced by BMY with a longer barrel M201 and direct fire M139 telescope firing new types of ammunition. Subsequently all M110 and most of the M107 were brought to this standard.
1978 adoption of the M110 A2 fitted with a double-maffle muzzle brake to fire charge 9. Most user countries adopted this configuration for their M107/M110.

Distinguishing features. Side. Five large roadwheels, front sprocket, nor idler neither return rollers. Flat sloped front. Unprotected mount for howitzer. Overhanging long barrel with double-baffle muzzle brake. Large spade at rear. Eight circular exhaust openings in right side. Front. Rounded hull front with sloped engine deck, driver left side. Sponsoned flat hull.
Vehicle can be retrofitted with a lightweight cover around the rear compartment to provide some weather protection for the crew.

Other details. Mounted at left rear of the hull is a hydraulically operated rammer and loading assembly which lifts the projectile from the rear or left side of the vehicle, positions and rams it into the chamber. Maximum range of the gun is 22.7 km, nuclear capable. There is a M115 sight (x4 degrees magnification, 10 degrees field of view) for indirect fire and the M116C telescope (x3 degrees magnification, 13 degrees field of view) for direct fire.
Only two rounds of artillery ammunition are carried on the vehicle, ammunition and the supplement crew are carried by a M548 tracked cargo carrier or by trucks.
(TABLE)

Role. In service with Bahrain (13 ex-Dutch), Denmark (36 CFE stock from USA), Greece (12 M107, 181 M110), Iran (38 delivered, status unknown), Israel (140 M107, 48 M110), Italy (36 M110 A2), Japan (201 A2), Jordan (102),

Pakistan (40), South Korea (100 M107, 99 M110), Spain (64 M110A2), Taiwan (75 A2), Turkey (36 M107, 219 M110A2), USA (669). Turkey is now the 175 mm ammunition producer.

Assessment. Proven self-propelled gun/howitzer. No armour protection for gun crew.

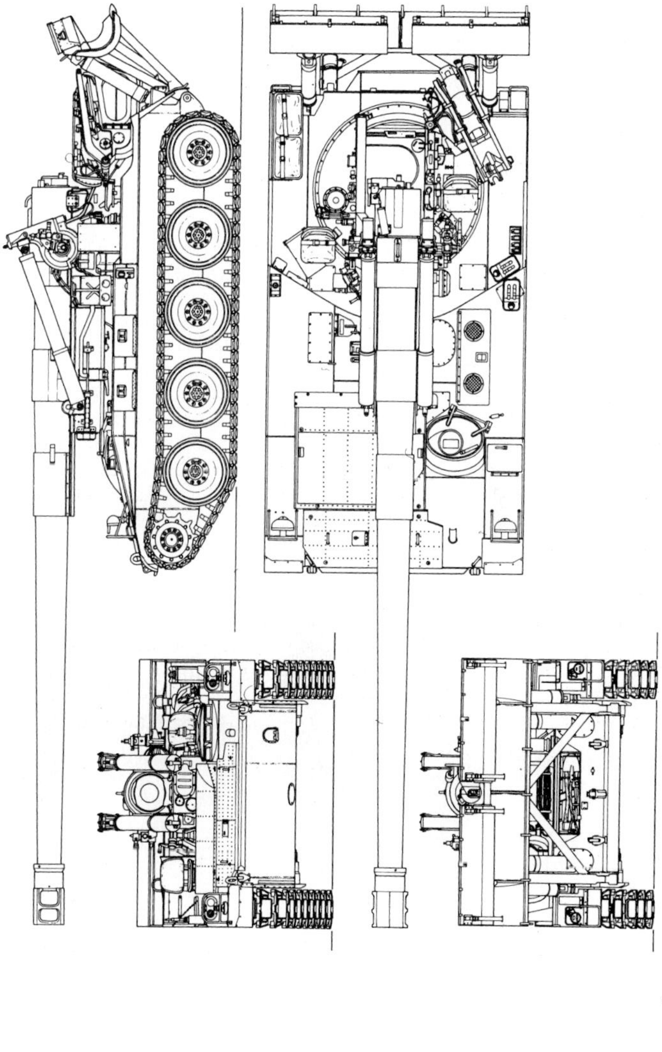

M110A2 AAH

M110A2 AAH

MLRS M270 armoured artillery rocket launcher
Multiple Launch Rocket System

Related Vehicle.

M4 ACCC

Non-serial versions.

XM5 GBCS-H (Ground-based Common Sensor-Heavy)
M987 with battlefield surveillance radar
AMTV (Armored Medical Treatment Vehicle) AAC
FAALS (Forward Area Armored Logistics System) AFARV
EFVS XM1070 (Electronic Fighting Vehicle System)
Armored Maintenance Vehicle (AMV)

Development and production.

1976 development start of a General Support Rocket System (GSRS).
1980 selection of Vought Corp. for series production.
1981 start of production.
1983 first vehicles issued to the troop.
1989 production start in Europe (Thyssen Henschel carrier integrator, Aerospatiale and Wegmann launcher integrators).
1991 delivery of two prototypes of EFVS by FMC.
1995 first five XM4 prototypes contracted.
1997 low-rate production of M-4 starts.

Distinguishing features. M987 carrier (modified M2 AIFV chassis) with six roadwheels in pairs closer together, two inner and two outer return rollers, sprocket front. Boxshaped operator's cabin on front end of chassis with sloped upper section. Large side doors. Three armoured shutters in front of windows. Engine compartment rear to front cabin. Chassis rear mounts big launcher box with two parallel sections of six tubes each.

Other details. The MLRS launcher is armed with twelve 227 mm rockets which can be launched within 50 seconds. Reload time is about ten minutes. The chassis has suspension lock-out blocks of 1st, 5th and 6th roadwheel pairs for loading and firing. The launch unit is hydraulically elevated and traversed under control of the on-board fire control system. Hand-held loading control unit. Ammunition is produced under a three stages program. The stage 1 warhead carries 644 double effect grenades of 38.7 mm diameter and 76.2 mm length consisting of a main hollow charge that can penetrate between 76 and 102 mm of steel and a casing with 200 fragmentations. With a range of 35 km each salvo covers an area of 15 hectares.
The stage 2 warhead contains 28 AT2 hollow charge anti-tank mines, each weighing 2.5 kg with a diameter of 102 mm and a height of 127 mm. A barrier of about 1 km width can be laid in less than two minutes at a distance up to 40 km.

The stage 3 warhead carries three terminally-guided sub-munitions with a range of 40 km.
(TABLE)
The MLRS launcher is also used to employ the Tactical Missile System (TACMS) as a LANCE replacement. One launcher carries two TACMS solid propellant fuel missiles (diameter 610 mm, length 3962 mm). The first two missile types use sub amunitions (1000 M74 bomblets in Block I) and precision guided submissiles (Block II) for use against moving armour. Range is about 150 km.
The MLRS system is also proposed as launcher vehicle for the ground launched anti-radar missile TACIT RAINBOW (2x6 missiles).
M4 has a large box-shaped superstructure housing different C3 installations depending of level of command including a 10 m telecoping mast.
XM5 is a prototyped IEW (Intelligence and Electronic Warfare) system. Current requirement is for 58 vehicles.
The EFVS will feature different version including one for the cooperation with J-STARS (Joint Surveillance and Target Attack Radar System). It is provided with a 20 m GBCS-H telscoping mast.

Role. In service with Bahrain (9), Denmark (12), France (56), Germany (154), Great Britain (71), Greece (18), Israel (48), Italy (22), Japan (36), Netherlands (22), Norway (12), Saudi Arabia (9), South Korea (29, 29 more on delivery), Taiwan (45 on order), Turkey (168 on order), USA (800), 46 M4 on production, total requirement is 439 vehicles..

Assessment. US standard hull, NATO standardised ammunition. Still not very outstanding range when compared with 155 mm base-bleed rounds. Cost-effectiveness of TGW doubtful.

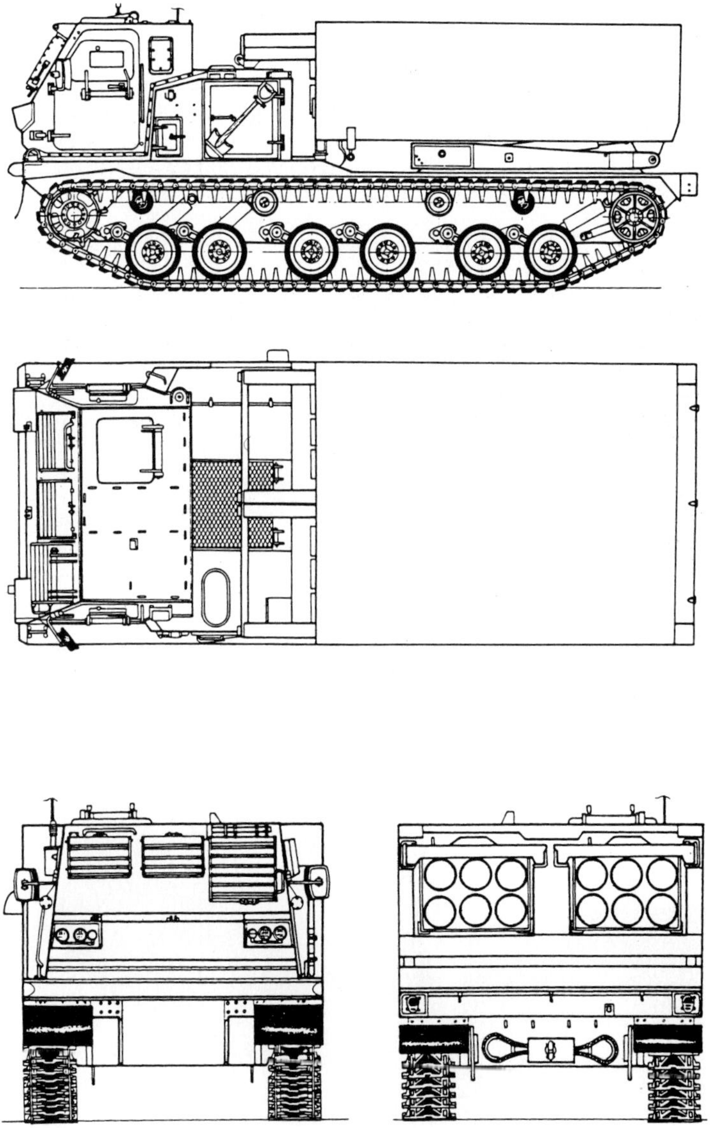

MLRS AARL

MLRS AARL

MLRS AARL (TACMS)

M4 ACCC

XM5 GBCS-H

XM1070 EFVS

M9 Armored Combat Earthmover (ACE)

Development and production.

1958 development of All-purpose Balastable Crawler (ABC) starts.
1975 four UET (Universal Engineer Tractor) prototypes completed.
1986 start of series production of M9 ACE, running until 1993 at BMY.

Distinguishing features. Side. Four roadwheels, first one acting as idler, no return rollers, sprocket rear. Dozer at front with large multi-purpose open-topped compartment behind it. Engine compartment rear right, driver's cab rear left. Front. Large two-piece dozer blade extending to both sides.

Other details. Dozer and dumper fittings hydraulically linked to front. The hull can be raised and lowered hydraulically. The vehicle is amphibious without preparation.
(TABLE)

Role. In service with the US Army (566).

Assessment. The project was handicapped by changing demands. Special vehicle to extend capability of armoured engineers, especially in route clearance/construction and improvement of fords.

M9 ACE

M728 Combat Engineer Vehicle (CEV)

Related version.

M1060 ROBAT (Robotic Breaching Assault Tank)

Development and production.

1959 T118 prototype on T95 MBT chassis, later T118E1 on M60.
1963 standardisation as M728.
1966 start of series production at Chrysler, Detroit, built until 1983.

Distinguishing features. M60 MBT hull. Dozer blade at front. M60A1 turret with short barreled gun and infra-red searchlight above mantlet. A-frame pivoted on sides of turret, lays back over rear in travelling condition. Commander's cupola with ADMG.

Other details. Main armament is a 165 mm demolition M135 gun based on British L9A1 gun. It has power elevation from -10 to +20 degrees. It fires the M123A1 HESH round of 30 kg up to ranges of 1000 m. A-frame and two-speed winch at the turret rear with a cable of 61 m and a diameter of 19 mm.
(TABLE)

Role. In service with Saudi Arabia, Singapore (8) and USA (291 and several ROBOTs).

Assessment. Versatile AEV with low mobility.

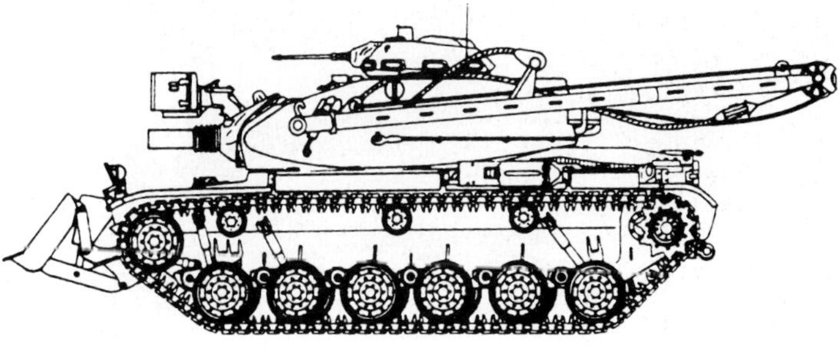

M9 ACE

M728 CEV

M1060 ROBAT

M47/M48/M60 AVLB

Models.

M47E AVLB (E)
M48A2 AVLB
M48A5 AVLB
A-26 AVLB (I)
M60A1 AVLB
M60 ALB with LEGUAN bridge

Related vehicle.

MICLIC mine explode carrier

Development and production.

1949 work begins on an AVLB based on the M46.
1956 tests of two M48 AVLB.
958/59 series production of AVLB on M48A2 chassis for export.
1964 production of AVLB on M60A1 hull.
1972 bridge laying system by ASTRA (A-26).
1976 conversion initiated of 203 US M48/M48A1 to A5 standard.
1994 prototype of AVLB-70 offered by GDLS for upgrade.
1996 Spanish contract for 12 LEGUAN AVLB on M-60 chassis.

Distinguishing features. Turretless MBT chassis with cupolas for commander and driver. Scissors bridge launched over front of vehicle, large rectangular footplate lowered to the ground prior to launching. Two track bridge is carried folded, large overhang front and rear.

Other details. Bridge is launched and retrieved hydraulically within three minutes. It can be recovered from both ends.
M48A5 AVLB has a weight of 55.76 t.
Local conversion in Spain of diesel-engined M-47 hulls.
A-26 is a converted M-47 hull with a similar scissor type bridge. Length of bridge is 22 m, width 4 m. It can carry 54 t. Launching or recovery time is about 6 minutes.
(TABLE)
The MICLIC is a field conversion for the transport of two launch boxes of mine clearance charges on top of the rear engine deck.

Role. In service with Israel, Italy (64 A-26), Singapore (8 M-60), Spain (25 M47, 12 M-60), Taiwan (12 M48A5), Turkey (10 ex-German M-48A2) and USA (650).

Assessment. Very large but effective vehicle for quick crossing of small obstacles in the combat zone. Capability insufficient for weight of M1 MBT.

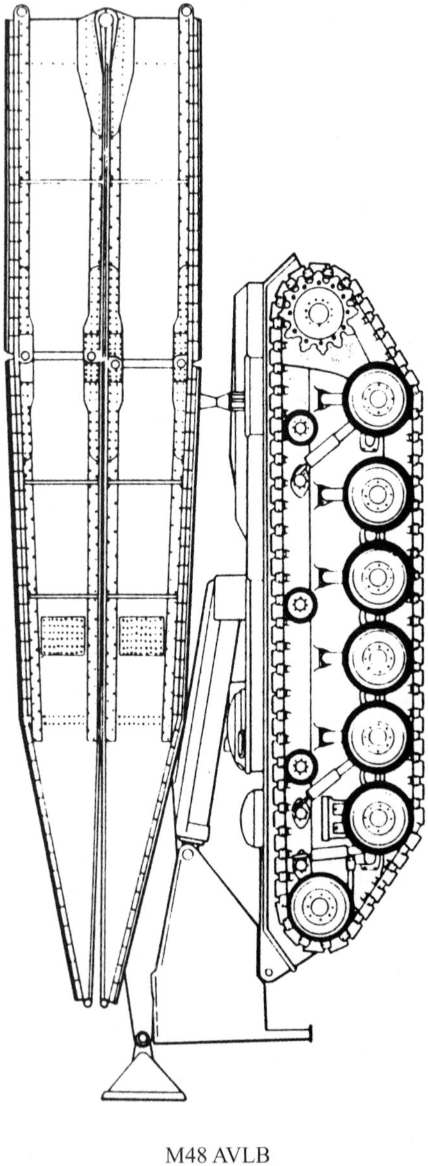

M48 AVLB

M60A1 AVLB

MICLIC

M88 armoured recovery vehicle

Models.

M88
M88 A1
M88 A2 IRV (Improved Recovery Vehicle) HERCULES
M48 ARV (Germany)
AB1 ARV (Jordan)

Non-serial version.

COV (Counter Obstacle Vehicle) AEV

Development and production.

1955 authorization given to produce three prototypes of medium ARV.
1959 standardization of T88 based on M48 A2 MBT chassis as M88.
1961 start of production running up to 1964 (1075).
1975 production resumes at BMY as M88 A1 diesel version.
1977 conversions of 878 US M88 to M88 A1 standard starts.
1986 1383 M88 A1 in US inventory. M88 AX with 772 kW engine issued.
1989 development of M88 A2 (M88E1) delayed due to budgetary reasons, 5 prototypes built.
1990 delivery of the first M48 ARV to the Turkish Army (150 required), development by Jung Jungenthal.
1995 conversion of IRVs starts.

Distinguishing features. Side. Six roadwheels, three top rollers, sprocket front. Large crew compartment at front with distinct step down to engine section at rear. Slightly sloped front with dozer blade. A-frame is pivoted on hull top folded back over rear when not in use. Crew compartment with large doors on both sides and flat cdr's cupola with externally mounted ADMG. Front. Bulky sponsoned hull with flat-sided superstructure. Two crew stations on upper glacis with three episcopes each.

Other details. Winch and hydraulic compartment separated from crew and engine contains the main hydraulic systems, the main winch, dozer blade actuating cylinders, hoist winch, bilge pumps and front fuel tank. The engine compartment also contains the auxiliary power unit and generator (7.5 kW), auxiliary hydraulics, jib erection cylinders, batteries and rear fuel tank.
The M88 is powered by the AVSI-1790-6A petrol engine with 599 kW, has a weight of 49.5 t, a fuel capacity of 1514 litres and a range of 350 km.
The M88 A1 has the AVDS-1790-2DR diesel engine, the Allison XT-1410-4 transmission and a new diesel personnel heating system.
(TABLE)
IRV has the following main improvements: 772 kW Teledyne engine, a stronger main winch with a constant pull of 635 kN, a 85.34 m cable, a rectangular

section boom with a capacity of 35 tons, upgraded torsion bars, shock absorbers and snubbers. For better protection the vehicle is provided with appliqué armour and ballistic suspension skirts.
M48 ARV is a German conversion of the M48 MBT using the components of the BERGEPANZER 2. Optional offered with either petrol or diesel engine.
AB1 is a Jordan conversion of M47 gun tanks to an ARV.

Role. In service with Austria (12), Bahrain (3), Egypt (43, 50 A2 on order), Greece (26 plus an option of 13), Israel, Jordan (20 plus 20 AB1 on order), Kuwait (74 A2), Morocco, Norway (4), Oman, Pakistan, Portugal (10), Saudi Arabia (43 plus an option of 11), South Korea, Spain (12), Sudan, Taiwan (22), Thailand (25, 6 A2 on delivery), Tunisia (4), Turkey (20 M48 ARV), USA (2988, 102 A2) and Yemen.

Assessment. Specialised ARV with a wide range of equipment. Good configuration and extensive standardization. A1 version unsuitable for recovery of tanks over 50 t.

M88A1 ARV

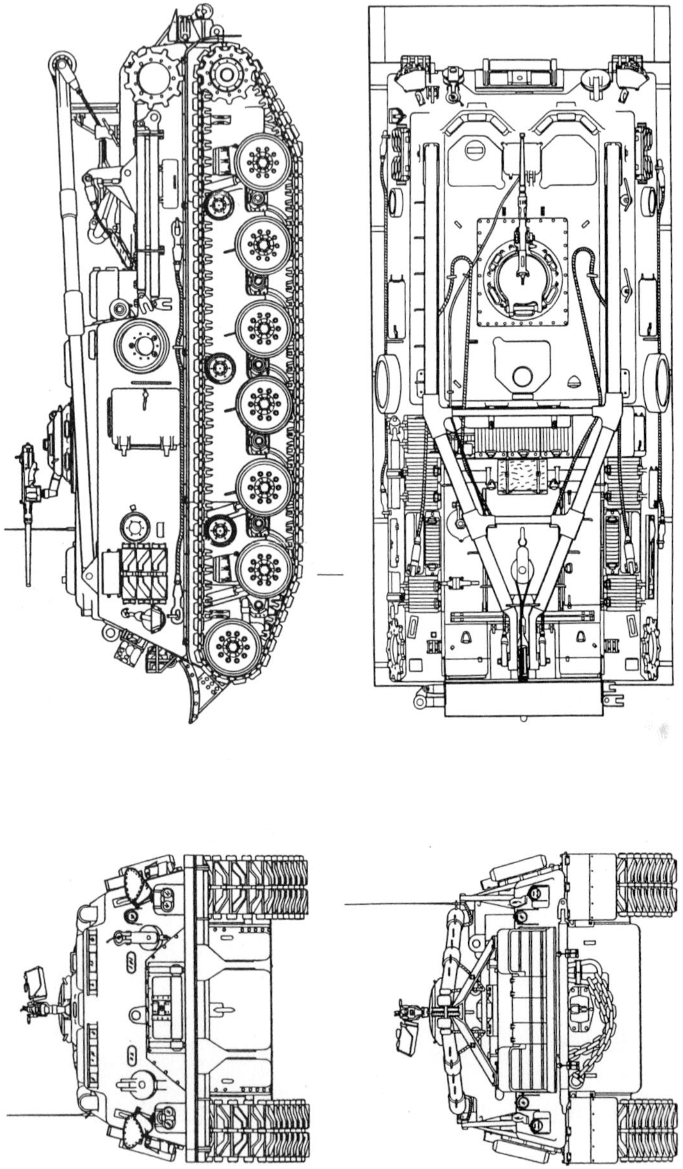

M88A1 ARV

M88A2 ARV

M48 ARV

M578 armoured recovery vehicle

Development and production.

1961 two prototypes built based on M107/M110 AAH chassis, the T119 Unarmored Wrecker and the T120 ARV.
1971 T120 standardized as M578 and in production (1018) until 1983.

Distinguishing features. Chassis identical to M110 AAH. Large square turret at rear third of vehicle length with crane extending out over vehicle front. Large spade mounted at vehicle rear, raised in travelling order.

Other details. (TABLE)
The US vehicles were upgraded by a SMA (Special Mission Adaption) which includes a boom extension of 1.219 m with a lift capacity of 6.8 t.

Role. In service with Bolivia, Brazil, Canada, Denmark, Egypt (43), Great Britain, Iran, Jordan, Morocco, Netherlands, Norway, Philippines, Saudi Arabia, Spain, Tunisia (6) and USA.

Assessment. Useful light ARV with minimal armour protection.

M578 ARV

BVP M80 armoured infantry fighting vehicle
Borbeno Vozilo Pesadije

Models.

BVP M80 (former M-980)
BVP M80A
BVP M80AK with 30 mm M86 cannon
BVP M80A KC / M80A KB ACCC
BVP M80A LT TDM
BVP M80A Sn AAV

Development and production.

1975 first appearance of M-980.
1984 BVP M80A exhibited for the first time in Cairo.

Distinguishing features. Side. Five "waffle" roadwheels (as Soviet light tank family), front sprocket, two track return rollers. Low, heavily sloped glacis, horizontal hull top, vertical rear. Three weapon ports each side. Low-profile turret slightly to rear of vehicle centre. Two ATGM on traversing launch rails on rear of turret. Crew compartment with three periscopes and one hatch each side. Front. Prominent leading edge with splashboard. Driver and commander left front, engine right front. Sponsoned hull with sloping sides. Turret with unarmoured slit mantlet and vision block apertures on left side. Ground and AD sight on the left side in front of gunner's hatch.

Other details. BVP M80A is of the same overall layout as the M-80 but slightly wider (+0.4 m) and heavier (+0.3 t).

Both tanks are armed with a HS 804 20 mm cannon and an 7.62 mm MG. Elevation of the cannon is -5 to +75 degrees. It has a cyclic rate of fire of 800 rpm and has a range of 2 km (1.5 km against aircraft targets). The M-80 has its MG coaxial in a parallel slit mantlet right of the main gun. The gunner is provided with a day sight (x1.1 magnification at a 7 degrees field of view) and a night sight (x7.3 magnification and 7 degrees field of view).
(TABLE)
The BVP M80A has the same armament but the MG is arranged above the 20 mm cannon. 1400 resp. 2000 rounds and four ATGM are carried. The BVP M80A is powered by a 4-stroke 10-cylinder direct-injection diesel engine. Fuel capacity is 510 litres.
Both tanks are amphibious, driven by the tracks at a maximum speed of more than 7 km/h. NBC system is fitted. The M80A versions are powered by the FAMOS 10V003 10 cylinder diesel engine developing 232 kW.
The BVP M80AK is armed with a 30 mm M86 cannon with weapon stabilization.
The command versions are for the company (KC) resp. battalion level (KB).

The TDM is armed with a turret with six ready-to-launch SAGGER ATGM as used in the BOV-1.
The ambulance version lacks the turret and can accomodate four stretcher patients.

Role. In service with Croatia (approx. 100), Slovenia (62) and the Yugoslav Army (512).

Assessment. A successful development on the line of the Soviet BMP family. Lightly armoured, but with good configuration and firepower.

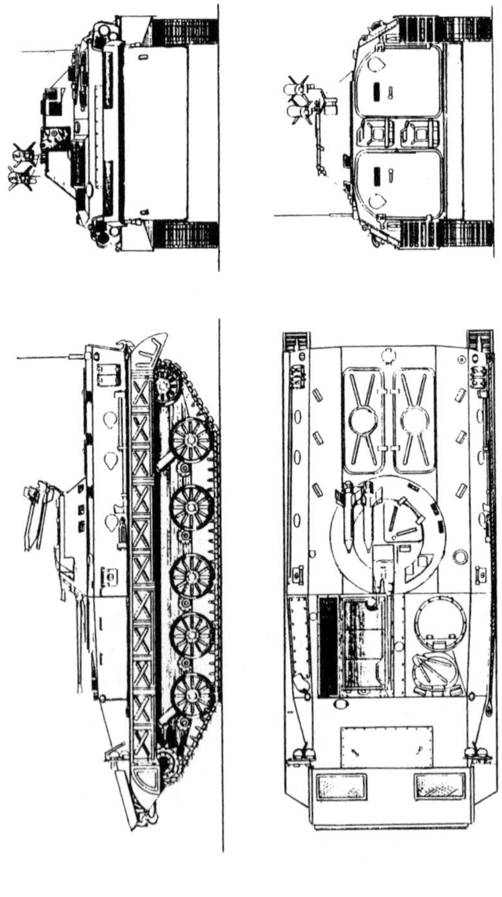

BVP M80A AIF

BVP M80A AIFV

M-60P armoured general purpose carrier

Wt 11 t L 5.02 W 2.77 H 1.86 103 kW 45 km/h 400 km Crew 3+10

Models.

M-60P APC
M-60PB TDG
M-60P K ACCC
M-60P Sn AAV

Development and production.

1965 first appearance during parade (initially called M-590).

Distinguishing features. Side. Five road wheels, three return rollers, front sprocket. Suspension extends each side of hull, upper part covered by skirt. High box-shaped superstructure with well sloped glacis. Vertical hull rear with two doors. Front. Front plate with driver left and co-driver on his right. Cupola right top with externally mounted ADMG, commander's cupola opposite. Troop compartment has vertical sides with top half sloping inwards. Weapon ports in sides and rear walls.

Other details. Main armament consists of a 12.7 mm ADMG and one 7.62 mm bow MG. Nine infantrymen sit on either side of the hull. The M-60P is not amphibious and lacks a NBC system.
The M-60PB has mounted on top of the left side hull two 82 mm recoilless rifles on a rotating mount with mechanical elevation and traverse. To the right is the loader's hatch. The effective range of the rifle is 1500 metres. The HEAT round penetrates 250 mm of armour, ten rounds are carried. It carries one crew member less.
The command versions have different radio layout.
The ambulance accomodates up to four stretchers.

Role. In service with the Croatian and Yugoslav Armies (112).

Assessment. Rather high, lightly armoured vehicle of low mobility. Technical standard comparable with US M113 APC.

M-60P APC

M-60PB TDG

M-60P Sn AAV

BOV armoured general purpose carrier
Borbeno Vozilo

L 5.59 W 2.52 110 kW 93 km/h 500 - 800 km Fuel 220 litres

Models.

BOV-M APC
BOV-1 TDM
BOV-3 AADG
BOV-30 AADG

Development and production.

1981 start of troop trials of new 4x4 vehicle family.
1983 first appearance of BOV-1 and BOV-3 in service.
1985 issue of BOV-30.

Distinguishing features. Side. Two large roadwheels each side which door/hatch on left side between. Angled nose with sloped glacis plate. APC with raised central crew compartment and almost vertical windscreen for commander and driver. Other versions with horizontal roof, turret arrangements central. Pointed rear. Front. Driver left, cdr right with single piece hatch cover. Hull sides slope inwards from just above wheel arches. AADG versions with turret sides protruding in hull sides.

Other details. APC has a pintle-mounted 12.7 mm MG and carries eight infantry soldiers. Three firing ports on either side of hull. Engine compartment rear (Deutz F 6L 413 F six cylinder diesel coupled to manual transmission with five forward and one reverse gears). Weight is 5.7 t, height 2.34 m.
BOV-1 with two launching pods containing three semi-automatic guided missiles (Yugoslav SAGGER C copy), missile range is 3 km. 7.62 mm MG between the pods. Weight is 8.1 t, height 2.8 m. The tank is operated by a crew of two.
BOV-3 with open topped turret for triple guns with 60 rounds drum magazines. 20 mm M-55 cannons (licence of Hispano-Suiza HS 804) in power operated turret, elevation -4.5 to +83 degrees. A total of 1500 rounds is carried. Rate of fire per barrel is approximately 700 rpm. Maximum anti-aircraft engagement range is 1.5 km. The vehicle's weight is 9.5 t, height 2.8 m, it is operated by a crew of four. J-171 optical sight is used. The gunner has a joystick for power control and fires the guns using a foot pedal.
BOV-30 has a larger enclosed turret armed with two 30 mm cannons on either side in external mountings. The gunner is seated in a raised cupola at the turret rear. Rate of fire is 600 rpm per barrel. Height is about 3 metres. It has a crew of three.
The vehicles have a central tyre pressure regulation system. They are non-amphibious and lack a NBC system.

Role. In service with Croatia, Slovenia (100+) and Yugoslavia (approx. 300).

Assessment. Cheap, mobile vehicle of limited cross-country capability. Weapons limited to day-fighting capability. Lightly armoured.

BOV-M APC

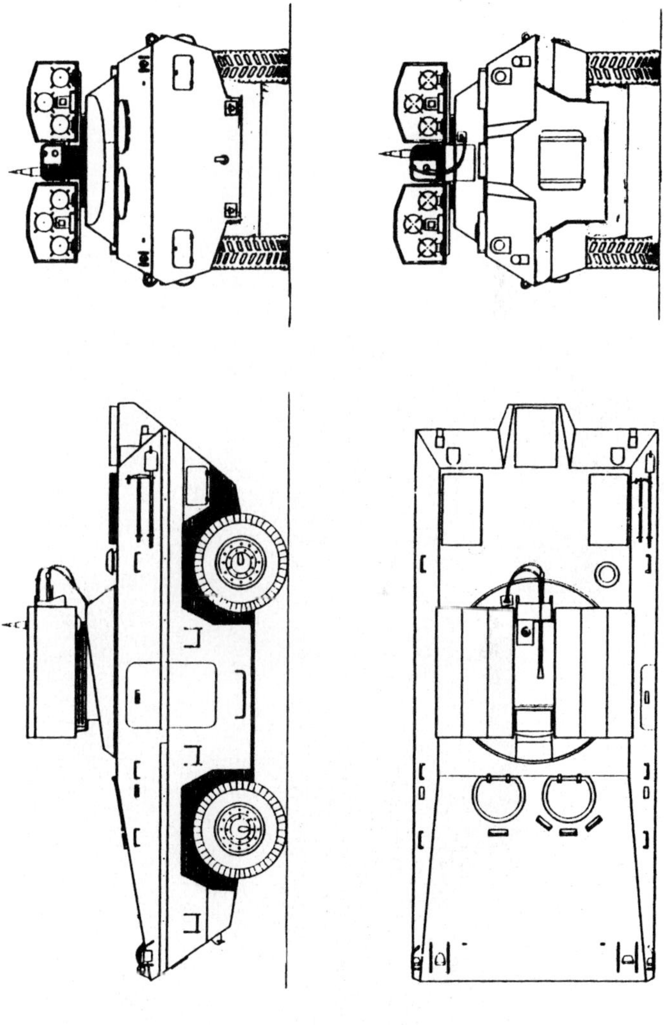

BOV-1 TDM

BOV-1 TDM

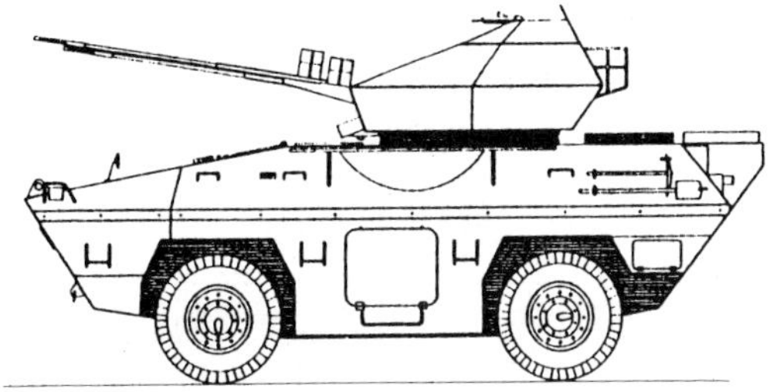

BOV-3 AADG

BOV-3 AADG

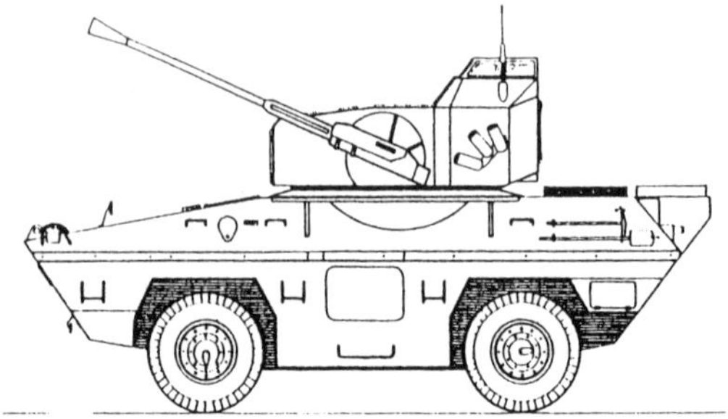

BOV-30 AADG

BOV-30 AADG

PART 2. TABLES

ARMOURED RECONNAISSANCE VEHICLES (WHEELED)

Characteristics	Unit	CASCAVEL 4
Country		BR
No of wheels/driven wheels		6x6
Crew	persons	3
Battle weight	t	13.4
Power-weight ratio	kW/t	11.64
Automotive performance		
Maximum speed	km/h	100
Road range	km	880
Fuel capacity	l	390
Fuel consumption	l/km	0.44
Vertical obstacle	m	0.6
Trench width	m	1.5
Water propulsion		-
Maximum water speed	km/h	-
Chassis		
Turning radius	m	9.5
Tyres	size	12.00x20
Suspension		CS/SEL
Wheelbase	m	2.34 and 1.41
Engine		
Model		
Manufacturer		Detroit Diesel
Type		6V-53N
Cylinders -no/arrangement		V-6
Capacity	l	522
Rating	kW	156
at peak power speed	rpm	2200
Cooling		liquid
Transmission		
Type of gearbox		Allison MT 643
No of speeds - forward/reverse		4/1

VBL	AMX-10 RC	SAGAIE	LUCHS
F	F	F	D
4x4	6x6	6x6	8x8
2/3	4	3	4
3.55	15.88	8.1	19.5
19.7	13.96	14.07	14.71
95	85	95	90
600	1000	700	800
	520	242	390
	0.52	0.32	0.49
0.5	0.7	0.8	0.6
	1.6	1.1	1.9
propellor	waterjets	2 hydrojets	propellors
4	7.2	9.5	10
5.77			11.5
9.00x16	14.00x20	11.00x16	14.00x20
HS/TB	HP	CS/HS	CS
2.45	1.55/1.55	1.63 + 1.22	1.4/2.26/1.4
XD 3T diesel	6F11 SRX	PRV	OM 403 A
Peugeot	Baudouin	Peugeot	Daimler-Benz
turbo-charged	turbocharged diesel	petrol	turbocharged
4	V6	V-6	V-10
	6.54		15.9
70	221	114	287
4150	1800	5500	2500
liquid	liquid	liquid	liquid
ZF 3HP 22	preselector TC	Panhard	ZF 4PW95H1
3/1	4/1	6/1	4/4

Characteristics	Unit	CASCAVEL 4
Electrics		
Generator	V/kW	24
Armament		
Main armament, model		EC-90
Manufacturer		ENGESA
Calibre	mm	90
Laying system		mechanical
Feed/loading system		manual
Rate of fire	rds/min	6
Muzzle velocities	m/sec	
Elevation arc	degrees	-8 +15
Traverse	degrees	360
Ammunition stowage	rounds	44
Fire control		OIP or Marconi
Secondary armament		
No/type/role		2/coax and ADMG
MG, calibre	mm	7.62 or 12.7
Ammunition stowage	rounds	2200
Smoke dischargers		2x3
NBC collective protection		none
Armour type		steel
Manufacturer		ENGESA
Dimensions		
Length	m	6.2
Hull length	m	5.2
Width	m	2.64
Height	m	2.68
Ground clearance	m	0.34

VBL	AMX-10 RC	SAGAIE	LUCHS
24	28/4.5	24	28/4.3
1 MG	MECA	F4	Rh202
	GIAT	GIAT	Rheinmetall
7.62	105	90	20
manual	elec-hyd	manual	elec-hyd
belt	manual	manual	belt
depends on model	8	8	1000
	HEAT 1120	1250 APFSDS	1050
-10 +50	-8 +20	-8 +15	-15 +80
180	360	360	360
3000	38	several options	500
none	COTAC		manual
MILAN ATGM (optional)	1 F1 ADMG	2 MG/ coax/AD	1 MG3 ADMG
	7.62	7.62	7.62
	4000	2000	800
6	2x2	2x2	2x4
optional	yes	optional	yes
steel	light alloy	steel	steel
Panhard	AMX	Panhard	Thyssen-Henschel
3.7	9.15	7.693	-
	6.36	5.098	7.743
2.02	2.95	2.495	2.98
2.14	2.68	2.254	2.905
0.37	0.21 - 0.6	0.294	0.406

Characteristics	Unit	CENTAURO
Country		I
No of wheels/driven wheels		8x8
Crew	persons	4
Battle weight	t	25
Power-weight ratio	kW/t	15.28
Automotive performance		
Maximum speed	km/h	105
Road range	km	800
Fuel capacity	l	540
Fuel consumption	l/km	0.68
Vertical obstacle	m	0.55
Trench width	m	1.55
Water propulsion		
Maximum water speed	km/h	-
Chassis		
Turning radius	m	9
Tyres	size	14.00x20
Suspension		HP
Wheelbase	m	1.6/1.45/1.45
Engine		
Model		MTCA
Manufacturer		IVECO
Type		turbocharged diesel
Cylinders -no/arrangement		V-6
Capacity	l	12.88
Rating	kW	382
at peak power speed	rpm	2300
Cooling		liquid
Steering		
Type of gearbox		ZF 5 HP 1500
No of speeds – forward/reverse		5/2
Electrics		
Generator	V/kW	28/15
Armament		
Main armament, model		105/52 LR

Type 87	ROOIKAT	VEC
J	ZA	E
6x6	8x8	6x6
5	4	5
15.0	27	13.75
14.95	15.56	16.58
100	120	96
500	1000	800
	500	400
	0.5	0.5
0.6	1	0.6
1.5	2	1.5
-	-	wheels
-	-	3
8.4	12.5	
14.00x20	14.00x20	13.00x20
CS	CS	HP
1.5/1.5	1.55/2.032/1.625	1.65/1.65
10 PBI	ADE	96 TEAZ34300
Isuzu		Pegaso
diesel	diesel	turbocharged diesel
V-10	V-10	6
14.022		11.94
224	420	228
2700		2200
liquid	liquid	liquid
	two front axles	front + rear axles
	automatic	ZF6HP500
6/1	6/1	6/1
28/		
KBA		M242 chain gun

Characteristics	Unit	**CENTAURO**
Manufacturer		OTO-Breda
Calibre	mm	105
Laying system		elec-hyd
Feed/loading system		manual
Rate of fire	rds/min	8
Muzzle velocities	m/sec	APFSDS 1650
Elevation arc	degrees	-6 +15
Traverse	degrees	360
Ammunition stowage	rounds	40
Fire control		TURMS
Secondary armament		
No/type/role		2 MG 42/59 AAD/coax
MG, calibre	mm	7.62
Ammunition stowage	rounds	2650
Smoke dischargers		2x4
NBC collective protection		yes
Armour type		steel
Manufacturer		IVECO/OTO Melara
Dimensions		
Length	m	8.55
Hull length	m	7.85
Width	m	2.95
Height	m	2.74
Ground clearance	m	0.42

Type 87	**ROOIKAT**	**VEC**
Oerlikon	LEW	Hughes
25	76	25
elec-hyd	electrical	electric
belt	manual	belt
550	6	100/200/500
1350	APFSDS 1600	APDS 1345
	-10 +20	-10 +50
360	360	360
	48	170 (turret)
manual	ESD IFCS	-
1 MG ADMG	1 coax MG	1 coax MG
	1 ADMG	
7.62	7.62	7.62
	3600	400 (turret)
2x3	2x4	2x3
no	yes	optional
steel	steel	aluminium
Komatsu	Sandock-Austral	ENASA
-	8.2	-
5.99	7	6.25
2.48	2.9	2.5
2.8	2.8	2.51
0.45	0.4	0.4

ARMOURED RECONNAISSANCE VEHICLES (TRACKED)

Characteristics	Unit	SCORPION
Country		GB
No of wheels/driven wheels		3
Crew	persons	7.94
Battle weight	t	17.87
Power-weight ratio	kW/t	0.36
Automotive performance		
Maximum speed	km/h	80
Road range	km	600
Fuel capacity	l	423
Fuel consumption	l/km	0.71
Vertical obstacle	m	0.5
Trench width	m	2.05
Water propulsion		tracks
Maximum water speed	km/h	6.5
Chassis		
Turning radius	m	front TB
Tyres	size	5/0
Suspension		single pin
Wheelbase	m	removable pads
Engine		
Model		J60
Manufacturer		Jaguar
Type		petrol
Cylinders -no/arrangement		6
Capacity	l	4.2
Rating	kW	142
at peak power speed	rpm	4500
Cooling		liquid
Transmission		
Type of gearbox		SCG TN 15
No of speeds - forward/reverse		7/7
		integral triple- differential mech.

Characteristics	Unit	SCORPION
Electrics		
Generator	V/kW	28/3.4
Armament		
Main armament, model		L23A1
Manufacturer		ROF
Calibre	mm	76
Laying system		manual
Feed/loading system		manual
Rate of fire	rds/min	8
Muzzle velocities	m/sec	APDS 1200
Elevation arc	degrees	-10 + 35
Traverse	degrees	360
Ammunition stowage	rounds	45
Fire control		-
Secondary armament		
No/type/role		1 L8A1 coax MG
MG, calibre	mm	7.62
Ammunition stowage	rounds	3000
Smoke dischargers		2x8
NBC collective protection		yes
Armour type		light alloy
Manufacturer		Alvis
Dimensions		
Length	m	-
Hull length	m	4.78
Width	m	2.235
Height	m	2.49
Ground clearance	m	0.37

MAIN BATTLE TANKS

Characteristics	Unit	TAM
Country		D/RA
Crew	persons	4
Battle weight	t	30
Power-weight ratio	kW/t	17.75
Nominal ground pressure	bars	0.79
Automotive performance		
Maximum speed	km/h	75
Road range	km	590
Fuel capacity	l	670
Fuel consumption	l/km	1.15
Vertical obstacle	m	1
Trench width	m	2.5
Shallow fording depth	m	1.5
Immersion depth	m	-
Water propulsion		-
Hull		
Sprocket		front
Suspension		TB
No of roadwheels/top rollers		6/3
Trach		end connectors removable pads
Engine		
Model		MB833Ka500
Manufacturer		MTU
Type		turbocharged diesel
Cylinders -no/arrangement		V-6
Capacity	l	22.4
Rating	kW	529
at speak power speed	rpm	2400
Cooling system		liquid
Transmission and steering		
Model		HSWL204 Renk
Type of gearbox		auto TC epicyclic 4/4
No of speeds - forward/reverse		hydrostatic
Type of steering		infinitely variable

LEOPARD 1 A5	**LEOPARD 2**	**AMX30B2**
D	D	F
4	4	4
42.4	55.15	37
14.39	20	13.31
0.89	0.83	0.87
65	72	65
560	550	520
985	1200	900
1.76	2.18	1.73
1.15	1.1	0.93
3	3	2.9
1.2	1	1.2
4	4	4
-	-	-
rear	rear	rear
TB	TB	TB
7/4	7/4	5/5
end connectors	end connectors	single pin
removable pads	removable pads	removable pads
MB838CaM500	MB873Ka501	HS110
MTU	MTU	Renault-Saviem
diesel	turbocharged diesel	diesel
V-10	V-12	HO-12
37.4	47.6	28.8
610	1103	495
2200	2600	2250
liquid	liquid	liquid
4HP250ZF	HSWL354 Renk	ENC200 SESM
auto TC	auto TC	auto TC
epicyclic	epcyclic	epicyclic
4/2	4/2	5/5
2-radius	hydrostatic	hydrostatic
regenerative	infinitely variable	

Characteristics	Unit	TAM
Electrics		
Generator	V/kW	28/9
Battery capacity	Ah	300
Auxiliary generator	kW	-
Main armament		
Model		Rh105-30
Manufacturer		Rheinmetall
Calibre	mm	105
Calibre length		L 51
Rate of fire	rds/min	8
Elevation arc	degrees	-7 +18
Ammunition stowage	rounds	50
Ammunition natures/MVs	m/sec	APDS 1470 HEAT 1170
Secondary armament		
Coax, model		FN
Calibre	mm	7.62
Cyclic rate	rds/min	900
Ammunition stowage	rounds	6000
ADMG, model		FN
Calibre	mm	7.62
Cyclic rate	rds/min	900
Ammunition stowage	rounds	(see coax)
Smoke dischargers		2x4
Laying, sighting and vision systems		
Powered mounting		elec-hyd
Stabilisation - primary/slaved		sight/gun
Rangefinder		ND-YAG laser
Ballistic computer		FCC for range, ammo, trunnion, tilt & lead
Gunner's primary sight		x8 sight

LEOPARD 1 A5	LEOPARD 2 A4	AMX30B2
28/9	28/20	28/9
400	500	400
-	-	-
L7A3	Rh-120	105F1
ROF	Rheinmetall	GIAT
105	120	105
51	L 44	L 56
8	8	8
-9 +20	-9 +20	-8 +20
60	42	47
APDS 1470	APFSDS 1650	HEAT 1000
APFSDS 1525	MP 1140	other 700
MG3	MG3A1	F2 (M693)
7.62	7.62	20
1200	1200	600
5500	4750	480
MG3	MG3A3	F1 (C1)
7.62	7.62	7.62
1200	1200	900
see coax	(see coax)	2150
2x4	2x8	2x2
elec-hyd	elec-hyd	elec-hyd
sight/gun	sight/gun	sight/gun
ND YAG laser	ND-YAG laser	MD-YAG laser
	Eltro	APX M550
yes	FLT2FCC	COTAC GIAT
	AEG-Telefunken	
EMES 18	EMES 15 (KAE)	M581 GIAT
	binocular (x12)	(x10)
	with LRF and	
	Zeiss thermal sight	

Characteristics	Unit	TAM
Cdr's vision system		all-around PERI-R12 (x2/x8)
Secondary sight		Zeiss TZF1A monocular 8x40
Manufacturer		TAMSE
Dimensions		
Length	m	8.23
Hull length	m	6.75
Width	m	3.29
Height	m	2.66
Ground clearance	m	0.45

Characteristics	Unit	LECLERC
Country		F
Crew	persons	3
Battle weight	t	54.5
Power-weight ratio	kW/t	20.2
Nominal ground pressure	bars	0.9
Automotive performance		
Maximum speed	km/h	71
Road range	km	550
Fuel capacity	l	1300
Fuel consumption	l/km	2.36
Vertical obstacle	m	1.25
Trench width	m	3
Shallow fording depth	m	1
Immersion depth	m	4
Water propulsion		-
Hull		
Sprocket		rear
Suspension		HP
No of roadwheels/top rollers		6/4
Track		end connectors removable pads
Engine		
Model		UDV 8X

LEOPARD 1 A5	LEOPARD 2 A4	AMX30B2
TRP (lenghtened) with light pipe	panoramic PERI-R17 (x2 and x8) with stab. line of sight	M496 Matra (x8)
TZF 1A	Leitz FERO Z18 monocular 8x40	M267.02 (x1 and x10), DIVT 13 light intensification
Krauss-Maffei, Krupp-MaK	Krauss-Maffei and MaK	ARE
9.54	9.61	9.48
7.09	7.66	6.59
3.41	3.7	3.1
2.58	2.78	2.86
0.44	0.5	0.44

CHALLENGER 2	ARIETE	MERKAVA MARK 3
GB	I	IL
4	4	4
62	52	61
14.23	18.3	14.47
0.96	0.9	0.96
56	65	55
600	550	500
1592	1200	1250
2.7	2.18	2.5
0.91	1.1	1
2.8	3	3.5
1.07	1.2	1.38
-	4	2.4
-	-	-
rear	rear	front
HP	TB	CS/HP
6/3	7/4	6/3
single pin	end connectors	single pin
removable pads	removable pads	no pads
CONDOR 12V1200	MTCA	AVDS 1790-9AR

Characteristics	Unit	**LECLERC**
Manufacturer		SACM
Type		hyperbar diesel
Cylinders -no/arrangement		V-8
Capacity	l	16.47
Rating	kW	1103
at peak power speed	rpm	2500
Cooling system		liquid
Transmission and steering		
Model		SESM ESM 500
Type of gearbox		auto hydraulic
No of speeds - forward/reverse		5/2
Type of steering		hydrostatic infinitely variable
Electrics		
Generator	V/kW	28/9
Battery capacity	Ah	500
Auxiliary generator	kW	9
Main armament		
Model		CN 120-26
Manufacturer		GIAT
Calibre	mm	120
Calibre length		L 52
Rate of fire	rds/min	10
Elevation arc	degrees	-8 +15
Ammunition stowage	rounds	40
Ammunition natures/MVs	m/sec	APFSDS 1790 MP 1100
Secondary armament		
Coax, model		GIAT
Calibre	mm	12.7
Cyclic rate	rds/min	600
Ammunition stowage	rounds	800
ADMG, model		GIAT
Calibre	mm	7.62
Cyclic rate	rds/min	900
Ammunition stowage	rounds	2000

CHALLENGER 2	ARIETE	MERKAVA MARK 3
Rolls-Royce	IVECO/FIAT	Teledyne Contin.
diesel	turbocharged diesel	diesel
V-12	V-12	V-12
26.1		29.3
882	956	882
2300	2300	2400
liquid	liquid	air
TN 37 David Brown	LSG-3000 ZF	CD850-6BX
auto TC	auto TC	Ashot (licence)
epicyclic	epicyclic	auto TC
		epicyclic
4/3	4/2	2/1
STN37	hydrostatic infi-	single radius
hydrostatic	nitely variable	regenerative
28/14	28/20	28/15.6
300	400	500
8.4	-	9
L11A5	Rh-120	
ROF	OTO-Breda	Israel Industries
120	120	120
L 55	L 44	L 44
8	8	8
-10 +20	-9 +20	-8 +20
48 - 52	42	50
APDS 1370	APFSDS 1650	
others 670	MP 1140	
L11A5	MG42/59	1 M240 MAG58
7.62	7.62	7.62
600/1000	1200	900
4000	2500	10000
L8A2	MG42/59	2
7.62	7.62	7.62
600/1000	1200	1200
850	(see coax)	(see coax)

Characteristics	Unit	**LECLERC**
Smoke dischargers		2x9
Laying, sighting and vision systems		
Powered mounting		electric
Stabilisation - primary/slaved		sight/gun
Rangefinder		ND-YAG laser Avimo
Ballistic computer		UCMH/PMF Serge Dassault
Gunner's primary sight		SAGEM HL 60 binocular (x2.5/x10) with LRF and Athos thermal sight
Cdr's vision system		
Secondary sight		-
Manufacturer		GIAT
Dimensions		
Length	m	9.87
Hull length	m	6.88
Width	m	3.71
Height	m	2.78
Ground clearance	m	0.5

Characteristics	Unit	**Type 74**
Country		J
Crew	persons	4
Battle weight	t	38
Power-weight ratio	kW/t	14.2
Nominal ground pressure	bars	0.86
Automotive performance		
Maximum speed	km/h	53
Road range	km	475
Fuel capacity	l	950
Fuel consumption	l/km	2

CHALLENGER 2	ARIETE	MERKAVA MARK 3
2x5	2x4	60 mm mortar
electrical	elec-hyd	electrical
gun/sight	sight/gun	sight/gun
ND-YAG laser	ND-YAG laser	ND-YAG laser
TLS2	Selenia	
digital No. 6 Mk1	COSMO	Elbit/El-Op
Marconi Space &	Marconi Italiana	
Defence 12/12P		
TLS No. 10 Mk1	OG-14L3 TURMS	periscopic (x12)
Barr & Stroud	monocular (x10)	
x10	with LRF and	
	Galileo thermal sight	
AFV No. 37 Mk5	panoramic SFIM	
Avimo	(x2.5 and x10) with	
(x1 and x10)	stab. line of sight	
AFV No. 80 Mk1	Galileo OG C-102	-
Vickers Instr.	monocular 8x	
ROF	IVECO/OTO-Melara	Ordnance Corps Fa.
11.55	9.67	8.78
8.327	7.59	7.6
3.518	3.6	3.7
2.95	2.5	2.76
0.5	0.44	0.53

Type 90	T-55	T-72 M1
J	RUS	RUS
3	4	3
50	36	43.5
20	11.85	13.22
0.89	0.81	0.9
70	48	60
400	400	460
1100	812	1190
2.75	2.03	2.59

Characteristics	Unit	**Type 74**
Vertical obstacle	m	1
Trench width	m	2.7
Shallow fording depth	m	1
Immersion depth	m	3
Water propulsion		-
Hull		
Sprocket		rear
Suspension		HP
No of roadwheels/top rollers		5/0
Track		end-connectors fixed pads
Engine		
Model		10F22WT
Manufacturer		Mitsubishi
Type		diesel
Cylinders -no/arrangement		V-10
Capacity	l	21.5
Rating	kW	529
at peak power speed	rpm	2200
Cooling system		air
Transmission and steering		
Model		MT75A Mitsubishi
Type of gearbox		semi-automatic epicyclic
No of speeds - forward/reverse		6/1
Type of steering		epicyclic
Electrics		
Generator	V/kW	
Battery capacity	Ah	
Auxiliary generator	kW	-
Main armament		
Model		1.7
Manufacturer		Japan Steel Works
Calibre	mm	105
Calibre length		L 51
Rate of fire	rds/min	8
Elevation arc	degrees	-12 +15

Type 90	T-55	T-72 M1
1.0	0.8	0.8
2.7	2.7	2.7
1	1.4	1.4
2	4.6	5.5
-	-	-
rear	rear	rear
TB/HP	TB	TB
6/3	5/0	6/3
end connectors	single pin	single pin
steel/rubber pads	no pads	no pads
10ZG32	W55	V-46-6
Mitsubishi		
supercharged diesel	diesel	diesel
V-10	V-12	V-12
21.5	38.88	38.8
1103	426	575
2400	2000	2000
liquid	liquid	liquid
auto TC	manual	manual
epicyclic		
4/2	5/1	7/1
hydrostatic rege-	clutch+brake	clutch+brake
nerative differential	fixed radius	fixed radius
28/20	28/5	28
500	280	
-	compressed air	compressed air
	starter	starter
Rh-120	D10T2S	2A46
Rheinmetall	state factory	
120	100	125
L 44	L 56	
8	6	5
-7 +15	-5 +18	-5 +18

Characteristics	Unit	**Type 74**
Ammunition stowage	rounds	55
Ammunition natures/MVs	m/sec	APDS 1470 HEP 760
Secondary armament		
Coax, model		Type 74
Calibre	mm	7.62
Cyclic rate	rds/min	1000
Ammunition stowage	rounds	4500
ADMG, model		M2
Calibre	mm	12.7
Cyclic rate	rds/min	500
Ammunition stowage	rounds	600
Smoke dischargers		2x3
Laying, sighting und vision systems		
Powered mounting		electrical
Stabilisiation - primary/slaved		gun
Rangefinder		Nihon Denki ruby laser
Ballistic computer		Mitsubishi analogue for 2 types of ammo
Gunner's primary sight		periscope J2 x1 and x8 with IR mode
Cdr's vision system		periscope J3 x1 and x8 with IR mode
Secondary sight		telecopic
Manufacturer		Mitsubishi
Dimensions		
Length	m	9.41
Hull length	m	6.7
Width	m	3.18
Height	m	2.67
Ground clearance	m	0.2-0.65

Type 90	T-55	T-72 M1
44	43	39
APFSDS 1650	HVAP 1250	HVAPFSDS 1615
MP 1140	AP 1000	HEATFS 1100
Type 74	PKT	PKT
7.62	7.62	7.62
1000	650	650
	3500	3000
M2	DShK38/46	NSVT
12.7	12.7	12.7
600	600	800
220	500	500
2x4	-	8
elec-hyd	hyd, traverse elec	electrical
sight/gun	gun	gun
ND-YAG laser	none	
	none	laser
	TSh2B22	yes
	x3.5 and x7	
Nikon	TPK U-2B	day TPD 2
	rotating head	night TPN 1-49-23
	periscope x5.5	
Fuji	TTSh16 (x4)	
	IR night sight	
Mitsubishi	state factories	state factories
9.755	9	9.35
7.5	6.45	6.67
3.43	3.27	3.46
3.045	2.4 (+ ADMG)	2.27 (+ ADMG)
0.2-0.6	0.425	0.47

Characteristics	Unit	T-80U
Country		RUS
Crew	persons	3
Battle weight	t	46
Power-weight ratio	kW/t	20
Nominal ground pressure	bars	0.92
Automotive performance		
Maximum speed	km/h	70
Road range	km	335
Fuel capacity	l	1090
Fuel consumption	l/km	3.25
Vertical obstacle	m	1.0
Trench width	m	2.85
Shallow fording depth	m	1.2
Immersion depth	m	5
Water propulsion		-
Hull		
Sprocket		rear
Suspension		TB
No of roadwheels/top rollers		6/4
Track		end connectors removable pads
Engine		
Model		GTD-1250
Manufacturer		
Type		gas turbine
Cylinders -no/arrangement		
Capacity	l	
Rating	kW	919
at peak power speed	rpm	
Cooling system		air
Transmission and steering		
Model		
Type of gearbox		manual
No of speeds - forward/reverse		4/1
Type of steering		radial
Electrics		
Generator	V/kW	27/20

Type 88	M60A3	M1A1 ABRAMS
ROK	USA	USA
4	4	4
51	52.6	57.15
17.3	10.48	19.3
0.86	0.87	0.96
65	48.2	66.8
500	480	465
1173	1420	1908
2.35	2.96	4.1
1	0.91	1.07
2.74	2.6	2.74
1.2	1.22	1.22
1.8	2.4	1.98
-	-	-
rear	rear	rear
TB/HP	TB	TB
6/3	6/3	6/2
end-connectors	end-connectors	end-connectors
fixed pads	removable pads	removable pads
MB871 Ka-501	AVDS 1790-2A	AGT 1500
MTU	Teledyne Contin.	Lycoming
turbocharged diesel	diesel	gas turbine
V-8	V-12	-
	29.34	
882	551	1100
2600	2400	3000
liquid	air	air
ZF LSG 3000	CD850-6A	XM100-3 GM Allison
	GM Allison	hydrokinetic
automatic with	auto TC	auto TC
mech lock-up	epicyclic	epicyclic
4/2	2/1	4/2
hydrostatic	single radius	hydrostatic
infinitively var.	regenerative	
	28/15.6	28/18.2

Characteristics	Unit	**T-80U**
Battery capacity	Ah	500
Auxiliary generator	kW	28
Main armament		
Model		2A46M-1
Manufacturer		Motovilikha
Calibre	mm	125
Calibre length		L 48
Rate of fire	rds/min	8
Elevation arc	degrees	-5 + 14
Ammunition stowage	rounds	45
Ammunition natures/MVs	m/sec	APFSDS 1750 MP 1050
Secondary armament		
Coax, model		PKT
Calibre	mm	7.62
Cyclic rate	rds/min	1000
Ammunition stowage	rounds	1250
ADMG, model		NSVT
Calibre	mm	12.7
Cyclic rate	rds/min	500
Ammunition stowage	rounds	500
Smoke dischargers		2x4
Laying, sighting and vision systems		
Powered mounting		electric
Stabilisation - primary/slaved		sight/gun
Rangefinder		ND-YAG laser
Ballistic computer		1V528
Gunner's primary sight		IG46 daysight monocular (x2.7/x12) with LRF
Cdr's vision system		PNK-4S daysight x5 and x7.5 light intensif. night sight
Secondary sight		-

Type 88	M60A3	M1A1 ABRAMS
	300	300
-	yes	-
M68A1	M68	Rh-120 (M256)
Watervliet	Chrysler	Watervliet (lic.)
105	105	120
L 51	L 51	L 44
8	8	8
-10 +20	-10 +20	-10 +20
47	63	40
APFSDS 1525	APFSDS 1525	APFSDS 1650
HEAT 1173	HEAT 1173	MP 1140
1 M60	M240 FN MAG58	M240 FN MAG58
1 external M60		
7.62	7.62	7.62
900	900	900
7200 + 1400	5950	12400
M2	M2	M2
12.7	12.7	12.7
500	500	500
2000	900	1000
2x6	2x6	2x6
elec-hyd	elec-hyd	elec-hyd
sight/gun	gun	sight/gun
CO2	Hughes ruby laser	ND-YAG
Comput. Device	M21 digital	digital Comp. Devices
GPTTS (x1/x3/x8) binocular thermal sight SFIM VS580-13 panoramic sight (x3, x10)	AN/VSG-2 + TTS (x8)	GPS (x9.5, x3) with thermal imaging light pipe to gunner's sight
Kollmorgen (x8)	M105D (x8)	Kollmorgen 939 (x8)

Characteristics	Unit	**T-80U**
Manufacturer		Omsk state factory
Dimensions		
Length	m	9.656
Hull length	m	7.0
Width	m	3.589
Height	m	2.2
Ground clearance	m	0.45

Type 88	M60A3	M1A1 ABRAMS
Hyundai	General Dynamics	General Dynamics
9.672	9.436	9.828
7.477	6.946	7.918
3.594	3.631	3.657
2.248 (+ ADMG)	3.27	2.886
0.46	0.45	0.432

TANK DESTROYERS, GUN

Characteristics	Unit	KÜRASSIER
Country		A
Crew	persons	3
Battle weight	t	17.5
Power-weight ratio	kW/t	12.6
Ground-pressure	bars	0.68
Automotive performance		
Maximum speed	km/h	68
Road range	km	520
Fuel capacity	l	400
Fuel consumption	l/km	0.77
Vertical obstacle	m	0.8
Trench width	m	2.4
Water propulsion		-
Maximum water speed/	km/h	-
fording depth		1
Running gear		
Sprocket		rear
Suspension		TB
No of roadwheels/top rollers		5/3
each side		end connector
Track		fixed pads
Engine		
Model		7FA
Manufacturer		Steyr
Type		turbocharged diesel
Cylinders -no/arrangement		6
Capacity	l	10
Rating	kW	221
at peak power speed	rpm	2300
Cooling		liquid
Transmission		
Type of gearbox		ZF S6-70/3 manual
No of speeds - forward/reverse		6/1
Steering		hydrostatic regenerative

Ikv 91

S
4
16.3
14.89
0.49

65
500
400
0.8
0.8
2.8
tracks
6.5
-

rear
TB
6/0
end connector
removable pads

TD 120 A
Volvo-Penta

turbocharged diesel
6

243
2200
liquid

HT740D
GM Allison
4/1
CLETRAC

Characteristics	Unit	**KÜRASSIER**
Electrics		
Armament		
Main armament, model		M57 Type D1504
Manufacturer		Fives-Cail Babcock
Calibre	mm	105
Laying system		elec-hyd
Feed/loading system		auto-loader
Rate of fire	rds/min	6
Muzzle velocities	m/sec	HVAP 1030
Elevation arc	degrees	-6 +13
Ammunition stowage	rounds	43
Traverse	degrees	360
Fire control		laser rangefinder
Secondary armament		
No/type/role		1 MG 42 coax
MG, calibre	mm	7.62
Ammunition stowage	rounds	2000
Smoke dischargers		2x3
NBC collective protection		yes
Armour type		steel
Manufacturer		Steyr-Daimler-Puch
Dimensions		
Length	m	7.763
Hull length	m	5.582
Width	m	2.5
Height	m	2.529
Ground clearance	m	0.4

Ikv 91

KV90S73
Bofors
90
elec-hyd
manual
8
HEAT 825

-10 +15
59
360
computerized

1 coax MG
1 ADMG
7.62
4250
2x6
yes
steel
Hägglund & Söner

8.84
6.41
3
2.32
0.37

TANK DESTROYERS, MISSILE

Characteristics	Unit	JAGUAR 1
Country		D
Running gear/sprocket		tracked/rear
Crew	persons	4
Battle weight	t	25.5
Power-weight ratio	kW/t	14.4
Ground-pressure	bars	0.7
Automotive performance		
Maximum speed	km/h	70
Road range	km	400
Fuel capacity	l	470
Fuel consumption	l/km	1.18
Vertical obstacle	m	0.75
Trench width	m	2
Water propulsion		-
Maximum water speed/	km/h	-
fording depth		1.8
Hull/chassis		
Suspension		TB
No of roadwheels/ top rollers each side		5/3
Track/tyres		end connectors removable pads
Engine		
Model		MB837A La500
Manufacturer		MTU
Type		diesel
Cylinders -no/arrangement		V-8
Capacity	l	29.9
Rating	kW	368
at peak power speed	rpm	2200
Cooling		liquid
Transmission		
Type of gearbox		Renk HSWL123 hydrostatic
No of speeds - forward/reverse		3/3
Steering		

VCAC MEPHISTO	BRDM-2 SPANDREL	Pvrbv 551
F	RUS	S
4x4	4x4	tracked/rear
4	2	4
13	7	9.7
13.01	14.71	10.3
-	-	0.4
92	100	41
770	750	300
300	290	240
0.39	0.39	0.8
0.6	0.4	0.61
1	1.25	1.5
by wheels	waterjet	-
7	10	-
-	-	0.9
TB	SEL	CS
-	-	6/2
14.00x20	13.00x18	single pin
2356HM71	GAZ-41	Model 2658E
Renault-Saviem		Ford
diesel	petrol	petrol
6	V-8	V-6
10.7	5.53	
169	103	100
220	3400	5200
liquid	liquid	liquid
semi-auto TC	manual	Mercedes W 4A-018
5/1	4/1	4/1
hyd servo	front	

Characteristics	Unit	**JAGUAR 1**
Electrics		
Generator	V/kW	
Armament		
Main armament, model		HLPB 04 (K3S)
ATGMS		HOT-2
Manufacturer		MBB
Missile diameter	mm	150
Missile length	m	1.275
Laying system		electric
Elevation arc	degrees	-20 +20
Traverse	degrees	140
Missile stowage	rounds	20
Sustained speed	m/s	250
Battle range (min-max)	m	75-4000
Secondary armament		
No/type/role		1 bow MG 1 ADMG
MG, calibre	mm	7.62
Ammunition stowage	rounds	3200
Smoke dischargers		8
NBC collective protection		yes
Armour type		steel
Manufacturer		Thyssen-Henschel
Dimensions		
Length	m	-
Hull length	m	6.61
Width	m	3.12
Height	m	1.98 (hull top)
Ground clearance	m	0.44

VCAC MEPHISTO	**BRDM-2 SPANDREL**	**Pvrbv 551**
24		24
MEPHISTO turret	9M113	pivot
HOT-2	AT-5 SPANDREL	TOW
Aerospatiale	state factory	Hughes
150	135	127
1.27	1.22	1.16
electrical	electric	manual
±10		
360		360
12	15	
250	185	312
75-4000	75-4000	65-3750
none	none	none
		none
2x4	-	
yes	yes	
steel	steel	steel
Renault-Saviem	state factory	Hägglund & Söner
-	-	-
5.98	5.7	4.81
2.49	2.35	2.54
2.06 (hull top)	2.68	2.54
0.4	0.43	0.33

ARMOURED INFANTRY FIGHTING VEHICLES and ARMOURED CARRIERS

Characteristics	Unit	**ASCOD**
Country		A/E
Crew	persons	3+8
Battle weight	t	25.24
Power-weight ratio	kW/t	12.98
Nominal ground pressure	bars	0.62
Automotive performance		
Maximum speed	km/h	70
Road range	km	600
Fuel capacity	l	650
Fuel consumption	l/km	1.08
Vertical obstacle	m	0.95
Trench width	m	2.5
Shallow fording depth	m	1.2
Amphibious capability		no
Water propulsion		-
Maximum water speed	km/h	-
Hull		
Sprocket		front
Suspension		TB
No of roadwheels/top rollers		7/3
Track		end-connectors removable pads
Engine		
Model		183 TE22 8
Manufacturer		MTU
Type		diesel
Cylinders -no/arrangement		V-8
Capacity	l	22.4
Rating	kW	441
at peak power speed	rpm	2300
Transmission		
Model		HSWL106 Renk
Type of gearbox		auto HM

4K7FA	**MARDER 1 A3**	**CV-9040**
A	D	SWE
2+8	2+7	3+8
14.8	34	22.8
15.9	12.98	17.7
0.55	0.9	0.53
70	70	70
520	500	300
360	625	525
0.69	1.25	1.75
0.8	1	1
2.1	2.5	2.4
1.0	1.5	1.2
-	no	no
-	-	-
-	-	-
front	front	front
TB	TB	TB
5/3	6/3	7/0
double pin, single pad	end-connectors removable pads	double pin
7FA	MB 833Ea-500	DS 14
Steyr	MTU	Scania
turbocharged diesel	diesel	diesel
6	V-6	V-8
9.98	22.4	14.2
235	441	404
2300	2200	2200
manual	HSWL194 Renk	Perkins X-300-5N
ZF S 6-80	auto TC	automatic TC

Characteristics	Unit	ASCOD
No of speeds - forward/reverse		6/6
Type of steering		hydromechanic infinitely variable
Electrics		
Generator	V/kW	28/9
Battery capacity	Ah	300
Armament		
Main armament, model		F
Manufacturer		Mauser
Calibre	mm	30
Feed/loading system		belt
Turret crew	persons	2
Rate of fire	rds/min	800
Muzzle velocity	m/sec	1095
Elevation arc	degrees	-10 +50
Ammunition stowage	rounds	402
Laying system		electrical
Secondary armament		
No/type/role		1 coax MG
Calibre	mm	7.62
Ammunition stowage	rounds	2900
ATGM - no, type		-
Smoke dischargers		12
Armour type		steel
Manufacturer		Steyr/S. Barbara
Dimensions		
Length	m	-
Hull length	m	6.22
Width	m	3.00
Height	m	2.65
Ground clearance	m	0.45

4K7FA	**MARDER 1 A3**	**CV-9040**
6/1	4/4	4/2
hydrostatic	hydrostatic	infinitely variable
split torque	infinitely variable	hydrostatic
24	28/9	28/13
180	300	204
M2 HB	Rh202	L70
Browning	Rheinmetall	Bofors
12.7	20	40
magazine	belt	auto-loader
1	2	2
550	1000	300
830	1050	1470
-15 +50	-17 +45	-8 +35
1000	2500	238
manual	elec-hydraulic	hydraulic/electric
none		
	1 coax MG	1 coaxial MG
	7.62	7.62
	5000	3000
	6 MILAN	-
4	6	2x3
steel	steel	laminate/steel
Steyr-Daimler-Puch	Thyssen-Henschel	Hägglund/Bofors
-	-	-
5.87	6.79	6.471
2.5	3.24	3.192
1.61	2.99	2.5
0.42	0.44	0.45

Characteristics	Unit	**WARRIOR**
Country		GB
Crew	persons	2+8
Battle weight	t	24.5
Power-weight ratio	kW/t	16.73
Nominal ground pressure	bars	68.0
Automotive performance		
Maximum speed	km/h	75
Road range	km	660
Fuel capacity	l	770
Fuel consumption	l/km	1.2
Vertical obstacle	m	0.75
Trench width	m	2.5
Shallow fording depth	m	1.3
Amphibious capability		-
Water propulsion		-
Maximum water speed	km/h	-
Hull		
Sprocket		Front
Suspension		TB
No of roadwheels/top rollers		6/3
Track		steel rubber pad
Engine		
Model		CV8 TCA
Manufacturer		Perkins
Type		diesel 4st
Cylinders -no/arrangement		V-8
Capacity	l	17.41
Rating	kW	410
at peak power speed	rpm	2300
Transmission		
Model		Allison DDA X300 4B
Type of gearbox		automatic
No of speeds - forward/reverse		4/2
Type of steering		hydrostatic

STORMER	DARDO	Type 73
GB	I	J
3+8	3+7	3+9
12.7	23	13.3
11.55	16.61	16.59
0.46	0.7	
80	70	60
600	500	300
405	620	450
0.68	1.24	1.5
0.46	0.85	0.7
1.75	2.5	2
1.1	1.5	-
no	no	yes
-	-	tracks
-	-	7
front	front	front
TB	TB	TB
6/2	6/3	5/0
single pin	end-connectors	single pin
removable pads	removable pads	fixed pads
T6.3544	8260	4ZF
Perkins	IVECO	Mitsubishi
diesel	diesel	diesel
6L	V-6	V-4
5.8	12.88	8.59
184	382	221
2600	2300	2200
T300	ZF LSG 1500	
Alvis		
semi-auto	auto TC	auto TC
		epicyclic
7/1	4/2	4/1
Alvis	hydrostatic	CLETRAC
	infinitely variable	

Characteristics	Unit	WARRIOR
Electrics		
Generator	V/kW	24
Battery capacity	Ah	
Armament		
Main armament, model		Rarden
Manufacturer		Royal Ordnance
Calibre	mm	30
Feed/loading system		belt
Turret crew	persons	2
Rate of fire	rds/min	90
Muzzle velocity	m/sec	APDS 1200
Elevation arc	degrees	-10 +45
Ammunition stowage	rounds	228
Laying system		electrical, manual elevation
Secondary armament		
No/type/role		1/M240/TMG M240 Chain Gun
Calibre	mm	7.62
Ammunition stowage	rounds	2200
ATGM - no, type		-
Smoke dischargers		2x4
Armour type		Aluminium/steel
Manufacturer		GKN Defence
Dimensions		
Length	m	-
Hull length	m	6.34
Width	m	3.034
Height	m	2.735
Ground clearance	m	0.4

STORMER	DARDO	Type 73
28/3	28/9	28/1
220	360	200
1 GPMG No 16	KBA Oerlikon	M2 HB Browning
7.62	25	12.7
belt	belt	belt
-	2	1
900	550	550
1000	1150	790
(external)	-10 +60	-10 +60
2000	200	600
manual	electric	manual
none		
	1 coax MG	1 ball mounted MG
	7.62	7.62
	700	4500
-	2 TOW	-
2x3	8	6
steel	aluminium	aluminium
Alvis	IVECO/OTO-Melara	Mitsubishi
-	-	-
5.33	6.71	5.8
2.69	3.0	2.8
2.27	2.64	2.2
0.425	0.4	0.4

Characteristics	Unit	BMP-1
Country		RUS
Crew	persons	3+8
Battle weight	t	13.9
Power-weight ratio	kW/t	15.87
Nominal ground pressure	bars	0.59
Automotive performance		
Maximum speed	km/h	80
Road range	km	500
Fuel capacity	l	460
Fuel consumption	l/km	0.92
Vertical obstacle	m	0.8
Trench width	m	2.2
Shallow fording depth	m	-
Amphibious capability		yes
Water propulsion		tracks
Maximum water speed	km/h	8
Hull		
Sprocket		front
Suspension		TB
No of roadwheels/top rollers		6/3
Track		single pin no pads
Engine		
Model		5 D 20
Manufacturer		state factory
Type		diesel
Cylinders -no/arrangement		V-6
Capacity	l	15.9
Rating	kW	221
at peak power speed	rpm	2600
Transmission		
Model		
Type of gearbox		manual
No of speeds - forward/reverse		5/1
Type of steering		clutch + brake

BMP-2	BMP-3	MT-LB
RUS	RUS	RUS
3+7	3+7	3+9
14.6	18.7	11.9
17.63	19.7	14.83
	0.6	0.45
80	70	60
500	600	500
460	690	450
0.92	1.15	0.9
0.7	0.8	0.7
2.5	2.5	2.7
-	-	-
yes	yes	yes
tracks	hydrojet	tracks
7	10	6
front	rear	rear
TB	TB	TB
6/3	6/3	6/0
double pin	double pin	single pin
		no pads
5D20	UTD29M	YaMZ 238V
state arsenals	state arsenals	
diesel	diesel	diesel
V-6	V-10	V-8
15.9		
257	368	176
	2600	2100
	hydromechanic	clutch
5/1	4/2	6/1
clutch + brake	clutch + brake	clutch + brake

Characteristics	Unit	BMP-1
Electrics		
Generator	V/kW	27
Battery capacity	Ah	280
Armament		
Main armament, model		2A28
Manufacturer		state factory
Calibre	mm	73
Feed/loading system		auto-loader
Turret crew	persons	1
Rate of fire	rds/min	8
Muzzle velocity	m/sec	435 to 700
Elevation arc	degrees	-4 +33
Ammunition stowage	rounds	40
Laying system		manual, traverse elec-hydr
Secondary armament		
No/type/role		1 coax MG PKT
Calibre	mm	7.62
Ammunition stowage	rounds	2000
ATGM - no, type		4 SAGGER
Smoke dischargers		-
Armour type		steel
Manufacturer		Leningrad plant
Dimensions		
Length	m	-
Hull length	m	6.74
Width	m	2.94
Height	m	1.92
Ground clearance	m	0.39

BMP-2	BMP-3	MT-LB
27	27	28
2A42	2A70	1 PKT MG
state arsenals	state arsenals	
30	100	7.62
belt	auto-loader	belt
2	2	1
200-300 or 500	10	650
	250	855
-5 +50	-6 +60	-5 +30
500	32 HE/8 9M117	2000
electric	electric	manual
		none
1/ADMG	1 coaxial	
7.62	30	
2000	305 HE/195 AP	
4/AT-5 Spandrel	8 9M117	-
2x3	2x3	-
laminate/steel	laminate/steel	steel
state arsenals	Kurganmashzavod	state factories
-	7.2	-
6.858	7.14	6.454
3.089	3.23	2.85
2.077	2.3	1.865
0.4	0.19-0.51	0.4

Characteristics	Unit	M2 BRADLEY
Country		USA
Crew	persons	3+7
Battle weight	t	22.59
Power-weight ratio	kW/t	22.13
Nominal ground pressure	bars	0.54
Automotive performance		
Maximum speed	km/h	66
Road range	km	483
Fuel capacity	l	662
Fuel consumption	l/km	1.37
Vertical obstacle	m	0.914
Trench width	m	2.54
Shallow fording depth	m	-
Amphibious capability		yes
Water propulsion		tracks
Maximum water speed	km/h	7.2
Hull		
Sprocket		front
Suspension		TB
No of roadwheels/top rollers		6/3
Track		single pin removable pads
Engine		
Model		VTA-903T
Manufacturer		Cummins
Type		turbocharged diesel
Cylinders -no/arrangement		V-8
Capacity	l	14.8
Rating	kW	368
at peak power speed	rpm	2600
Transmission		
Model		HMPT-500 General Electric
Type of gearbox		automatic
No of speeds - forward/reverse		3/1
Type of steering		integral infinitely variable hydrostatic

AIFV	M113 A2	M-980
USA	USA	YU
3+7	2+11	2+8
13.7	11.34	13
14.17	13.94	9.05
0.66	0.56	0.64
61	67.59	60
490	483	500
416	360	
30.36	0.75	
0.64	0.61	0.8
1.63	1.68	2.2
-	-	-
yes	yes	yes
tracks	tracks	tracks
6.3	5.8	7.5
front	front	front
TB	TB	TB
5/0	5/0	5/2
single pin	single pin	single pin
removable pads	fixed pads	no pads
6V-53T	6V-53N	HS 115-2
Detroit Diesel	Detroit Diesel	SNECMA
turbocharged diesel	diesel	turbocharged diesel
V-6	V-6	V-8
5.21	5.21	8.21
194	158	191
2800	2800	3000
TX-100-1A	TX-100-1	
Allison		
torque converter	torque converter	preselector TC
3/1	3/1	4/1
controlled	controlled	
differential mech	differential mech	

Characteristics	Unit	**M2 BRADLEY**
Electrics		
Generator	V/kW	28/5.3
Battery capacity	Ah	200
Armament		
Main armament, model		1 M242 chain gun
Manufacturer		Hughes
Calibre	mm	25
Feed/loading system		belt
Turret crew	persons	2
Rate of fire	rds/min	100/200
Muzzle velocity	m/sec	1345 APDS
Elevation arc	degrees	-10 +60
Ammunition stowage	rounds	1340
Laying system		elec-hydraulic
Secondary armament		
No/type/role		1 coax MG M240
Calibre	mm	7.62
Ammunition stowage	rounds	1340
ATGM - no, type		7 TOW-2
Smoke dischargers		2x4
Armour type		compound
Manufacturer		FMC
Dimensions		
Length	m	-
Hull length	m	6.453
Width	m	3.2
Height	m	2.972
Ground clearance	m	0.43

AIFV	M113 A2	M-980
28/2.8	28/2.8	28
100	100	
KBA	Ms HB	HS804
Oerlikon-Bührle	Browning	Hispano-Suiza
25	12.7	20
belt	belt	drum magazines
1	-	1
570	550	800
1100	790	AP 850
-10 + 60	(external)	-5 +75
324	2000	400
elec-hydraulic	manual	manual
	none	
1 coax MG FN MAG		1 MG53
7.62		7.92
1840		2250
-	-	4 SAGGER
6	optional	-
compound	aluminium	steel
FMC	FMC	FAMOS
-	-	-
5.258	4.863	6.4
2.819	2.686	2.59
2.619	1.85 (hull top)	2.3
0.43	0.43	0.4

ARMOURED CARRIERS (WHEELED)

Characteristics	Unit	PANDUR
Country		A
No of wheels/driven wheels		6x6
Crew	persons	2+8
Battle weight	t	11.9
Power-weight ratio	kW/t	16.1
Automotive performance		
Maximum speed	km/h	100
Road range	km	700
Fuel capacity	l	275
Fuel consumption	l/km	0.39
Vertical obstacle	m	0.5
Trench width	m	1.1
Amphibious capability		no
Water propulsion		-
Maximum water speed	km/h	-
Chassis		
Turning radius		8.5
Tyres		12.5R20
Suspension		CS/TB
Wheelbase		1.53 + 1.53
Engine		
Model		WD 612.95
Manufacturer		Steyr
Type		turbocharged diesel
Cylinders -no/arrangement		V-6
Capacity	l	6.59
Rating	kW	191
at peak power speed	rpm	2400
Cooling system		liquid
Steering		2 front axles
Gearbox		
Model		MT 653 DR Allison
Type		automatic
No of speeds - forward/reverse		5/1
Electrics		
Generator	V/kW	28/6
Battery capacity	Ah	200

EE-11 URUTU 5	GRIZZLY	FUCHS
BR	CDN	D
6x6	6x6	6x6
13	3+6	2+10
14	10.5	17
13.66	15.06	13.84
105	102	105
750	603	800
380	204	390
0.51	0.34	0.49
0.6	0.38	0.6
1	1.2	1.06
yes	yes	yes
propellor	2 propellors	2 propellors
10	7	10
10	7.3	8.5
14.50x20	11.00x16	14.00x20
CS/rear SEL	CS/rear SEL	CS
2.09/1.41	2.04+1.04	1.75/2.05
6V 53T	6V-53	OM402A
Detroit Diesel	Detroit Diesel	Daimler-Benz
turbocharged diesel	diesel	turbocharged diesel
V-6	V-6	V-8
5.22	5.22	12.76
191	158	235
2800	1800	2500
liquid	liquid	liquid
servo hydraulic	servo hydraulic	servo hydraulic
front axle	front axle	
Allison MT 643	MT653TC	ZF 6HP500
torque converter	GM Allison	automatic
4/1	5/1	6/1
	24/2.4	28/5

Characteristics	Unit	**PANDUR**
Armament		
Main armament, model		M2 HB
Manufacturer		Hughes
Calibre	mm	12.7
Feed/loading system		belt
Turret crew	persons	1
Rate of fire	rds/min	500
Muzzle velocity	m/sec	1000
Elevation arc	degrees	-15 +30
Ammunition stowage	rounds	500
Laying system		manual
Secondary armament		
No/type/role		1 coax MG
Calibre	mm	7.62
Ammunition stowage	rounds	1500
ATGM - no, type		-
Smoke dischargers		optional
Armour type		steel
Manufacturer		Steyr-Daimler-Puch
Dimensions		
Length	m	-
Hull length	m	5.697
Width	m	2.5
Height	m	1.81
Ground clearance	m	0.42

EE-11 URUTU 5	GRIZZLY	FUCHS
TMG/M2 HB	M2 turret	1 MG 3
Browning	Cadillac Gage	Rheinmetall
12.7	12.7	7.62
belt	belt	belt
1	1	(external)
550	550	1200
790	790	
-7 + 50	-8 + 60	-
1200	1000	1000
manual	manual	manual
		none
optional	1 TMG coaxial	
	7.62	
	4400	
-	-	-
2x2	2x4	6
steel	steel	steel
ENGESA	GM Ontario	Thyssen-Henschel
-	-	-
6.1	5.968	6.76
2.65	2.53	2.98
2.72	2.53	2.3
0.38	0.392	0.406

Characteristics	Unit	**PASI**
Country		SF
No of wheels/driven wheels		6x6
Crew	persons	2+10
Battle weight	t	15
Power-weight ratio	kW/t	11.76
Automotive performance		
Maximum speed	km/h	100
Road range	km	800
Fuel capacity	l	300
Fuel consumption	l/km	0.375
Vertical obstacle	m	0.6
Trench width	m	1
Amphibious capability		yes
Water propulsion		2 propellors
Maximum water speed	km/h	10
Chassis		
Turning radius		
Tyres		14.00x20
Suspension		LS
Wheelbase		2.37
Engine		
Model		611 DSBJ
Manufacturer		Valmet
Type		turbocharged diesel
Cylinders -no/arrangement		6
Capacity	l	6.6
Rating	kW	176
at peak power speed	rpm	2500
Cooling system		liquid
Steering		front/rear
Gearbox		
Model		MT 643
Type		Allison
No of speeds - forward/reverse		4/1
Electrics		
Generator	V/kW	24
Battery capacity	Ah	

VAB-VTT	VCR/TT	SAXON
F	F	GB
4x4	6x6	4x4
2+10	3+9	12
13	7.9	11.66
12.44	13.496	10.53
92	90	96
1000	700	480
300	242	153
0.3	0.35	0.32
0.5	0.8	0.4
	1.1	-
yes	yes	1.12 fording
waterjet	wheels	-
7	4	-
9		7.52
14.00x20	11.00x16	12.00x20
TB	CS/HP	SEL
3.00	1.66/1.425	3.07
MIDS 06.20.45	PRV V6	500
Renault (MAN)	Peugeot	Bedford
turbocharged diesel	petrol	diesel
6	V-6	6
9.84	2.75	8.2
162	107	123
2200	5500	2200
liquid	liquid	liquid
servo hydraulic	front axle	servo hydraulic
Renault VI	Panhard	AT 540
Transelfluide	manual	Allison
5/1	6/1	4/1
28/1.8	24	24/0.8

Characteristics	Unit	**PASI**
Armament		
Main armament, model		1 MG
Manufacturer		
Calibre	mm	7.62
Feed/loading system		belt
Turret crew	persons	-
Rate of fire	rds/min	1200
Muzzle velocity	m/sec	
Elevation arc	degrees	-
Ammunition stowage	rounds	1000
Laying system		manual
Secondary armament		none
No/type/role		
Calibre	mm	
Ammunition stowage	rounds	
ATGM - no, type		
Smoke dischargers		4
Armour type		steel
Manufacturer		Oy SISU
Dimensions		
Length	m	-
Hull length	m	7.35
Width	m	2.9
Height	m	2.3
Ground clearance	m	0.4

VAB-VTT	VCR/TT	SAXON
1 MG	different	1 M240 chain gun
		Hughes
12.7 or 7.62		7.62
belt		belt
-		1
-		
-		
(external)		-10 +55
different		800
manual		manual
		none
none	none	
MILAN optional		-
-		-
steel		steel
Renault/Creusot-L.		GKN Sankey
-	-	-
5.98	4.875	5.17
2.5	2.5	2.49
2.06 (hull top)	2.13 (hull)	2.86
0.4	0.375 (6 wheels)	0.41

Characteristics	Unit	**6614**
Country		I
No of wheels/driven wheels		4x4
Crew	persons	1+10
Battle weight	t	8.5
Power-weight ratio	kW/t	13.84
Automotive performance		
Maximum speed	km/h	100
Road range	km	700
Fuel capacity	l	142
Fuel consumption	l/km	0.2
Vertical obstacle	m	0.4
Trench width	m	
Amphibious capability		yes
Water propulsion		wheels
Maximum water speed	km/h	4.5
Chassis		
Turning radius		6.8
Tyres		10.00x20
Suspension		CS
Wheelbase		2.8
Engine		
Model		8062.24
Manufacturer		Fiat
Type		supercharged diesel
Cylinders -no/arrangement		6
Capacity	l	5.5
Rating	kW	118
at peak power speed	rpm	3200
Cooling system		liquid
Steering		servo hydraulic
Gearbox		
Model		
Type		Fiat
No of speeds - forward/reverse		5/1
Electrics		
Generator	V/kW	24
Battery capacity	Ah	

Timoney Mk 6	BTR-70	BTR-80
IR	RUS	RUS
4x4	8x8	8x8
3+9	2+9	2+8
9.98	11.5	13.6
14.74	14.71	14.04
95	80	80
1000	500	600
300	350	290
0.3	0.7	0.48
0.4	0.5	0.5
	2.0	2
yes	yes	yes
wheels	1 hydrojet	1water jet
4.8	10	9
6.1	12.6	13
11.00x20	13.00x18	13.00x18
CS	TB	TB
2.87	1.4+1.78.1.4	1.4+1.78+1.4
4.53T	ZMZ-4905 (two)	AZ-7403
Detroit Diesel		
turbocharged diesel	petrol	turbocharged diesel
4	8	V-8
3.44		
147	88 each	191
1600		
liquid		liquid
servo hydraulic	2 front axles	2 front axles
AT540TC		
automatic	manual	manual
4/1	4/1	5/1
24/2.4	24	
	85	

Characteristics	Unit	**6614**
Armament		
Main armament, model		M2 HB
Manufacturer		Browning
Calibre	mm	12.7
Feed/loading system		belt
Turret crew	persons	1
Rate of fire	rds/min	550
Muzzle velocity	m/sec	790
Elevation arc	degrees	(externally)
Ammunition stowage	rounds	1000
Laying system		manual
Secondary armament		none
No/type/role		
Calibre	mm	
Ammunition stowage	rounds	
ATGM -no, type		-
Smoke dischargers		-
Armour type		steel
Manufacturer		Fiat
Dimensions		
Length	m	-
Hull length	m	5.86
Width	m	2.5
Height	m	2.18
Ground clearance	m	0.37

Characteristics	Unit	**RATEL 20**
Country		ZA
No of wheels/driven wheels		6x6
Crew	persons	11
Battle weight	t	18.5
Power-weight ratio	kW/t	11.21
Automotive performance		
Maximum speed	km/h	105
Road range	km	1000
Fuel capacity	l	430
Fuel consumption	l/km	0.43
Vertical obstacle	m	0.6

Timoney Mk 6	BTR-70	BTR-80
TLi 127 turret	KPV	KPVT
Creusot-Loire	state arsenals	
12.7	14.5	14.5
belt	belt	belt
1	1	1
550	600	600
790	945	945
-10 +50	-5 +30	-4 +60
	500	500
manual	manual	manual
none		
	1 TMG	1 TMG PKT
	7.62	7.62
	2000	2000
-	none	-
2x4	none	6
steel	steel	steel
Technology Investm.	Gorki Plant	State arsenals
-	-	-
4.95	7.65	7.65
2.41	2.8	2.9
2.75	2.35	2.35
0.41	0.475	0.475

BMR-600	LAV-25
E	CDN
6x6	8x8
3+10	3+6
14.0	12.88
16.28	15.7
100	100
1000	670
400	300
0.4	0.45
0.6	0.5

Characteristics	Unit	**RATEL 20**
Trench width	m	1.15
Amphibious capability		no, 1.2 m
Water propulsion		
Maximum water speed	km/h	
Chassis		
Turning radius		7.95
Tyres		14.00x20
Suspension		CS
Wheelbase	m	2.809 + 1.4
Engine		
Model		D 3256 BTXF
Manufacturer		
Type		turbocharged diesel
Cylinders -no/arrangement		6
Capacity	l	
Rating	kW	207
at peak power speek	rpm	2200
Cooling system		liquid
Steering		front
Gearbox		
Model		automatic
Type		
No of speeds - forward/reverse		6/2
Electrics		
Generator	V/kW	24
Battery capacity	Ah	100
Armament		
Main armament, model		M693
Manufacturer		GIAT/ARMSCOR
Calibre	mm	20
Feed/loading system		belt
Turret crew	persons	2
Rate of fire	rds/min	740
Muzzle velocity	m/sec	
Elevation arc	degrees	-8 +38
Ammunition stowage	rounds	1200
Laying system		manual

BMR-600	**LAV-25**
1.6	2.06
yes	yes
2 water jets	2 screws
9	10.5
7.5	7.7
13.00x20	11.00x16
HP	CS/TB
1.65	1.1 + 1.35 + 1.04
9157/8	6V-53T
Pegaso	GM Detroit Diesel
turbocharged diesel	turbocharged diesel
6	V-6
11.94	5.22
228	202
2200	2100
liquid	liquid
front/rear	2 front axles
automatic	MT 653 DR Allison
ZF 6 HP 500	automatic
6/1	5/1
24	24
M2 HB	M242
Browning	Hughes
12.7	25
	belt
-	2
550	100/200/500
830	
-15 + 60	-10 +60
	630
manual	hydraulic

Characteristics	Unit	**RATEL 20**
Secondary armament No/type/role		2
Calibre	mm	7.62
Ammunition stowage	rounds	6000
STGM - no, type		0
Smoke dischargers		2x2
Armour type		steel
Manufacturer		Sandock-Austral
Dimensions		
Length	m	-
Hull length	m	7.212
Width	m	2.516
Height	m	2.915
Ground clearance	m	0.34

BMR-600	LAV-25
	1 coax MG
	7.62
	1620
-	-
none	2x4
steel	steel
Empresa	Diesel Div. GM
-	-
6.15	6.393
2.5	2.5
2.36	2.69
0.4	0.5

ARMOURED AIRBORNE VEHICLES

Characteristics	Unit	WIESEL (TOW)
Country		D
Running gear/sprocket		tracked/front
Crew	persons	3
Battle weight	t	2.8
Power-weight ratio	kW/t	22.86
Ground-pressure	bars	0.35
Automotive performance		
Maximum speed	km/h	80
Road range	km	200
Fuel capacity	l	80
Fuel consumption	l/km	0.4
Vertical obstacle	m	0.4
Trench width	m	1.2
Water propulsion		-
Maximum water speed/ fording depth	km/h	-
Hull/chassis		
Suspension		TB
No of roadwheels/top rollers each side		3/1
Track/tyres		endless belt
Engine		
Model		Type 069, 2 DE
Manufacturer		VW
Type		turbocharged diesel
Cylinders -no/arrangement		5
Capacity	l	2000
Rating	kW	64
at peak power speed	rpm	4500
Cooling		liquid
Steering		Cletrac rev. gear
Type of gearbox		ZF 3 HP 22
No of speeds - forward/reverse		6/2

SO-120 (2S9)	BMD-1
RUS	RUS
tracked/rear	tracked/rear
4	3+4
8	6.7
22	26.34
0.5	0.598
60	60
500	320
300	300
0.6	0.94
0.6	0.8
2	1.6
2 water jets	2 water jets
9	10
	-
HP	HP
6/5	5/4
end connector	end connectors
5D-20	2D-20
diesel	diesel
V-6	V-6
176	176
2600	2600
liquid	liquid
clutch/brake	clutch/brake
5/1	5/1

Characteristics	Unit	**WIESEL (TOW)**
Electrics		
Generator	V/kW	24
Armament		
ATGMS		TOW-2
Manufacturer		Hughes
Missile diameter	mm	152
Missile length	m	1.174
Laying system		manual
Elevation arc	degrees	±10
Traverse	degrees	±45
Missile stowage	rounds	7
Sustained speed	m/s	312
Battle range (min-max)	m	60-3750
Main armament (excl. ATGMS)		none
Calibre	mm	
Battle range	m	
Ammunition stowage	rds	
Muzzle velocity	m/s	
Secondary armament		none
No/type/role		
MG, calibre	mm	
Ammunition stowage	rounds	
Smoke dischargers		-
NBC collective protection		-
Armour type		steel
Manufacturer		Krupp MaK
Dimensions		
Length	m	-
Hull length	m	3.265
Width	m	1.82
Height	m	1.87
Ground clearance	m	0.3

SO-120 (2S9)	**BMD-1**
24	24
	AT-4 Spigot
120	~115
	~0.9
	manual
-4 +80	-5 +20
±35	360
	3
	185
	10-2500
	73
8800	1300
60	40
	425 (boosted to 700 Muzzle velocity)
none	
	2 MG/bow/coax
	7.62
	2000
-	-
yes	yes
steel	steel
State arsenals	State factories
-	
6.02	5.4
2.63	2.63
2.3 (raised)	1.97 (elevated)
0.45	0.45 (elev.)

ARMOURED ARTILLERY VEHICLES

Characteristics	Unit	M-77 DANA
Country		CS
Detachment	persons	5
Battle weight	t	29
Sprocket		8x8 wheeled
Engine rating/	kW	254
at peak power speed	rpm	2200
Power-weight ratio	kW/t	8.76
Nominal ground pressure	bars	
Maximum speed	km/h	80
Road range	km	740
Fuel capacity	l	500
Auxiliary generator		none
Amphibious capability	m or km/h	1.4
Water propulsion		-
Armament		
Calibre	mm	152
Calibre length		
Rate of fire - normal	rds/min	5
Rate of fire - intense	rds/t sec	-
Maximum muzzle velocity	m/sec	
Maximum range - normal	km	20
with rocket assistance		24
Elevation arc	degrees	-3 +60
On-gun ammunition	rounds	60
Loading		hydraulic
Top traverse		hydraulic
ADMG/calibre	mm	1/12.7
Armour type		steel
Manufacturer		State factories
Dimensions		
Length	m	10.4
Hull length	m	8.87
Width	m	2.97 (turret)
Height	m	3.525
Ground clearance	m	

PzH 2000	GCT	AS90B
D	F	GB
5	4	5
55.33	42	42
front	rear	front
735	529	485
2000	2000	2800
13.3	12.6	11.55
0.98	0.9	0.75
60	60	55
420	450	430
970	970	640
5	5	yes
2.1	2.1	1.5
-	-	-
155	155	155
52	40	39
12	8	6
3/9.2	6/45	3/10
1000	810	
30	24	24.7
40	30	32
-2.5 +65	-5 +66	-5 +70
60	42	48
automatic	automatic	flick rammer
electric	hydraulic	electric
1/7.62	1/7.62	1/12.7
steel	steel	steel
Wegmann	ARE	VSEL
11.67	10.25	9.7
7.3	6.7	7
3.58	3.15	3.3
3.46	3.25	3
0.44	0.42	0.41

Characteristics	Unit	**Type 75**	**2S1**	**2S3**
Country		J	RUS	RUS
Detachment	persons	6	4	4+2
Battle weight	t	25.3	15.7	27.5
Chassis		front	front	front
Engine rating/	kW	331	176	382
at peak power speed	rpm	2200	2100	
Power-weight ratio	kW/t	13.08	11.24	14.16
Nominal ground pressure	bars	0.64	0.49	0.5
Maximum speed	km/h	47	60	62
Road range	km	300	500	500
Fuel capacity	l	650	550	850
Auxiliary generator		-		
Amphibious capability	m or km/h	1.3	4.5	1
Water propulsion		-	tracks	-
Armament				
Calibre	mm	155	122	152
Calibre length	L	26		
Rate of fire - normal	rds/min	3	5	4
Rate of fire - intense	rds/sec	6/60		
Maximum muzzle velocity	m/sec	720	726	655
Maximum range - normal	km	18.5	15.26	18.5
with rocket assistance			21.9	24
Elevation arc	degrees	-5 +65	-3 +70	-5 +65
On-gun ammunition	rounds	28	40	46
Loading		semi-automatic	manual, assisted	
Top traverse		hydraulic	elec-hydr	electric
ADMG/calibre	mm	1/12.7	-	1/7.62
Armour type		light alloy	steel	steel
Manufacturer		Mitsu-bishi HI	State factory	State factories
Dimensions				
Length	m	7.79	-	7.765
Hull length	m	6.64	7.26	7.8
Width	m	3.09	2.85	3.25
Height	m	2.55	2.725	3.05
Ground clearance	m	0.4	0.4	0.45

2S19	RHINO G6	Bandkanon 1A	M109 A2	M110A2
RUS	ZA	S	USA	USA
5+2	6	5	6	5
42	46	53	24.95	28.35
rear	6x6	front	front	front
574	386	176 + 221	298	298
2000		3750/38000	2300	2300
13.66	8.39	7.49	11.94	10.5
0.92	-	0.85	0.85	0.76
60	90	28	56.3	54.7
500	600	230	349	523
	700	1445	511	984
16	yes		-	no
1.2	1	1	1	1.07
-	-	-	-	-
152	155	155	155	203
45	45	50	39	39
7	4	15	4	0.5
	3/21	-	3/20	1/60
828		865	684	711
24.7	30.8	25.6	23.5	24
28.9	39		-	29.1
-3 +68	-5 +75	-3 +38	-3 +75	-2 +65
50	47	14	36	2
semi-automatic	hydraulic	automatic	manual	hydraulic
electric	hydraulic	manual	hydraulic	hydraulic
1/12.7	1/12.7	1/7.62	1/12.7	none
steel	steel	steel	light alloy	steel
Ural-transmash	Armscor	Bofors	BMY	BMY
11.92	10.335	11	9.12	10.73
	9.2	6.55	6.19	5.72
3.584	3.4	3.37	3.1	3.15
2.985	3.8	3.85	3.28	2.743
0.435	0.45	0.37	0.45	0.393

ARMOURED ARTILLERY ROCKET LAUNCHERS

Characteristics	Unit	Type 75
Country		J
Detachment	persons	3
Battle weight	t	16.5
Running gear/sprocket		tracked/front
Engine rating/	kW	221
at peak power speed	rpm	2200
Power-weight ratio	kW/t	13.37
Nominal ground pressure	bars	0.63
Auxiliary generator	kW	-
Automotive performance		
Maximum speed	km/h	53
Road range	km	300
Fuel capacity	l	
Fuel consumption	l/100 km	
Vertical obstacle	m	0.7
Trench width	m	2.1
Amphibious capability	m or km/h	
Water propulsion		-
Rocket		
Type		
Calibre	mm	130
Length	m	1.856
Weight	kg	43
On-launcher missiles		30
Range zone	km	max 15
Manufacturer		Aerospace Div. Nissan Motor Comp.
Dimensions		
Length	m	-
Hull length	m	5.78
Width	m	2.8
Height	m	2.67
Ground clearance	m	0.4

MLRS

USA
3
25.15
tracked/front
368
2400
14.62
0.6
-

64
483
617
1.28
1
2.29
1.1
-

M-77 (and others)
227
3.94
310
2x6
30+
LTV Missiles and
Electr. Group

-
6.972
2.972
2.617 (travelling)
0.43

ARMOURED AIR DEFENCE GUNS

Characteristics	Unit	GEPARD
Country		D
Detachment	persons	3
Battle weight	t	46.3
Running gear/sprocket		tracked/rear
Engine rating/	kW	610
at peak power speed	rpm	2200
Power-weight ratio	kW/t	13.2
Nominal ground pressure	bars	0.97
Auxiliary generator	kW	66
Automotive performance		
Maximum speed	km/h	65
Road range	km	490
Fuel capacity	l	985
Fuel consumption	l/100 km	2.01
Vertical obstacle	m	1.15
Trench width	m	3
Amphibious capability	m or km/h	2.25
Water propulsion		-
Armament		
Model		KDAL/R04
Manufacturer		Oerlikon
Numer of barrels		2
Calibre	mm	35
Calibre length	L/	90
Feed		
Muzzle velocity	m/sec	1175
Rate of fire/barrel	rds/min	550
On-gun ammunition	rds	680
Mounting		elec-hydr Mounting
Elevation arc		-10 +85
Surveillance/ acquisition radar		Siemens MPDR12 coherent pulse Doppler, range 15 km
Tracking radar		Siemens ALBIS monopulse Doppler range 15 km
Computer		Siemens analogue
Manufacturer		Krauss-Maffei

2S6

RUS
4
34
tracked/rear
522
2000
15.35

59

65
500

1.0
2.0
-
-

2A38M3

2
30

linked
960
2500
1904
electric
-6 +80
1RL-144M E-band
range 18 km

same system J-band
tracking range 13 km

1A26
state factories

Characteristics	Unit	**GEPARD**
Dimensions		
Length	m	7.73
Hull length	m	6.85
Width	m	3.37
Height	m	4.03
Ground clearance	m	0.44

2S6

-
7.93
3.236
4.02 (with radar)
0.45

ARMOURED AIR DEFENCE MISSILE LAUNCHERS

Characteristics	Unit	FlaRakPz ROLAND
Country		D
Detachment	persons	3
Battle weight	t	32.5
Running gear/sprocket		tracked/front
Engine rating/	kW	441
at peak power speed	rpm	2200
Power-weight ratio	kW/t	13.57
Nominal ground pressure	bars	0.92
Auxiliary generator	kW	25
Automotive performance		
Maximum speed	km/h	70
Road range	km	365
Fuel capacity	l	640
Fuel consumption	l/100 km	175
Vertical obstacle	m	1
Trench width	m	2.5
Amphibious capability	m or km/h	1
Water propulsion		-
ADMS		
Type		ROLAND 2
Missile length/weight	m/kg	2.4/63
Maximum speed	Mach	1.6
On-launcher missiles		10
Range zone	km	0.5-6.3
Fire control system		
All-weather		yes
Guidance		optical + radar
Surveillance/ acquisition radar		Thomson-CSF D-band coherent pulse Doppler with IFF range 16 km
Tracking radar		Siemens coherent monopulse range 16 km

AMX30R	BUK-M1 (SA-11)	SA-13 GOPHER	TOR-M1 (SA-15)
F	RUS	RUS	RUS
3	4	3	3
33	32.34	13	34
tracked/rear	tracked/rear	tracked/front	tracked/rear
529	552	176	552
2400	2000	2100	
16.04	17.07	13.57	16.24
0.78			
45	75		75
65	60	55	60
520	500	450	450
970		450	450
186		1	1
0.93	1	0.7	0.7
2.9	2	2.7	2.7
1.5	-	6	6
-		tracks	tracks
ROLAND 1/2	9M38M1	9M37M	9M37M
2.4/63	5.55/690	2.2/55	2.2/55
1.6	3	1.5	1.5
10	4	4	4
0.5-6.3	1-35	0.5-8	0.5-8
no/yes	yes	no	no
optical/opt.+radar	pulse Doppler	passive IR	passive IR
Thomson-CSF D-band pulse Doppler w. IFF range 16 km	seperate vehicle	optical	optical
Siemens monopulse range 16 km	H/I-band	range only	range only

Characteristics	Unit	FlaRakPz ROLAND
Computer		digital
Manufacturer		Euromissile MBB
Dimensions		
Length	m	-
Hull length	m	6.915
Width	m	3.24
Height	m	2.92
Ground clearance	m	0.44

AMX30R	BUK-M1 (SA-11)	SA-13 GOPHER	TOR-M1 (SA-15)
digital Euromissile Aerospatiale			
-	-	-	-
6.59	9.3	6.6	6.6
3.1	3.25	2.9	2.9
3.26	7.72 (firing)	3.8 (firing)	3.8 (firing)
0.45	0.45	0.4	0.4

ARMOURED ENGINEER VEHICLES

Characteristics	Unit	DACHS
Country		D
Crew	persons	3
Battle weight	t	43
Power-weigh ratio	kW/t	14.19
Nominal ground pressure	bars	0.92
Automotive performance		
Maximum speed	km/h	62
Road range	km	650
Fuel capacity	l	1410
Fuel consumption	l/km	2.17
Vertical obstacle	m	1.15
Trench width	m	2.5
Shallow fording depth	m	1
Immersion depth	m	4
Maximum water speed	km/h	-
Water propulsion		-
Running gear		
Sprocket		rear
Suspension		TB
No of roadwheels/top rollers		7/4
Track		end-connectors removable pads
Engine		
Model		MB 838 Ca.M500
Manufacturer		MTU
Type		multifuel diesel
Cylinders -no/arrangement		V-10
Capacity	l	37.4
Rating	kW	610
at peak power speed	rpm	2200
Cooling		liquid
Transmission		
Model		4HP250 ZF
Type of gearbox		auto TC epicyclic
No of speeds - forward/reverse		4/2

FV180 CET	IMR-2M1	M728 CEV	M9 ACE
GB	RUS	USA	USA
2	2	4	1
17.7	50	53.2	14.5
13.5	12.54	10.37	14.96
0.46	0.76	0.89	0.64
56	60	48	48
480	500	480	320
430	812	1420	507
0.9	2.03	2.96	1.58
0.6	0.8	0.76	0.457
2.05	2.7	2.5	1.574
1.8	1.2	1.22	1.83
amphibious		4	amphibious
9		-	4.8
hydrojet		-	tracks
front	rear	rear	front
TB	TB	TB	HP
5/0	5/3	6/3	4/0
pin	single-pin	end-connector	end-connector
removable pads		fixed pads	removable pads
C6TFR	V-55	AVDS 1790-2A	V903C
Rolls-Royce	State arsenals	Teledyne-Continental	Cummins
diesel	diesel	diesel	diesel
6	V-12	V-12	8
12.2	38.8	29.34	
235	618	551	217
2100	2000	2400	2600
liquid	liquid	air	liquid
TN26 Self-changing gears		CD850-6 GM Allison	Caterpillar
epicyclic with TC	manual	epicyclic with TC	epicyclic
4/4	5/1	2/1	6/2

Characteristics	Unit	**DACHS**
Type of steering		2-radius regenerative
Electrics		
Generator	V/kW	28/9
Battery capacity	Ah	400
Armament		
Main armament, model		1 ADMG
Calibre	mm	7.62
Ammunition stowage	rounds	4250
Turret traverse		-
Secondary armament		none
Coax - calibre	mm	
Ammunition stowage	rounds	
ADMG - calibre		
Ammunition stowage	rounds	
Operating characteristics		
Crane jib - lift	t	-
Hydraulic dozer blade - capacity	cubic m	270/h
Maximum winch pull	kN	350
Anchor holding power	kN	-
NBC collective		yes
Armour type		steel
Manufacturer		MaK
Dimensions		
Length	m	9.01
Hull length	m	
Width	m	3.25
Height	m	2.57
Ground clearance	m	0.44

FV180 CET	**IMR-2MI**	**M728 CEV**	**M9 ACE**
clutch/brake	clutch/brake	fixed radius regenerative	
28/3.4	24	28/15.6	
200		300	
none	none		none
		petard	
		165	
		30	
		hydraulic	
		7.62	
		2000	
		12.7	
		600	
-	2	7.9	-
1.7	0.17	1.9	2
133		113	113
80 (rocket projection to 100 m)	-	-	-
yes	yes	yes	yes
light alloy	steel	steel	light alloy
ROF	State arsenals	Chrysler Detroit	Caterpillar
7.3	9.32	9.3	3.2
5.334		8.91	2.79
2.921	3.5	3.7	3.2
2.83	3.42	3.2	2.36
0.457	0.46	0.381	0.457

ARMOURED VEHICLES LAUNCHED BRIDGE

Characteristics	Unit	**BIBER**
Country		D
Crew	persons	2
Battle weight	t	45.2
Engine rating	kW	610
Power-weight ratio	kW/t	13.5
Nominal ground pressure	bars	0.97
Maximum speed	km/h	65
Road range	km	490
Fuel capacity	l	985
Type of bridge		free nose
Laid length/		22
width of bridge	m	4
Load capacity (MLC)	class (US tons)	60
Weight of bridge	t	9.8
Maximum silhouette		
height on laying	m	3.6
Dimensions		
Length	m	11.82
Hull length	m	10.56
Width	m	4
Height	m	3.55
Ground clearance	m	0.42

AMX30PP	MTU-20	MTU-72	BRÜPZ 68	M60 A1 AVLB
F	RUS	RUS	CH	USA
3	2	2	2	2
42.5	37	40	44.6	55.7
515	426	618	485	551
12.1	11.53	15.45	10.88	9.9
1.0	0.9	0.808	1.03	0.9
60	50	60	55	48
470	450	500	350	450
970	960	1000	855	1420
scissors	deck + 2 folding end ramps	deck + 2 folding end ramps	deck	scissors
22	20/	20/	18.23/	19.2//
3.95	3.3	3.3	3.8	4
50	50	50	60	60
8.5	7	6.4	6.6	14
11	6	7.56	4	10
11.4	11.64	11.64	20.1	11.28
6.7		6.86		8.648
3.95	3.3	3.46	4	4
4.29	3.4	3.38	3.39	3.9
0.4	0.425	0.492	0.4	0.36

ARMOURED RECOVERY VEHICLES

Characteristics	Unit	GREIF
Country		A
Crew	persons	4
Battle weight	t	19.8
Engine rating	kW	235
Power-weight ratio	kW/t	11.88
Nominal ground pressure	bars	0.75
Maximum speed	km/h	63
Road range	km	625
Fuel capacity	l	500
Hydraulic dozer blade/spade		yes
Main winch		
Maximum pull	kN	200
Maximum paying out speed	m/min	
Length of rope	m	95
Auxiliary winch		none
Pull/rope length	kN/m	
Crane jib		
Maximum lift	t	6.5
Traverse	-	234
Immersion	m	-
Armament - MG	no/calibre	1/12.7
Maximum trailing load	t	20
Dimensions		
Length	m	6.27
Hull length	m	5.58
Width	m	2.5
Height	m	2.38
Ground clearance	m	0.4

BERGEPANZER 2	BERGEPANZER 3	AMX30D
D	D	F
4	3	4
40.8	54.5	40
610	1100	515
15.3	20.22	12.9
0.87	0.83	0.87
65	68	65
850	650	565
1620	1620	1050
yes	yes	yes
350	343	350
72	85	20
90	180	90
none		
	6.5/245	3.5/120
10	30	12
270	270	240
4	4	4
2/7.62	1/7.62	1/7.62
55	61	50
7.57	9.06	7.53
7.09	6.97	6.59
3.25	3.42	3.15
2.7	2.92	2.65
0.44	0.51	0.45

Characteristics	Unit	**DNG**
Country		F
Crew	persons	3
Battle weight	t	60
Engine rating	kW	1103
Power-weight ratio	kW/t	18.38
Nominal ground pressure	bars	0.82
Maximum speed	km/h	70
Road range	km	550
Fuel capacity	l	1500
Hydraulic dozer blade/spade		yes
Main winch		
Maximum pull	kN	350
Maximum paying out speed	m/min	85
Length of cable	m	180
Auxiliary winch		
Pull/cable length	kN/m	13.5/230
Crane jib		
Maximum lift	t	30
Traverse		240
Immersion	m	4
Armament - MG	no/calibre	1/12.7
Maximum trailing road	t	61
Dimensions		
Length	m	9.15
Hull length	m	6.88
Width	m	3.42
Height	m	3.16
Ground clearance	m	0.485

CHIEFTAIN	CHARRV	Type 78
GB	GB	J
4	3	4
55.64	62	38
537	882	529
9.65	14.23	14.98
0.96	0.99	0.85
42.2	59	53
400	400	300
955	1500	950
yes	yes	yes
300	510	372
13.73	60	30
122	150	60
30/295	15/300	-
2.85	6.5	20
270	360	270
-	1.0	1
1/7.62	1/7.62	1/12.7
65	67.5	50-
8.57	9.64	7.95
7.52	8.33	
3.53	3.55	3.38
3.43	3.96	2.4
0.5	0.5	0.4

Characteristics	Unit	**T-55T**
Country		RUS
Crew	persons	4
Battle weight	t	32
Engine rating	kW	427
Power-weight ratio	kW/t	13.34
Nominal ground pressure	bars	0.8
Maximum speed	km/h	50
Road range	km	500
Fuel capacity	l	960
Hydraulic dozer blade/spade		no
Main winch		
Maximum pull	kN	440
Maximum paying out speed	m/min	
Length of cable	m	
Auxiliary winch		
Pull/cable length	kN/m	80/400
Crane jib		
Maximum lift	t	1.5
Traverse		
Immersion	m	5.5
Armament - MG	no/calibre	1/7.62
Maximum trailing road	t	42
Dimensions		
Length	m	7.47
Hull length	m	6.45
Width	m	3.27
Height	m	1.89
Ground clearance	m	0.43

BREM-1	ENTP-PZ 65	M578	M88 A1
RUS	CH	USA	USA
3	3	3	4
41	38	24.5	50.8
574	485	313	551
14.34	12.77	12.8	10.86
	0.86	0.71	0.76
60	55	60	42
480	450	725	483
	905	905	1514
yes	yes	yes	yes
250	216	272	410
0.2	0.4	58	52
200	120	70	70
5/400	6/240	-	-
19	15	15	22.68
360	0	360	0
5.5	-	-	-
1/12.7	1/7.5	1/12.7	1/12.7
+40	48	30	50
7.1	7.6	6.426	8.267
	6.88	5.588	
3.46	3.15	3.149	3.428
2.4	3.25	3.416	3.225
	0.45	0.44	0.43